Hindu Perspectives on Evolution

Offering new insights into the contemporary creationist–evolution debates, this book looks at parallel debates within the Hindu cultural-religious traditions of India. To provide background for appreciating contemporary Hindu responses to Darwinism, the book examines ancient Vedic and classical Hindu or Dharmic views regarding the origin of the universe. Of special note are the diverse evolutionary views ranging from ideas of material evolution to devolution or manifestation of matter from a higher spiritual reality. At the same time, traditional Dharmic schools developed a number of creationist models of the universe employing sophisticated versions of the design argument often spiced with uniquely Hindu elements.

These traditional perspectives laid the groundwork for the rich variety of responses to Darwinism when it first became known in India. The book highlights the significance of the colonial context of India's encounter with the West. The oppressiveness of British colonialism resulted in considerable ambivalence towards modern science on the part of Hindu intellectuals grappling with the challenges of modernization, secularization, and westernization. Given the enormous prestige of modern science, Hindus could not and did not wish to ignore it, recognizing at the same time its corrosive potential to destroy traditional ideals and values. Their ambivalence manifested itself in the complementary trends of scientizing traditional spiritual ideas and of spiritualizing science. These developments led to notions of spiritual evolution transcending or completing Darwinian evolution, as well as to creationist rejections of Darwinism as non-scientific. Analysing critically the question of compatibility between traditional Dharmic theories of knowledge and the epistemological assumptions underlying contemporary scientific methodology, the book raises broad questions regarding the frequently alleged harmony of the Hindu Dharmic traditions with modern science, and with Darwinian evolution in particular.

C. Mackenzie Brown is Jennie Farris Railey King Professor of Religion at Trinity University, USA. His research interests include the relationship of Hinduism and modern science, with particular focus on creationism and evolution.

Routledge Hindu Studies Series

Series Editor: Gavin Flood, Oxford Centre for Hindu Studies

Former Series Editor: Francis X. Clooney, SJ, Harvard University

The *Routledge Hindu Studies Series*, in association with the Oxford Centre for Hindu Studies, intends the publication of constructive Hindu theological, philosophical and ethical projects aimed at bringing Hindu traditions into dialogue with contemporary trends in scholarship and contemporary society. The series invites original, high quality, research level work on religion, culture and society of Hindus living in India and abroad. Proposals for annotated translations of important primary sources and studies in the history of the Hindu religious traditions will also be considered.

Epistemologies and the Limitations of Philosophical Inquiry
Doctrine in Mādhva Vedānta
Deepak Sarma

A Hindu Critique of Buddhist Epistemology
Kumarila on perception: the 'determination of perception' chapter of Kumarilabhatta's *Slokarvarttika*
Translation and commentary John Taber

Samkara's Advaita Vedanta
A way of teaching
Jacqueline Hirst

Attending Kṛṣṇa's Image
Caitanya Vaiṣṇava Mūrti-sevā as devotional truth
Kenneth Russell Valpey

Advaita Vedānta and Vaiṣṇavism
The philosophy of Madhusūdana Sarasvatī
Sanjukta Gupta

Classical Sāṃkhya and Yoga
An Indian metaphysics of experience
Mikel Burley

Self-surrender (Prapatti) to God in Śrīvaiṣṇavism
Tamil cats and Sanskrit monkeys
Srilata Raman

The Caitanya Vaiṣṇava Vedānta of Jīva Gosvāmī
When knowledge meets devotion
Ravi M. Gupta

Gender and Narrative in the Mahābhārata
Edited by Simon Brodbeck and Brian Black

Yoga in the Modern World
Contemporary perspectives
Edited by Mark Singleton and Jean Byrne

Consciousness in Indian Philosophy
The Advaita doctrine of 'awareness only'
Sthaneshwar Timalsina

Desire and Motivation in Indian Philosophy
Christopher G. Framarin

Women in the Hindu Tradition
Rules, roles and exceptions
Mandakranta Bose

Religion, Narrative and Public Imagination in South Asia
Past and place in the Sanskrit Mahābhārata
James Hegarty

Interpreting Devotion
The poetry and legacy of a female Bhakti Saint of India
Karen Pechilis

Hindu Perspectives on Evolution
Darwin, Dharma, and design
C. Mackenzie Brown

Hindu Perspectives on Evolution

Darwin, Dharma, and design

C. Mackenzie Brown

Routledge
Taylor & Francis Group
LONDON AND NEW YORK

First published 2012
by Routledge
2 Park Square, Milton Park, Abingdon, Oxfordshire OX14 4RN

Simultaneously published in the USA and Canada
by Routledge
711 Third Avenue, New York, NY 10017

First issued in paperback 2015

Routledge is an imprint of the Taylor & Francis Group, an informa business

British Library Cataloguing in Publication Data
A catalogue record for this book is available from the British Library

Library of Congress Cataloging in Publication Data
Brown, Cheever Mackenzie.
Hindu perspectives on evolution : Darwin, Dharma, and design /
C. Mackenzie Brown.
p. cm. -- (Routledge Hindu studies series)
Includes bibliographical references and index.
1. Evolution (Biology)--Religious aspects--Hinduism. 2. Hinduism--India.
I. Title.
BL263.B745 2011
294.5'24--dc23
2011030243

ISBN13: 978-1-138-11936-9 (pbk)
ISBN13: 978-0-415-77970-8 (hbk)

Typeset in Times New Roman
by GreenGate Publishing Services, Tonbridge, Kent

For Andrew O. Fort

Contents

Tables and figures

Tables

Figures

Preface

Early in my career when I was researching the mythology of Hindu goddesses in the medieval Sanskrit texts known as Purāṇas, I would occasionally come upon modern commentaries that equated some of the highly destructive armaments of one or another warrior goddess with modern atomic weaponry. That was my introduction to what I later came to realize was part of a widespread contemporary phenomenon, the scientizing of the ancient Hindu or Vedic tradition. This book examines one aspect of that phenomenon relating specifically to modern evolutionary theory.

The immediate inspiration for this study was the anti-Darwinian criticism of the Hindu creationist A.C. Bhaktivedanta Swami Prabhupada. My interest in Prabhupada led to my first publication in the Hinduism and science discourse, on "Hindu and Christian Creationism: 'Transposed Passages' in the Geological Book of Life" (2002). I soon began to wonder what other approaches modern Hindus have taken to evolution, and how these Hindu perspectives might compare with the Christian spectrum of responses, from young-Earth creationism and Intelligent Design to divinely guided evolutionary theories. As this study shows, Hindus have developed their own vast and distinctive array of responses.

This work is the culmination of over a decade of research. During that time I have published a number of articles (listed in the Bibliography) deriving from that research. Parts of those articles, in turn, have been utilized and incorporated into the larger framework of this book, with significant revisions. Of special note, the three chapters focusing on design (3, 4, and 5) were utilized in writing "The Design Argument in Classical Hindu Thought" (2008). The chapter on Vivekananda (11) was revised and expanded upon for the book chapter "Vivekananda and the Scientific Legitimation of Advaita Vedānta" (2011). Parts II and III of this book, on Colonial and Post-Colonial developments, were greatly abstracted and revised for the essay "Hindu Responses to Darwinism: Assimilation and Rejection in a Colonial and Post-Colonial Context" (2010).

All translations from Sanskrit are my own unless otherwise indicated. On occasion, after a translated passage, I cite both a Sanskrit text and another scholar's English translation (e.g. *BSB* 1.1.2; Thibaut 1890: 15–16). In these instances, the translations are my own based on the Sanskrit texts cited, and the references to another scholar's translation are provided to allow those who do not read Sanskrit to find the cited passages in their larger context. Translations citing a bilingual (Sanskrit and English) edition of a text are my own.

Many colleagues and friends have provided insightful comments and critiques of various parts of the book. I would especially like to thank Professors Andrew O. Fort of Texas Christian University and Jeffrey J. Kripal of Rice University for their interest and encouragement from the very beginning. I also wish to acknowledge several fruitful conversations with Dr. Meera Nanda, independent scholar, and with Michael A. Cremo, member of the International Society for Krishna Consciousness. Professors J.E. Llewellyn of Missouri State University and Robert N. Minor of the University of Kansas deserve my thanks for their critical feedback on individual chapters. For help in formulating and testing the questions on my international survey of Hindus on evolution, I am grateful to Bala Vishwanathan, San Antonio physician and member of the Hindu Temple of San Antonio. Professor Diane Saphire, statistician and Director of Institutional Research at Trinity, was of considerable help in the formatting of the survey and in helping me to interpret the results. I wish to thank my son Nathan Brown, who showed me how to increase responses to my survey through the use of Facebook. I am also indebted to my sister, Deborah Reason, and wife, Dr. Charleen Moore, for reviewing the entire manuscript for general readability. For assistance in formatting citations and the bibliography I am indebted to Benjamin Harris, bibliographer in the Trinity University Library. Maria G. McWilliams, in charge of Interlibrary Loans at Trinity, was extremely helpful in obtaining obscure articles and rare books.

I am particularly thankful to Trinity University for a year's academic leave in 2006–2007, and for providing me with a reduced teaching load through the Jennie Farris Railey King Professorship in Religion. I gratefully acknowledge the Advaita Ashrama Trust for permission to quote extensively from *The Complete Works of Swami Vivekananda*. My gratitude is also due to the Sri Aurobindo Ashram Trust for permission to quote extensively from Sri Aurobindo's *The Life Divine*.

1 Introduction
Design and Darwin in the Hindu Dharma traditions

The ongoing controversy over creationism and evolution in the United States reflects the deep tension between Christian conservatives who espouse various forms of biblical literalism and secular society at large. This tension also exists between Christian and Christian, between those same conservatives and more liberal thinkers who grant God the capacity to utilize evolutionary processes in fulfilling his creative intentions for the world. These tensions are but part of a larger rift between those subscribing to time-honored religious orientations rooted in reverential deference to traditional authority and post-Enlightenment thinkers who tend to question all authorities and whose skeptical perspective helped to create not only modern secular societies but modern science as well. Similar tensions can be found within other cultural and religious communities as seen in the case of Islam, especially with the rise of anti-evolutionary creationism in Turkey. Central to the debates are the challenges raised by modern evolutionary sciences to traditional understandings of scripture, God's action in the world, human nature, and ethics.

This book aims to provide new insights into the contemporary creationist–evolution debates by looking at similar controversies in a radically different context: the Hindu cultural-religious traditions of India, which I collectively call the Hindu Dharma traditions. This usage of the term *dharma* accords with its contemporary Hindu interpretation as a rough analogue for what in the West is called *religion*. As Rabindranath Tagore elaborates: "In the Sanskrit language, religion goes by the name *dharma*, which in the derivative meaning implies the principle of relationship that holds us firm, and in its technical sense means the virtue of a thing, the essential quality of it" (1931: 141–2). The term *dharma* has thus come to mean the eternal principles and divine order thought by Hindus to embody the essential and final truths about the nature of the cosmos and our responsibilities within it.[1] But as we shall see, there is no single Hindu Dharmic perspective regarding modern science, and indeed there is an array of often conflicting responses to Darwinian evolution—from Vedic Creationism to Vedāntic (Advaitic) Evolutionism.

Much has been written on the Hindu experience of an oppressive colonialism with its hegemonic secularizing and modernizing proclivities. And in the last thirty years scholarly attention has increasingly focused on one major aspect of those proclivities, the introduction of modern science into India and the resulting interactions of religion and science in a Hindu cultural and nationalist context. But relatively little work has been done on the specific set of issues arising from the Hindu encounter with Darwinism and the related topics of creationism and design. This book situates the Hindu perspectives on these problems within their own historical, political, and cultural contexts, as well as presenting a critical analysis of these perspectives. Such an analysis raises broad questions regarding the frequently alleged harmony of the Hindu Dharma traditions with modern science.

The argument from design, invoking especially the perceived order in the movement of heavenly bodies and the complex structure of organic beings, is one of the most intuitively appealing of the rational proofs for God, or at least for divine providence. The argument not only has a long and resurgent history, appearing already in the *Timaeus* and *Phaedo* of Plato and manifesting today in the guise of Intelligent Design, but its intuitive appeal has also been appreciated and appropriated in many different cultural contexts.

In the Indian context, we find already in Vatsyayana's *Kāma Sātra* (ca. second century CE) a bare-bones presentation of the astronomical version of the design argument employed against the Materialists' assertion that postmortem existence is doubtful and thus religious acts are fruitless.[2] Vatsyayana offers several reasons for assuming the existence of afterlife rewards, including the assertion that "the constellations, moon, sun, stars, and the circle of the planets are seen to act for the sake of the world as if they thought about it first" (Vatsyayana 2002: 10). The argument here points not to a supreme superintending intelligence, but only to the apparently providential actions of various celestial agents. Yet the critical design element is present: the idea of conscious intent rather than randomly acting natural forces being responsible for the beneficent arrangement of the heavens. Vatsyayana leaves to the reader to divine exactly how such superintending providence supports the notion of postmortem existence. His general thrust is clear enough, though, in his rejection of naturalistic and materialistic denials of supernatural causation.

And more recently, let us consider the following statement from the nineteenth-century Hindu revivalist Dayananda Saraswati (1824–1883), founder of the Arya Samaj:

> When a man sees an object, he becomes conscious of two things, first, the perception of the object as it is; second, the idea of its maker from observing its plan [*racanā*]. For instance, a man comes across a beautiful ornament in a forest. On examining it, he finds it to be made of gold by some intelligent artist. In like manner, the wondrous design [*vividha-racanā*] of the world of many kinds of creatures proves the existence of its author, the Supreme Ruler of all the creation.
>
> (1970: 217–18)[3]

This passage, from Dayananda's discussion of creation in his *Satyārth Prakāś* (*Light of Truth*), first published in Hindi in 1883, readily brings to mind William Paley's famous opening passage of his *Natural Theology* that appeared more than eight decades earlier. In introducing the design argument, Paley asks us to suppose that, in crossing a heath, he stumbles against a rock, and later finds a watch on the ground. In the former instance, he tells us, he might well assume the rock had been there forever, but in the latter, he would conclude that the watch had *not* rested there eternally, as it gives clear evidence of having been designed, "its several parts … framed and put together for a purpose," and thus to have had a beginning and a maker (Paley 2006: 7).

Dayananda's example of the gold ornament clearly differs from Paley's watch, appealing implicitly to the aesthetic quality of the jewelry rather than to its mechanical ingenuity. The skillful craftsmanship of the ornament elicits the wonder of its artistic design. Such aesthetic emphasis is useful for explaining, by way of analogy, the beautiful designs visible in the universe, like the starry heavens.[4] At the same time the passage above, with its reference to the "many kinds of creatures," hints at the more familiar version—the adaptive-mechanical variant—of the design argument that attempts to account for the clever and complex adjustment of organisms to their environments promoting survival and reproduction. Indeed, Dayananda's presentation of the ornament analogy follows a long passage that portrays the marvelous design of the human body.[5]

The lynchpin of the design argument in all its versions is analogical. So like an artifact is the universe, or some aspect of it, that it, too, must have been designed by a conscious, intending creator. In the adaptive-mechanical variant, just as a complex machine whose interrelated parts subserving some purpose must have been devised and assembled by a knowledgeable mechanic, so the complex "works of nature" including the human body, which give "every indication of contrivance, every manifestation of design, which existed in the watch," must have had an artificer, an intending mind directing a skilled hand, in this case God (Paley 2006: 16). Paley illustrates the analogy using a telescope and the human eye: the eye must be an artifact of an intelligent creator, given its intricate working parts that together work to produce a focused image like a telescope. Dayananda utilizes the same illustration of the eye (without reference to the telescope), but with a classical Hindu karmic twist:

> … consider what skill and wisdom have been displayed in the construction of the eye. Great anatomists and doctors have been studying it up to the present time, yet their collective knowledge of it is imperfect. No one can know *thoroughly* how God fashioned it, and what beauties He has endowed it with. The creation of the Sun, the Moon, etc., and the upholding of these is the work of God, and He it is that dispenses to the souls the fruits of their actions, and none else.
>
> (Quoted in Singh 1971: pt. I, pp. 134–5)

One may wonder if Dayananda was not simply mimicking Paley's argument as he perchance heard it from the mouths of Christian missionaries during British colonial rule in India, and then gratuitously grafting indigenous notions onto an argument derived essentially from European and especially British natural theology. The answer, as we shall see throughout this book, is a decided no. Whatever specific Western examples Dayananda may have incorporated into his discussion of creation, his design argument is firmly rooted in a rich Hindu philosophical and theological heritage.

His substitution of a gold ornament for a watch, for instance, derives from a classical Hindu illustration of the design argument found already in the writings of the Vedāntic philosopher Śaṅkara.[6] And Dayananda's teleological reflections on cosmology with their karmic twist are indebted to the classical school of the Nyāya-Vaiśeṣika, or Logical Atomists. According to the school, at least in its later theistic forms, atoms are eternal but require an omniscient and intelligent creator to set them in motion and to combine them into appropriate dyads and triads that come to constitute the gross, sensible elements of the universe. Such a process, the Logical Atomists insist, cannot take place randomly. Dayananda reiterates this view in a passage just preceding his reference to the skillful construction of the eye:

> Earth or clay has not the power to shape itself into a jar. Matter cannot assume gross forms of itself, for it is devoid of knowledge requisite for the purpose. Nor can any human soul accomplish the result, for no man has yet been able even to make a single hair. No one has ever been able to grasp the atoms and to fashion anything out of them by any means. Nor is there anyone in the present time, who can unite even two *trisrenus* [atomic triads]. This shows that God alone has the power to fashion the world.
>
> (Quoted in ibid.: pt. I, p. 134)

The Logical Atomists further argue, like Dayananda centuries later, that God alone can assemble the gross elements into animate bodies appropriate for each soul according to its karmic situation at the renewal of a cosmic cycle.

Two major challenges to the design argument have gradually emerged over the centuries. The first is speculative with ancient roots; the other, empirically based and modern. The speculative challenge has taken two distinct but interrelated approaches: on the one hand examining logical difficulties inherent in the basic analogical inference of the design argument itself, and on the other exploring negative implications of the argument for the nature of God, including perhaps most forcefully the problem of theodicy. The eighteenth-century Scottish philosopher David Hume, best known of the early-modern philosophers to challenge the design argument on logical and theological grounds, is often accredited as the thinker most responsible for the decline in popularity of the argument in the West. The Dharma traditions have their own Hume, the Vedāntic theologian Rāmānuja, who anticipated many of Hume's arguments by seven centuries.

In India and the West many remained unconvinced by such reasoning, for until Charles Darwin there was no empirical, viable explanation for the evident design observed especially in organisms. The apparent explanatory power of the design argument—even if somewhat attenuated by the philosophers—to account for the complexity of life and the adaptation of means (such as integrated nervous systems and eyeballs) to ends (vision) short-circuited for most believers any need for a better explanation. This was especially the case when the only alternatives seemed to be either the chance assembly of thousands of specific parts in a unique configuration that just happens to turn out to serve some advantageous function, or a "Lamarckian" evolutionary theory of gradual adaptation of parts through use and disuse and the bequeathing of such acquired characteristics to successive generations. Paley himself provided empirical and theoretical arguments against both of these alternatives. Darwin proposed a third alternative, one that never occurred to Paley: the gradual adaptation of means to end and of animals to habitats by the process of natural selection working on minute, chance variations over immense periods of time, thus combining elements of randomness and necessity in an inefficient, often brutal process (Gould 1993: 143–6).[7] Darwinian evolution, with its dismissal of any grand purpose and rendering divine intention irrelevant, presents what is probably the most stubborn and even overwhelming challenge to all notions of conscious or intelligent design. Design arguments, in whatever cultural configurations they appear, are grounded for the most part in fundamental assumptions about the empirical world and how it works that are gradually and some would say inevitably corroded by Darwin's "dangerous idea."[8]

Paley died four years before Darwin was born and thus avoided having to face this dangerous idea. Dayananda, however, lived during the time of Darwin, and Darwinian ideas (often in Spencerian guise) began filtering into India soon after the publication of his *On the Origin of Species* in 1859. While Dayananda had only vague knowledge of Darwin's ideas, the general notion of organic evolution caught his attention late in life. Sensing its incompatibility with his religious views and his notions of design, he dismissed it in a manner reminiscent of contemporary creationist misunderstandings of the theory: "If man descended from monkeys, how is it that process had come to an end and monkeys no longer evolve into men?" (quoted in G.R. Garg 1984: 70).

Dayananda's rejection of Darwin may surprise those who believe or insist that Hinduism is more compatible with modern science, including evolution, than the Abrahamic traditions. Was Dayananda perhaps exceptional in his anti-evolutionism? On one hand, Pratik Chakrabarti tells us that Darwinism "was widely contested in nineteenth-century India" (2004: 175). On the other, Dermot H. Killingley observes: "It was easier for a Hindu to combine modern knowledge with adherence to tradition than for a Muslim, or even, as might appear from the controversies in the West over Darwin, for a Christian" (1995: 179). And Deepak Kumar argues that "modern science was not seen as an alien import. Darwinism, for instance, was imported readily and the theological issues at its heart did not cause a ripple in India" (2010: 677). My own reading of the historical situation is that most Hindu intellectuals in the nineteenth century, who almost universally granted modern science enormous prestige, wished to avoid direct confrontation with such a major scientific theory as Darwinism. A principal argument of this book is that, while those Hindus who denied and continue to deny Darwinism outright were and are relatively few, the apparent historical ease of assimilating modern science and evolutionary theory on the part of Hindus is misleading—indeed, an illusion.

For instance, a contemporary Hindu follower of the Vedānta affirms the harmony of his tradition and modern science, noting that, while many perceive a three-way tension or conflict among philosophy, theology, and modern science, "[t]here is no such paranoia in Hinduism, where philosophy and theology are one and the same, and where the arrival of challenges in the form of scientific enquiry is welcome for they serve only to confirm spiritual truths." But then, tellingly, this Vedāntic apologist adds: "Vedanta emphasises not the physical evolution of man as propounded by Darwin, but a spiritual evolution of man" (Joshi 1997: 116). Conflict is thus often avoided by sidestepping crucial issues, such as the implications of natural selection for the notion of God's benevolent action and purposeful creation of all species.

A close examination of traditional Dharmic world views reveals a variety of assumptions that create ambiguous orientations regarding Darwinian evolution.[9] One oft-cited factor supposedly demonstrating the compatibility of Hindu and Darwinian perspectives is the Hindu notion that all sentient beings have souls that in the process of spiritual development transmigrate through a series of animal bodies. Human beings once occupied ape bodies, for instance, thereby denying the absolute uniqueness of humankind espoused by the Abrahamic traditions. As David L. Gosling notes, it seems

> that Darwin's theory was readily assimilated by educated Indians, and that they failed to see any threat to their religious beliefs comparable to that felt by the Victorians. Assuming that the crux of the matter in England was the idea of a common ancestry for human beings and animals, then it is not hard to see why educated Indians reacted differently, since for them the theory of reincarnation pre-supposed such a belief.
>
> (1976: 15)

At the same time, traditional Hindu texts often emphasize the exceptional status of human beings in a "Great Karmic Chain of Being," a notion that assumes the fixity of species according to their essential karmic natures and endowing human beings alone with the spiritual capacity for liberation.[10] Dayananda is true to this latter tradition when he writes:

> Of all living beings man alone has been created with a mind … By a peculiar arrangement of the molecules of the organs of the human body God has made them fit for being used as instruments for acquiring knowledge. Man alone, and no other animal, therefore, is capable of knowing right and wrong …
>
> (1958: 42)

While souls may occupy successively a huge variety of animal bodies in their karmic sojourn, such spiritual evolution is quite distinct from organic evolution, despite frequent comingling of the two by those wanting to reconcile Hinduism and modern science.

A second factor that might seem to render the Hindu Dharma traditions hospitable towards organic evolution is their acceptance of enormous cycles of time more consonant with the Darwinian time scale than the six thousand years or so of traditional biblical, Mosaic chronology. But the Dharmic time scale is far too large, at least when it comes to the history of terrestrial life. Dayananda, for instance, places the creation of humankind some 1,960,852,976 years ago, at which time, he claims, God created all animal species, including thousands of fully adult men and women, and instilled souls into them, after which they reproduced in the usual sexual manner.[11] Such Ancient Hominid Creationism, like Young Earth Creationism, has its own share of problems with Darwinian evolution, according to which modern humans emerged only 100,000 to 200,000 years ago. Further, the traditional Hindu view of the cyclical nature of the universe contrasts with the linear view generally assumed by modern science and by Darwinism in particular with its notion of a unique and largely unrepeatable history of life's evolution. This difference may not be absolute, for modern science has toyed with the idea of a limitless series of cosmic Big Bangs. But in the scientific view, successive cycles—if there are such—start from scratch, going uphill from chaos to cosmos, versus the traditional Hindu scheme of gradual degeneration following the zenith of physical and moral perfection manifest at the start of every eon.

One final factor to mention here—a factor somewhat in tension with the degenerative view of time—is the notion found in several Dharma traditions that the physical cosmos has evolved from some primal element like ether or from an original material substance such as Prakṛti (Primordial Nature).[12] We may note in this regard that many Hindu cosmogonic myths stress organic rather than mechanistic models of cosmogenesis. The organic models incorporate fluid, vegetative, and sexual motifs that would seemingly lend themselves to notions of gradual development guided by internal or innate potential, rather than to the idea of external manipulation of raw materials by an external artisan, craftsman, or mechanic imposing his sense of order upon those materials from above.[13] Metaphorically, for instance, the motif of a gradually evolving grand Cosmic Egg would seem potentially harmonious with a Darwinian perspective. Notions of cosmic, stellar, and planetary evolution in the Dharmic perspectives, however, do not necessarily extend into the realm of organic evolution. This is especially the case when design concepts are emphasized, as in Dayananda's world view. While rejecting Darwinism, he accepted an evolutionary perspective on the inorganic portion of the cosmos, arguing as he so often did that this was just one more example of the discoveries of modern science that had been anticipated by the ancient Hindu or Aryan sages. As he, or close disciples, wrote in the *Arya Magazine* of 1882:

> Thus the highest speculations of modern science dovetail very wonderfully with the ancient Aryan belief which asserted that all [the] universe has come into existence out of one primal element, the Akas (ether). True, this theory of our ancient rishis [seers] when first known to the Western world was ridiculed and set down as another example of the diseased imagination of the

> Asiatics. But the arguments above clearly show that far from being a subject of contempt, this ancient theory of evolution of matter is amply justified by the researches of modern science.
>
> (Quoted in Thistlethwayte 1998: 82–3)[14]

Dayananda notwithstanding, Hindu intellectuals have developed a variety of strategies to avoid outright conflict with Darwinism. Foremost among these is recourse to an essentially non-Darwinian version of evolution—often Lamarckian or Spencerian in nature—combined with some sort of integration with traditional Hindu evolutionary notions, including the radical subordination of organic to spiritual evolution. By such strategies, many Hindus find that "Darwinism" seems not to be a problem. But Dayananda's case reminds us that not all Hindus accept Darwin even in a minimal or reduced sense, suggesting a far greater incompatibility between a robust Darwinism and the Dharma traditions than is often acknowledged. The emergence of Vedic Creationism in America among the Hare Krishna devotees and their filing of an amicus curiae brief in Cobb County, Georgia, in support of a Christian Creationist effort to have high school biology textbooks include a "warning label" about evolution is simply another manifestation of this tension (Nanda 2006).

Dayananda epitomizes in many ways the situation of Hindu intellectual elites under British rule in the nineteenth and early twentieth centuries. Caught up in the encounter of tradition with modernity refracted through issues of class, religious, and national identity, they were fascinated by Western technology while simultaneously being squeezed by an increasingly oppressive colonial imperialism abetted by that same technology and allied with an aggressive Protestant proselytism that frequently disparaged all things Hindu. While child marriage, suttee, widow remarriage, caste, the worship of "idols," and the assimilation of Western technology were among the most prominent social-religious issues of the day, ideas about evolution and design had their own role to play in the complex cultural negotiations between colonizers and colonized. By the end of the nineteenth century, we find Hindu religious concerns becoming intertwined with escalating nationalist sentiments; it is clear that issues of religion and science cannot be separated from social and political interests, especially in a colonial context.

While conservative and even moderate Hindu views on the relation of science and religion, and on evolution, manifest many of the same concerns as Christian fundamentalist perspectives, a closer parallel is with the Islamic tradition, in part because of a shared colonial past. Taner Edis in his *An Illusion of Harmony: Science and Religion in Islam* provides a number of characteristics describing the efforts of various Muslim intellectuals as well as the Muslim populace at large to come to terms with modern science over the last two centuries. These characteristics, fully understandable only within their colonial context, include: (1) regarding Islam as the most scientific and rational religion; (2) interpreting the Qur'an as fully harmonious with modern science, and even as anticipating by centuries various modern scientific discoveries; (3) viewing the science–religion hostility in the West as an artifact of the authoritarian medieval Catholic Church; (4) desiring to assimilate Western technology as a means of catching up with the West and the modern world in general; (5) combining pro-technology attitudes with pseudo-scientific ideas and resistance to Darwinism; (6) tending to interpret science superficially as an assemblage of facts, without appreciation of its underlying theoretical frameworks and conceptual schemes, imposing instead overarching and rather vague metaphysical systems; (7) inclining strongly towards the "certainty" of revealed truth or metaphysical reasoning while exhibiting considerable wariness towards skepticism that produces merely tentative conclusions; and (8) favoring epistemological ideals that emphasize moral values, personal trust, and holistic approaches to nature rather than analytical and systematic investigations of impersonal laws of nature anchored in empirical reality testing. All these characteristics, as we shall see, readily apply to the Hindu case as well when appropriately translated into a Dharmic context.

Many of the above characteristics are also seen in the Buddhism and science discourse over the last century and a half. In some regards, the encounter of Hinduism with modern science exhibits even closer parallels with the Buddhist than the Islamic experience. This is hardly surprising given the common Indian origin of the two traditions and their many shared assumptions—however modified these became as the traditions diverged. Such assumptions include the vast expanse of the cycles of cosmic time and especially the "law" of karma and rebirth. In his *Buddhism and Science: A Guide for the Perplexed*, Donald Lopez notes a number of frequent elements or tropes in the Buddhism and science discourse which, in their general structure (but stated in terms of Buddhist teachings and ideas), are shared by Hindu commentators on science. These include the notions that: (1) Buddhism is the most rational and scientific religion; (2) the Buddha's teachings are fundamentally in harmony with, and affirmed by, the discoveries of modern science; (3) the omniscient Buddha, with his enlightened eye, knew scientific truths millennia before they were discovered by modern scientists; (4) Christianity, as opposed to Buddhism, is an adversary of science; (5) Buddhism welcomes all new scientific discoveries; (6) modern science deals with the outer world while Buddhism focuses on the inner; (7) karma and reincarnation are better explanations for the diversity of life than Darwinian evolution; and (8) Darwinian evolution is incomplete as it fails to account for consciousness and especially compassion or altruism.[15]

Clearly modern science, and especially evolutionary theory, have posed major challenges to a wide variety of religious traditions. But the challenges are not uniform. The Buddhism and science discourse is distinct from the Islamic and Christian engagements with science due to Buddhism's non-theistic perspective, and specifically its antipathy to any notions of an extra-cosmic creator responsible for designing the universe—an antipathy generally shared by the non-theistic strands of Hinduism manifest in the Neo-Vedānta or Neo-Advaita movements. Another distinctive element in the Hindu and Buddhist discourses with modern science that sets them apart from both the Islamic and Christian is their assimilation and adaptation of various Theosophical motifs and its fundamental ideal of a scientific and scientized religion. This ideal, born in part as a reaction against traditional Christian dogmas that seemed to be crumbling with the onslaught of modern science, drew much of its own inspiration from the "wisdom traditions" of the "East," especially Buddhism and Advaita Vedānta. The Theosophist proclamation of complete harmony between the eastern wisdom traditions and the findings of modern science was a siren song to those of the East who sorely felt the oppression of Western imperialism and Christian evangelization.

To provide background for a comprehensive understanding of the Hindu Dharmic perspectives on evolution and creation that have developed during colonial and post-colonial times, I examine in Part I the ancient and classical texts relevant to our inquiry. We shall begin with pertinent cosmological motifs and philosophical speculations in the ancient Vedic, Upaniṣadic, and epic literature, noting an array of theistic-creationist and non-theistic-evolutionist themes. We shall continue with the general cosmological trends of the early philosophical schools, and conclude with the elaboration of these trends in the sophisticated arguments of certain classical writers belonging to the so-called orthodox perspectives, or *darśanas*, of Vedānta and Nyāya-Vaiśeṣika. These orthodox perspectives were to provide Hindus with frequent inspiration in facing the challenges of British colonialism and modern science.

Part II focuses on the colonial period during which, in the first few decades of British rule, the classical design arguments were frequently called upon to defend against the combined onslaught of Protestant evangelism, European skeptical rationalism and scientific materialism, and by the end of the nineteenth century against Darwinism itself. At the same time, there was a growing distrust of the design argument as leading to Deism and even atheism, accompanied by a Romanticist reaction emphasizing direct or intuitive experience of God. I also examine in Part II the variety of attempts in the late nineteenth and early twentieth centuries to accommodate Darwinism within traditional Dharmic perspectives, especially the teachings of Advaita Vedānta.

Part III looks at post-colonial developments in India and the diasporic Hindu communities around the world. We shall observe a revival of Hindu design arguments, with their increasing scientific sophistication utilizing the latest discoveries in molecular biology and quantum mechanics to bolster ancient Hindu anti-evolutionary perspectives. I also call attention to the interrelationship between Christian and Hindu creationist strategies in challenging Darwinian evolution. Part III concludes with an analysis of contemporary Hindu attitudes towards evolution and related issues, as ascertained by an online survey of Hindus around the world, conducted between October 2009 and January 2011. The survey, while non-random, when interpreted in light of this study as a whole, reveals a number of ongoing critical issues regarding the compatibility between traditional Dharmic theories of knowledge and the epistemological assumptions underlying contemporary scientific methodology. These issues raise a number of challenging and troublesome questions for those proposing an alleged concord of Hinduism, the eternal Dharma, and modern science with its non-teleological, evolutionary model of the cosmos and of life itself.

Part I

The classical background

2 Creationism and evolutionism in the ancient and classical traditions

How did the ancient Hindus suppose the universe came into being? Was it the result of a conscious and wise creator? If so, where did the matter now constituting the world come from? Did the creator evolve matter out of his (or her) own body or mind, or work with pre-existing materials that he (or she) manipulated and assembled into the vast universe in which we find ourselves? Alternatively, is the universe self-existent, transforming itself—without assistance from any superintending divine being—from a primordial state, primal substance, or set of elements into the ordered structures which we see around us? Is the universe some kind of organism with conscious purpose? Or is the empirical realm fundamentally unconscious, but with an innate capacity for self-transformation, perhaps to meet the needs of spiritual entities existing alongside material nature? Or finally, is a self-evolving physical universe all there is, with consciousness manifesting as an emergent quality of matter?

Affirmative answers to most of these inquiries appear in the ancient Vedic, Upaniṣadic, and epic texts, as well as in the classical philosophical-theological schools (*darśanas*) rooted in and descended from those texts. Even the concept of a material universe devoid of any higher spiritual reality, although overwhelmingly rejected by the later Hindu Dharmic traditions, appears in various forms in some of the earliest Indian religious literature and constantly reappears in the classical period as the lowest and most disreputable of all possible world views. Not only do we encounter diverse answers to basic cosmogonic questions in the ancient and classical traditions, we find throughout that different viewpoints are frequently juxtaposed, affirmed, transformed, and synthesized.

This integrationist approach is manifest right from the start of Hindu reflections on origins:

> Vedic cosmogony is very complex and elusive ... The cosmogonic myths, metaphors, and paradigms may vary and merge with one another even within one single text, such as *Ṛgveda* X, 129. Here, the cosmogonic process is characterized as birth or germination, but also as transformation and emanation ... Numerous other ideas are found in other texts; metaphors of birth and biological formation occur side by side with conceptions that present the world as an artefact.
>
> (Halbfass 1992: 29)

And in *Ṛg Veda* X.81, the creator Viśvakarman (All-Maker) is variously portrayed as a sculptor fashioning the raw material of the world, as a smith producing heaven and earth and welding them together, as a carpenter carving out wood to create the sky and earth, and as a sacrificial offering whose body becomes the universe. In the next hymn (X.82) Viśvakarman is the primeval germ or embryo placed in the primordial waters and from which the cosmos eventually emerges (Doniger 1981: 34–6).

This juxtaposition of dualistic and monistic motifs in which accounts of the distinctness of creator and creation stand side by side with affirmations of their ultimate oneness or unity is characteristic of the presystematic thought not only of the Vedas, but also of the Upaniṣads and epics. Soon incorporated into these dualistic and monistic cosmological speculations, beginning especially with the Upaniṣads, are lists of various ingredients or entities, such as the five great elements ether, air, fire, water, and earth, that constitute the universe. Two basic enumerative arrangements can be discerned: "horizontal" schemes listing discrete, nonevolutionary elements, eternally irreducible, yet capable of being combined mechanistically to produce the universe, and "vertical" schemes describing successive stages of cosmic emanation or evolution out of some primordial substance or being. There are also hybrid schemes involving discrete, horizontal elements at one level, which then evolve of themselves, through various combinations, into the cosmos (Halbfass 1992: 48, 50).

An important development following the Vedic quest for origins was the increasingly unchallenged assumption that there was no beginning, at least no absolute beginning, to the cosmos. The early "fascination with the 'beginning,' ... was replaced by, or transformed into, a preoccupation with schemes of recurrent evolution and involution and with the permanent foundations of such regular and cyclical processes" (ibid.: 47). Such a cyclical approach came to dominate all cosmological speculations, whether evolutionary or creationist in outlook. This cyclical perspective reoriented cosmological into soteriological concerns: what is humankind's relationship to reality, knowing which, we can transcend the eternally cycling wheel of existence? Cosmology was important as it provided the framework in which human beings are to work out their ultimate destiny. Three cosmological models eventually came to predominate.

Three fundamental cosmological approaches

The early dualistic and monistic perspectives were variously integrated with the two enumerative schemes, horizontal and vertical, to produce three fundamental approaches to cosmogonic and cosmological issues relevant to our concerns: (1) theistic creationism, (2) descending evolutionism, and (3) ascending evolutionism. Theistic creationism generally assumes a horizontal scheme in which an independent and distinct creator God sets into motion or brings together and assembles diverse eternal or pre-existing entities and elements to construct the universe. Descending evolutionism, what might also be termed devolutionism or involutionism, begins with some sort of universal mind or consciousness, from which emanates, manifests, or appears all of material creation—the latter at times regarded as illusory. Ascending evolutionism views matter or nature as the ultimate source of the physical cosmos, which commonly included various mental or psychic factors regarded as distinct from a non-material and eternal conscious principle. These three fundamental approaches illustrate what came to be known as the doctrines of production (*ārambha*), manifestation (*vivarta*), and transformation (*pariṇāma*). They are well represented in the later classical Vaiśeṣika, Advaita Vedānta, and Sāṃkhya schools, respectively (cf. Halbfass 1988: 354; 1992: 48). We should also note that ascending evolutionism had an extreme materialist form that regarded even life and consciousness as emergent properties of matter. This robust materialist form of ascending evolutionism, espousing an evolutionary naturalism, is embodied in the Cārvāka or Lokāyata philosophy. It rejected the soteriological concerns of the other classical schools, and its teachings survive today only in the accounts of its opponents.

Theistic creationism

An early Upaniṣadic horizontal listing of elements within a theistic-creationist perspective appears in *Śvetāśvatara Upaniṣad* 6.1–10. Here five elements, apparently pre-existing, are enumerated: earth, water, heat, wind, and sky. These are identified with the work of creation and said to unfold (*vivartate*) under the control of God (6.2).[1] He is described as the efficient cause of combinations (*samyoga-nimitta-hetu*), seemingly referring to his combining of the various elements in making the universe (6.5) (see Figure 2.1a). Alternative views of the ultimate cause of the cosmos as either the inherent nature (*svabhāva*) of matter or time (*kāla*) are specifically rejected (6.1). Further, this creator, the Supreme Lord (Maheśvara) who is other than this cosmos, is said to be free of motivated action (*kārya*) and of any sense organs or limbs (*karaṇa*) (6.7–8). Rather, the activity or working of his wisdom and strength is simply inherent (*svābhāvikī*) (6.8). Similarly, God is said to conceal himself, like a spider, with threads derived from matter (*pradhāna*), by his own nature (*svabhāva*) (6.10). The *Śvetāśvatara*, in applying the notion of *svabhāva* both to matter and to God, points to an ongoing controversy over whether an innate capacity for creation belongs only to some supernatural being, or instead inheres within matter or its elements.

Theistic creationism, derived in part from the Vedic myths of the universe as artifact, readily invites elaborations of the design argument—as well as the less renowned causal or cosmological argument. But the *Śvetāśvatara*'s denial of purpose on the part of God and of his having organs and limbs was to create considerable difficulties for later advocates of the design argument. Any teleological notions of purpose attributed to God suggest that he is lacking something he needs to accomplish. And if he has no sense organs, and no hands or feet, then how can he, as a disembodied spirit, produce anything, since all the producers we know in everyday experience do have bodies?

Śvetāśvatara Upaniṣad 1.2 provides us with another and longer list of alternative views regarding the ultimate cause of the universe. In addition to time and

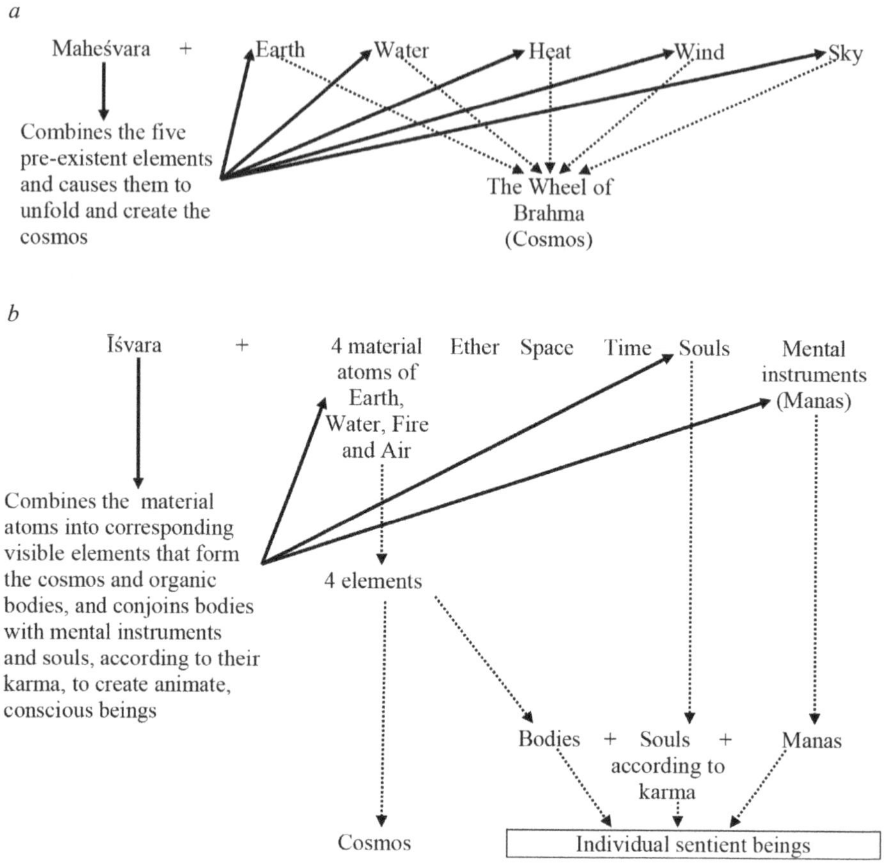

Figure 2.1 Diagrams of theistic creationism

a An early theistic-creationist scheme from *Śvetāśvatara Upaniṣad* 6.1–10

b Theistic-creationist scheme of the Vaiśeṣika (Atomists)

inherent nature, *Śvetāśvatara* 1.2 includes other mostly naturalistic causes such as necessity or destiny (*niyati*), chance (*yadṛcchā*), the elements (*bhūtas*), the womb (*yoni*), the person or soul (*puruṣa*), or some combination of these. The *Śvetāśvatara* discounts all of them, insisting that God governs all causes (1.3). A similar list of alternative, naturalistic causes is given in *Mahābhārata* 12.224.50–1: human or personal effort (*puruṣa-kāra*; *pauruṣam karma*), fate (*daiva*), the inherent nature (*svabhāva*) of the elements, or all of these forces combined. Such views are dismissed as misleading or dangerous, and opposed to the truth that God/Brahman creates all things and bestows on each its distinctive properties in each new creation (12.224.16–17).

Another possible example of early theistic creationism occurs in the teachings of the sage Asita Devala (*Mahābhārata* 12.267.4ff). According to Wilhelm Halbfass, Asita presents the eternal and unchanging great elements "as the material out of which the Lord produces this whole world and all beings in it" (1992: 51).[2] Yet the text of the Critical Edition represents time, not the Lord, as producing the five elements impelled by its own nature (*bhāva-pracodita*). Further, Asita insists that there is nothing higher or beyond these five, and that they, along with time, are powerful by their inherent nature (*svabhāva*) (12.267.4; cf. 12.267.6), thereby representing an ascending evolutionary or materialist viewpoint, although beginning with a horizontal scheme of the elements. The ease with which various cosmogonies can be reinterpreted by redefining basic terms or embedding one scheme within another is characteristic of these early texts.

The Vaiśeṣika, often cited as the prime example of the horizontal approach among the classical schools, proposes nine distinct and eternal substances (*dravya*): four material atoms of earth, water, fire, and air, plus ether, space, time, souls (*ātman*), and mental instruments (*manas*). Originally the school seems to have been close to the Materialists, requiring no supernatural but only mechanistic causes to bring about the combination of material substances with their inherent properties to produce the phenomenal world. Natural causes, such as the heaviness or fluidity of various elemental entities, were apparently regarded as sufficient to initiate the movements that resulted in the production of the universe (Frauwallner 1973: vol. 2, p. 57; Smart 1964: 156). During the course of the first millennium CE, with Indian culture as a whole becoming increasingly concerned with the ideal of deliverance from the round of rebirth, notions of karma were introduced to help comprehend not only the fate of individuals caught up in *saṃsāra*, but also to explain the start-up process at the beginning of each cosmic cycle. Finally a theistic element was added to account for the workings of the insentient karmic force: an overseeing Lord assembles atoms to create the elements of the cosmos, as well as organic bodies, which he then conjoins with mental instruments and souls according to their karma (see Figure 2.1b). The theistic Vaiśeṣika produced some of the most sophisticated defenses of the design argument in Hindu philosophical-theological writings.

The Vaiśeṣika subscribes to the notion that the production of effects, such as the universe, is a genuine innovation and that the effect is not latent within the cause (*asatkārya-vāda*), in contrast to the notion that nothing arises out of nothing, that the effect lies within its cause (*satkārya-vāda*), a viewpoint espoused by the Sāṃkhya. This latter view is already expressed in *Bhagavad-Gītā* 2.16, which affirms that effects are not different from their cause (*satkārya-vāda*). Halbfass points out, however, that the *asatkārya* perspective is not "a genuine alternative," as it "recognizes only temporarily 'new' configurations of preexistent constituents, such as the atoms. It is just a different type of substantialism and eternalism" (1992: 28; cf. 55–8). Both perspectives are thus quite distinct from the biblical notion of creation out of nothing. Nonetheless, the *satkārya* perspective frequently has been employed by advocates of evolutionary views of the universe, while the *asatkārya* perspective, with its emphasis on innovative, immediate production rather than gradual transformation, readily harmonizes—in a theistic context—with the notion of a constructed universe and a teleological cosmology.

Descending evolutionism

Vertical evolutionary perspectives seem to have been dominant in the early Vedic and Upaniṣadic literature (Halbfass 1992: 49–50). Frequently, though, the vertical is intermixed with the horizontal, as in the *Aitareya Upaniṣad* (3.1.3), where the earliest extant list of what came to be known as the five great elements (*mahābhūtas*) is embedded in a devolutionary framework. Earth (*pṛthivī*), wind (*vāyu*), space (*ākāśa*), water (*apas*), and light (*jyotīṃṣi*) are presented horizontally, that is, not as emanations one from another, for they all derive or devolve as a group from the intelligent Brahman. These things, mingled in their fine forms, are the seeds of all animate beings, the whole established in and guided by intelligence (*prajñā*) (see Figure 2.2a).

One of the earliest and strictly vertical lists occurs in the *Chāndogya Upaniṣad* (6.2.2–4), where the original being (*sat*)—a conscious entity since it "desired" to become many—first emitted fire (*tejas*), which in turn desired to increase itself and emitted water (*apas*), which in similar manner emitted food (*annam*). Another early vertical scheme that expands the *Chāndogya*'s list to include the classical great elements is found in *Taittirīya Upaniṣad* (2.1.1): "From the Self (*ātman*) arose space, from space air, from air fire, from fire water, from water earth, from earth herbs, from herbs food, from food the person" (see Figure 2.2b).

The vertical schemes above presuppose that the ultimate source from which all emanates is conscious and intelligent, and thus represent descending evolution. The monistic Advaita Vedānta exemplifies an extreme, idealist elaboration of this perspective, in which consciousness is the sole reality, the material realm being only a manifestation of consciousness. The ultimate reality, the one Brahman, is both material and efficient cause of the universe, illustrated by the famous Upaniṣadic metaphor of the spider spinning out its web (*Muṇḍaka Upaniṣad* 1.1.7). But in Advaita, since Brahman is not only eternal but also unchanging, cosmic evolution is apparent only. Thus the idea of real transformation or evolution (*pariṇāma*) is rejected in favor of the notion of phenomenal or apparent manifestation (*vivarta*). At the same time, and critical for our purposes, Advaita supposes that there are two levels of truth regarding the world. The world may be only an appearance from the higher perspective, like a snake falsely perceived in a rope, but from the lower level, the manifest variety and multiplicity of the everyday world can be granted a tentative reality. And so from this lower perspective, it is reasonable to see the physical universe not so much as a self-manifesting entity but as the product, the fabrication, of a designing intelligence.

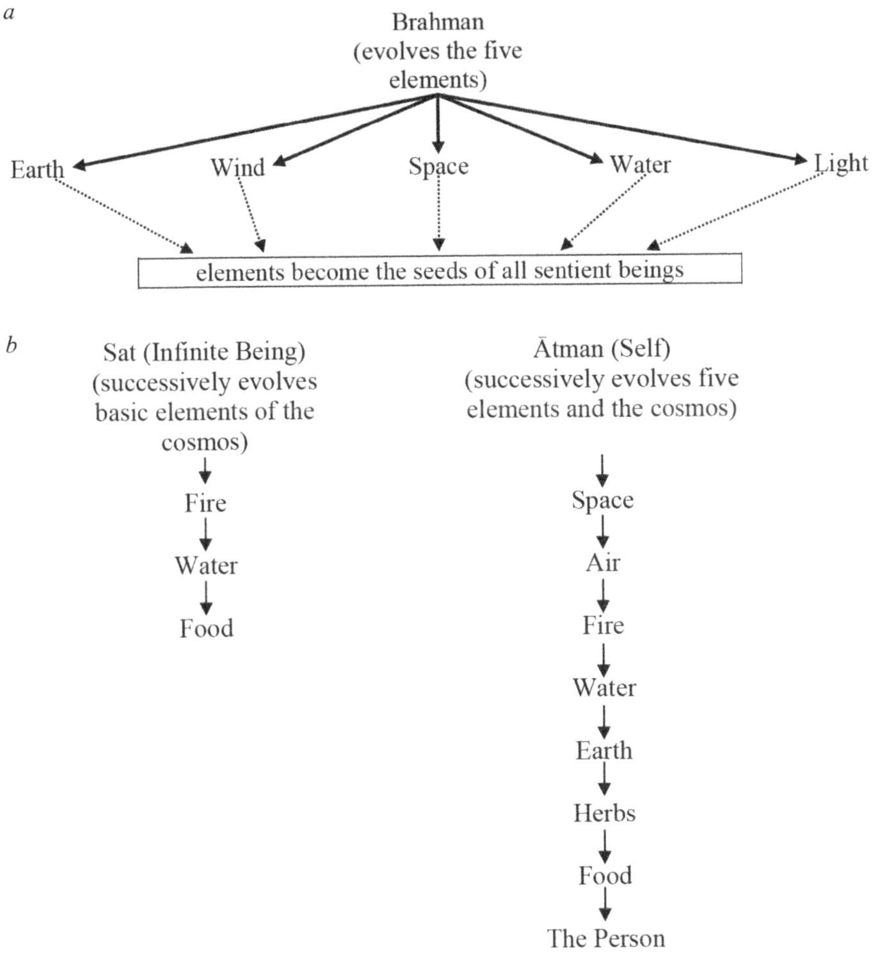

Figure 2.2 Diagrams of descending evolutionism

a A mixed horizontal and descending evolutionary scheme from *Aitareya Upaniṣad* 3.1.3

b Descending evolutionary schemes from the *Chāndogya Upaniṣad* 6.2.2-4 and *Taittirīya Upaniṣad* 2.1.1 (forerunners of idealist *vivarta* doctrine of Advaita Vedānta)

Ascending evolutionism

The dualistic Sāṃkhya model

Genuine transformation (*pariṇāma*), of an ascending type, is embraced in the Sāṃkhya. The Sāṃkhya, at least in its earliest classical form as found in Īśvarakṛṣṇa's *Sāṃkhya-Kārikā* (ca. fifth century CE), was atheistic or non-theistic, but this did not exclude a spiritual aspect belonging to the universe.[3] It exemplifies a form of ascending evolutionism, but a compromised form that rejects a robust materialism as it does not see consciousness as an emergent property of matter. Rather, Sāṃkhya affirms an ontological dualism, professing the ultimacy and eternity of both consciousness or spirit (Puruṣa), and matter (Prakṛti). Puruṣa, like the Brahman of Advaita, is unchanging and conscious, but plural, consisting of many souls (*puruṣas*) with no overarching Great Puruṣa or Supreme Being. Prakṛti, Primordial Nature, is insentient but active, evolving out of itself the physical universe in successive stages, beginning with the psycho-physical entities of intellect, egoism, and mind, followed by five sense organs and five motor organs, and concluding with the five subtle and five gross material elements. Succeeding this primary evolution is a secondary evolution in which individual entities like trees and animals appear. The psycho-physical entities are not in themselves conscious but simply instruments by which, or in which, consciousness of the *puruṣas* becomes reflected in individual sentient beings (see Figure 2.3a).

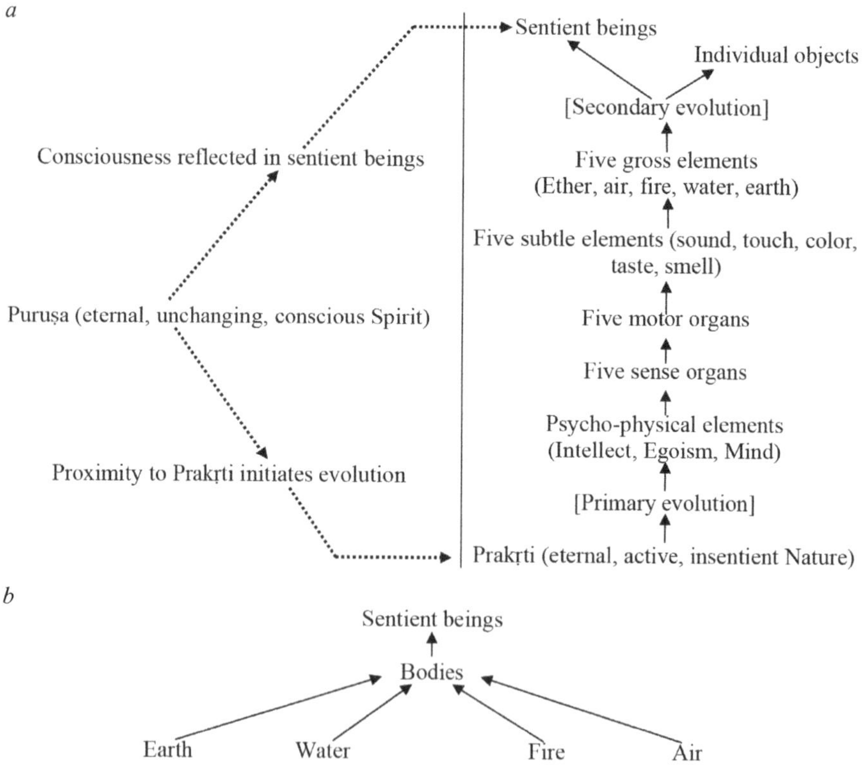

Figure 2.3 Diagrams of ascending evolutionism

a Dualistic, ascending evolutionary scheme of the Sāṃkhya.

b The mixed horizontal and ascending evolutionary scheme of the Cārvāka

Sāṃkhya intriguingly retains a teleological perspective regarding Nature by assuming that the purpose of material evolution is to provide souls with experiences of pleasure and pain leading eventually to liberation from the cycle of rebirth. But the *puruṣas* are not the contriver of the body or the world, for that is the work of Prakṛti. The Sāṃkhya teleological argument is used rather to prove the existence of the soul that is the *beneficiary* of design. As Mysore Hiriyanna notes: "The Sāṃkhya concludes from the presence in nature of means adapted to the accomplishment of ends, not to God as their author, but to the self for whom it supposes them to exist. It thus accepts design, but denies a conscious designer" (Hiriyanna 1949: 124–5).[4] In different words, Sāṃkhya accepts guided evolution, but denies a conscious guide.

The evolution of Primordial Nature into the universe is initiated by the disequilibrium of the three constituent factors of Prakṛti known as *guṇas*. Exactly how the *guṇas* become disequilibriated is not made clear, although it is suggested that the proximity of the unchanging Puruṣa is the cause, like a non-moving magnet attracting iron filings. The early Sāṃkhya seems to have assumed that Prakṛti "operates of itself (*svabhāvataḥ*)" and thus could bring about of its own accord the evolved universe (Frauwallner 1973: vol. 1, p. 318; cf. vol. 2, p. 61). A similar view regarding the innate capacity of matter to evolve of itself is found in the *Mahābhārata* (12.215.15–17), where the demon Prahrāda declares that all sorts of entities roll forth into existence and return by inherent nature (*svabhāva*), without any intention or supervision on the part of persons (*puruṣārtha*), suggestive of an evolutionary naturalism.

The evolutionary naturalism of the Cārvāka/Lokāyata

The Materialists, Cārvākas or Lokāyatas, are worth our serious consideration due to their importance in the development of the "orthodox" schools. As Halbfass notes, "although they [the Cārvākas] were rejected, they were part of traditional Hindu self-understanding" (1988: 361). We shall see, for instance, the significant role Cārvākas played, as foils, in articulations of design arguments in both Advaita Vedānta and Nyāya-Vaiśeṣika. And as the Nobel-prize-winning economist Amartya Sen has observed, "A defeated argument that refuses to be obliterated can remain very alive" (2005: 6).

Regarding the historical roots of Materialism, Hiriyanna observes that in the *Mahābhārata* we find a proto-Sāṃkhyan trend of thought that, while recognizing permanent souls, nonetheless resembles the doctrine of "naturalism" or *svabhāva-vāda* "in endowing matter with practically all the power necessary to unfold the whole universe out of itself" (1932: 106). Other thinkers dispensed with the idea of souls altogether. One of the earliest descriptions of this viewpoint is found in Buddhist literature, which summarizes the teaching of one Ajita Keśakambalī, a contemporary of the Buddha. Ajita denies that there is any afterlife and thus any karmic effects extending beyond this life. Human beings do not spring into life spontaneously but only from mothers and fathers. Humans are constituted of the four elements earth, water, fire, and air, and, upon death, the various earthy, fluid, warm, and breathy aspects of persons dissolve back into their respective elements. Their senses and mind dissipate into space, and thus they become completely annihilated (Davids 1990: 48). While the elements are listed in a horizontal fashion, collectively they are capable of evolutionary transformations, bringing about even consciousness from mere matter.

Similar ideas are put forth in the *Mahābhārata*. The sage Ājagara, for instance, declares that "the origin, growth, decay, and death of beings is without efficient or external cause (*animitta*)," as they "arise and roll forward just by their inherent nature (*svabhāva*) on which they depend" (12.172.10–11). Also in the *Mahābhārata* we occasionally find ascending evolutionary schemes where the elements by themselves evolve one from another. The sage Bhṛgu relates an ancient, divinely revealed cosmogonic theory according to which, long ago, before there were stars, moon, sun, or wind, there was only silent space. Then arose water, and in a complicated evolutionary succession, wind, air, fire, and finally earth, in or from which all kinds of animals were born (12.176.9–17).[5] Such materialist views in the epic are often expressed not by sages but by demons. Thus, the demon Prahrāda declares that all sorts of entities roll forth into existence and return by inherent nature (*svabhāva*), without any intention or supervision on the part of persons (12.215.15–17).

Given the frequent occurrence of the term *svabhāva* in early Indian cosmogonic speculations, we need to clarify and distinguish its various meanings. We have seen it applied to God, who is thus freed from any necessity of purpose or design, since it is simply his inherent nature to create. We have also noted its frequent ascription to matter or its elements, suggestive of a naturalistic process of cosmogenesis. In this materialist context two rather distinct interpretations of *svabhāva* emerge.

The pronominal adjective *sva* denotes what belongs to oneself or to an entity, what is innate or peculiar to it, while *bhāva* may designate something existing, a state of being, and an innate property or inherent nature—overlapping in this sense with *sva*. Accordingly, the compound *svabhāva* invokes the idea of an inborn essence or identity peculiar to an entity, eternal and unchanging (see Halbfass 1988: 334, 337; 1992: 54). This essentialist interpretation of *svabhāva* works against an evolutionary perspective inasmuch as individual entities, including species, are regarded as unchanging ideal types with unique characteristics (and leaving open the possibility that a creator god instilled these characteristics into every entity from the beginning). Indeed, twentieth-century Hindu critiques of Darwinism, as we shall see, invoke the idea of the inherent nature of species.[6]

But *bhāva* also has a dynamic sense, of becoming. In this case, *svabhāva* takes on the meaning of self-becoming and self-transformation, and when applied to the universe or its physical elements, indicates the inherent capacity of the world or nature to transform itself. The claim that nature is endowed with the power of self-transformation, along with the assertion that matter is the sole reality, constitutes what Kiwal Krishan Mittal (1974: 16) characterizes as "evolutionary naturalism."

For a fuller understanding of evolutionary naturalism in the ancient Indian context, we need to turn back to the *Śvetāśvatara*'s list (in 1.2) of naturalistic causes of the universe, and to consider what distinguishes the doctrine of inherent nature (*svabhāva*) from necessity (*niyati*) and chance (*yadṛcchā*). *Niyati*, derived from the verb *niyam*, to control, restrain, regulate and direct, denotes not only the idea of a fixed and inevitable order—and thus is equivalent to *daiva* or fate—but also the notion that this order is guided towards some preordained end. *Niyati* in both its deterministic and teleological aspects thereby contrasts with the idea of chance or accidentalism (*yadṛcchā*; also known as *animitta* and *ākasmika*). The *svabhāva-vāda* agrees with accidentalism in rejecting any notion that nature reveals some underlying divine power or intervening supernatural being, but rejects the idea that the world is lawless, the product of mere accident or chance. The world "is self-determined, not undetermined" (Hiriyanna 1932: 104). As Mittal summarizes the distinguishing characteristics of the *svabhāva* doctrine, which he identifies with Materialism: "Rejecting both teleology (*niyati*) and the rule of chance (*Yadṛcchā*) and insisting on its second fundamental dogma,[7] 'the reign of Law,' materialism adopts 'Evolutionary-Naturalism' (*svabhāvavāda*) as its cosmological theory" (1974: 16; cf. 52).[8]

One of the most useful surviving synopses of the Cārvāka is found in the *Sarva-Darśana-Saṅgraha*, composed by the fourteenth century philosopher Mādhava, an Advaita Vedāntin. Let us look in some detail at Mādhava's summary to clarify the basic teachings of the Cārvāka, keeping in mind that Mādhava's presentation in the final analysis is that of an opponent of the system.

In his *Sarva-Darśana-Saṅgraha* Mādhava summarizes all the major philosophical schools, or *darśanas*, of his time. He presents them in hierarchical order, proceeding from the lowly Cārvākas, or Lokāyatas as he also calls them, followed by the other "heretical" schools of Buddhists and Jains, up to the apex of Advaita. Despite his overall Advaitic agenda, Mādhava treats each school relatively sympathetically in its own right, its alleged weaknesses brought out in discussion of the following school. He sees the Materialist rejection of the notion that final happiness comes from the Supreme Lord as hard to eradicate, given the popularity of hedonistic inclinations among all sentient beings. He introduces the alleged founder of the Materialist school, Cārvāka, as the "crown-jewel of the naysayers (*nāstikas*)" (Mādhava 1981: 3). In general, the term *nāstika* refers to schools that deny the Veda and Vedic way of life, including Buddhism and Jainism. The Cārvākas, according to Mādhava, fit this definition perfectly, for they disparage the Vedas as the "incoherent prattling of rogues" and Vedic sacrifices as simply a means of livelihood for the priests lacking intelligence and manliness (ibid.: 5–6; cf. 11–12).

The dissenting Cārvākas go much further than Buddhists and Jains in denying all things religious—thus earning their place at the bottom of Mādhava's *darśana* ladder. They first deny any independent existence of the soul apart from the body and its material elements: "The four elements beginning with earth are the sole primary substances (*tattvas*), from which alone, when transformed (*pariṇata*) into a unified body, consciousness (*caitanya*) arises (*upajāyate*)" (ibid.: 3) (see Figure 2.3b). The Cārvākas thus subscribe to the same notion of evolutionary transformation (*pariṇāma*) taught by the Sāṃkhya, but unlike the latter include consciousness itself as an evolute of matter. The Cārvākas liken the emergent evolution of consciousness from the four elements (earth, water, fire, and air) to the production of intoxicating powers in sugary liquids by the process of fermentation.[9] The Cārvākas next deny any other-worldly reward, for "when these elements constituting the body become disrupted, then consciousness itself perishes" (ibid.). At this point, Mādhava has the Cārvākas cite a supposedly confirming passage from the *Bṛhadāraṇyaka Upaniṣad* (2.4.12): "Arising as a mass of understanding (*vijñāna*) from just these elements, one vanishes back into them; once dead, there is no more consciousness (*saṃjñā*)." The Cārvāka conclusion follows immediately: "Therefore the soul (*ātman*) is simply the body characterized by consciousness (*caitanya*), for there is no evidence (*pramāṇa*) supporting the existence of the soul separate from the body" (Mādhava 1981: 4).

Mādhava portrays the Cārvākas as selfishly hedonistic, seeking pleasure (*sukha*), the sole end of man (*puruṣārtha*), in the embrace of women and the like. Such disparagement seems unjustified, an easy way of smearing the Cārvāka viewpoint as a whole (Mittal 1974: 54–60). Much of the acrimony against the Cārvākas' alleged immorality stems from their denial of karma, the unseen moral force (*adṛṣṭa*) of merit and demerit overseeing the process of rebirth. Without the restraining influence of karma, so the complaint goes, how can one be moral? The Cārvākas reject karma on the basis that it cannot be perceived, only inferred (Mādhava 1981: 10-11). They are frequently criticized for disallowing all means of knowledge (*pramāṇas*) except for perception. But as Hiriyanna points out, the Cārvākas seem only to have rejected inferences thought by others to establish the supposed existence of such non-empirical entities as God and the afterlife (1932: 188). The Cārvāka attitudes towards inference and their rejection of Vedic revelation as mere prattling serve to highlight epistemological issues that are crucial to major themes of this book.

The Cārvāka denial of karma elicits the objection that diversity in the world cannot then be explained, for diversity in the absence of the unseen force would be accidental or uncaused (*ākasmika*). Mādhava immediately presents the Cārvāka rejoinder: "That is not correct, for diversity can arise by inherent nature (*svabhāva*) alone" (Mādhava 1981: 11). This denial of accidentalism is supported by the maxim: "Fire is hot, water cold, and wind induces a middling sensation; by whom has this variety been produced? It has arisen from the inherent nature (*svabhāva*) of these things" (ibid.). The maxim, while emphasizing the idea of innate characteristics, does not deny the evolutionary aspects of *svabhāva*, for as Mādhava stated earlier, the Cārvāka regards the four elements as a group, transformed into the body, as having the inherent capacity to produce consciousness. *Svabhāva* and *pariṇāma* are complementary. The basic point of the maxim is to affirm causation through properties within matter itself, not by unseen causes like karma. Accordingly, the Cārvāka as a whole represents a strong form of naturalistic ascending evolutionism.

Aside from its importance for the classical schools, the Cārvāka perspective is especially significant in the modern context as it readily accommodates a Darwinian or Neo-Darwinian[10] understanding of the universe with little or no need of theological or philosophical negotiation. As N. S. Dravid notes, the

Cārvāka perspective "is more or less the modern scientific view of the purely physico-chemical structure of the human body" (see Dravid's commentary in Udayana 1996: 78). Other versions of the self-transformative principle—including less vigorous forms of evolutionary materialism like the Sāṃkhya—as well as the theistic-mechanistic viewpoint, however much refined and updated in modern times, have inevitably encountered tensions in grappling with a robust Neo-Darwinian perspective, tensions that may not be capable of resolution. The Cārvāka provides an important alternative to such views when it is not misleadingly characterized as exclusively accidentalist or determinist, embracing as far as we can ascertain from the extant texts a combination of chance and necessity, to use Jacques Monod's famous characterization of Darwinian evolution. The ancient Cārvākas have apparently received the same groundless critique as modern evolutionists when spuriously credited with the preaching of accidentalism in a universe that is clearly not merely accidental in its order and structure.

The three cosmological models and three classical thinkers

We now turn to three classical thinkers, Śaṅkara (eighth century CE), Udayana (tenth–eleventh century CE), and Rāmānuja (eleventh–twelfth century CE), who present us with detailed arguments for and against the three cosmological models in their various versions outlined above. Theistic creationism, with its promotion of the design argument, is presented in all three of these classical philosophers, although not with equal exuberance.

Śaṅkara, representative of Advaita Vedānta, espouses the extreme, idealist form of descending evolutionism (*vivarta-vāda*), in which the eternal, unitary, and unchanging Brahman appears to become the manifold and changing world of everyday experience. Yet in dealing with this everyday level of reality, however illusory or misleading it may be, Śaṅkara sees fit to introduce the design argument as a means to counter rival schools, especially the non-theistic Sāṃkhya and the early Vaiśeṣika or Atomist school, also non-theistic, as they fail, in his view, to account for the wondrous order and arrangement seen in the universe. Most importantly, in Śaṅkara's view, these other schools fail to grasp the ultimate oneness of reality. When difficult questions arise within Śaṅkara's discussion of theistic design, for instance regarding the problem of evil in the world, we find him taking final recourse to his non-theistic idealism in which the world is ultimately only an appearance.

Given Śaṅkara's thorough treatment of the Sāṃkhya model of dualistic, ascending evolutionism, I will incorporate discussion of the Sāṃkhya cosmological scheme into my examination of Śaṅkara's critique of it. The Sāṃkhya, while eventually losing its vitality as an independent school (Larson 1969: 163), remains important as its model of evolution was largely subsumed in and adapted to later monistic and theistic approaches, and it continues to function in today's debates about evolution and creation.

Udayana, a philosopher of the late theistic Vaiśeṣika conjoined with the Nyāya school of Logic, is a theistic creationist to the core. Taking into account the sorts of critiques of non-theistic schools offered by Śaṅkara, Udayana presents within the Hindu tradition the most elaborated rational arguments for the existence of God, utilizing various cosmological and teleological approaches. In the process he takes aim at such non-theistic and atheistic schools as the Buddhists, Jains, and Materialists. We shall also consider other Nyāya-Vaiśeṣika theologians, whose views anticipate or complement Udayana's.

Rāmānuja, a theologian of the Viśiṣṭādvaita, a theistic school of Vedānta opposed to the monistic idealism of Śaṅkara, follows Śaṅkara in utilizing the design argument against non-theistic opponents. But ultimately Rāmānuja is quite uncomfortable with such a rational approach to God and provides a thorough critique of all notions of divine design that occasionally reverberate in more recent theological discussions.

In dealing with these three thinkers we shall pay close attention not only to their arguments regarding design and evolution, including the implications of their traditional cosmological views for Darwinian evolution, but also to their underlying epistemological assumptions. For in the end, any resolution of the current tensions existing between Dharmic and Darwinian viewpoints will involve a thorough grappling with the traditional epistemological perspectives that have continued into the modern period and that underlie the ongoing controversies.

3 Manifestation and apparent design in the Advaita Vedānta of Śaṅkara

The Advaita Vedāntin philosopher Śaṅkara (ca. 700–750 CE) presents an extreme, idealist form of descending evolutionism, in which the material realm is merely an apparent manifestation (*vivarta*) of consciousness.[1] Ultimately, for Śaṅkara, there is only one reality, the unitary or non-dual (*advaita*) and unchanging consciousness, Brahman, identical with the innermost essence (Ātman) of all sentient beings. The physical cosmos, despite its seeming multiplicity and constant change, is not different from Brahman. It is like the false impression of a snake superimposed on a rope, to use a renowned Advaita analogy. The world is a modification of Brahman but only in the sense that the diverse objects of the world are figments of nescience or imagination.

As Brahman is not a sensible object, it cannot be seen, and, prior to realization of one's identity with it, can only be known from the scriptures of Vedānta. These include the Upaniṣads, the *Bhagavad-Gītā*, and the *Brahma-Sūtras* of Bādarāyaṇa (also known as *Vedānta-Sūtras*), a compendium of aphorisms (*sūtras*) systemizing the teachings of the Upaniṣads. But as Śaṅkara points out in his commentary (*bhāṣya*) on the *Brahma-Sūtras*, inference can be used to ascertain and confirm the meaning of the scriptures as long as it does not contradict them (Śaṅkara 1983a [*Brahmasutra Bhashya*, henceforth abbreviated *BSB*]: 1.1.2; Thibaut 1890: 17).[2]

Basing himself on these Vedānta texts, Śaṅkara affirms that Brahman is both the material and efficient cause of all that is. He explains the relation between Brahman and the world as that of gold to gold ornaments, or clay to earthen pots, citing well-known metaphors from the *Chāndogya Upaniṣad* (6.1.4–5). He adds that Brahman is not only the gold and clay, the material cause (*prakṛti*, *upadāna-kāraṇa*) of the universe, but also the goldsmith and potter, the efficient cause (*nimitta-kāraṇa*) (*BSB* 1.4.23; Thibaut 1890: 283–6).

Given that the metaphor of artisan and artifact underlies the design argument and presupposes a dualistic model of theistic creationism, we may ask how it can be reconciled with Śaṅkara's descending evolutionism. Such devolution is better suggested by another Upaniṣadic analogy that Śaṅkara soon provides (*BSB* 1.4.27; Thibaut 1890: 288), that of a spider emitting and retracting its threads (*Muṇḍaka Upaniṣad* 1.1.7). This view accords with his understanding of *Bhagavad-Gītā* 2.16, which affirms that what truly exists cannot arise from non-existence, or restated, that effects are not different from their cause (*satkārya-vāda*). Thus, the world is not different from Brahman.

But both the artisan and spider metaphors imply that the universe is fully real, not merely an appearance reflected in or on Brahman. Later in his commentary (*BSB*, introductory comment to 2.1.1; Thibaut 1890: 290), Śaṅkara proposes yet another analogy for Brahman's relation to the world: Brahman, in the guise of the omniscient Lord of all (*sarvajña sarveśvara*), is like a magician (*māyāvī*) who through his mysterious creative energy, Māyā, maintains the illusion of the world. The magician simile suggests that the world is indeed merely an illusory manifestation.

Analogies, of course, are limited and should not be taken too literally. Thus, while the metaphors of artistic production, emission-retraction, and magical deception seem to imply very different cosmological models, Śaṅkara presents them as simple illustrations of the three phases of the cosmic cycle. The artisan represents the creation of the world, the magician sustaining his illusion symbolizes the maintenance of the world, and the spider reabsorbing all it emits embodies dissolution.[3]

The correlation of the three types of analogies with the three phases of the cosmic cycle, however, is only one way to relate them. The simile of the magician immediately evokes the famous Advaita idea of two levels of knowledge, the higher level of ultimate truth and the lower level of mundane experience. Herein lies the resolution to the apparent contradictions in the notion of Brahman as both changeless, on one hand, and as efficient and material cause of the world, on the other. On the highest level of truth, this world is illusory, an apparent manifestation only. On the lower, the empirical world can reasonably be interpreted as a supposedly real devolution from the Ultimate, or alternatively as the product of design of an overseeing, creative Lord, Īśvara, himself a manifestation of Brahman's magical power of Māyā. Accordingly, as Ninian Smart concludes, "natural theology, inappropriate at the higher level, reappears at the lower" (1964: 100).

We may still ask why Śaṅkara should concern himself with natural theology at all, since he clearly subordinates inference and other forms of reasoning to scripture, the ultimate authority. The answer is that not everyone agrees about the meaning of scripture, and some reject scripture altogether. In such cases, recourse to reasoning becomes requisite, for reasoning can assist seekers of the truth who might otherwise be misled by false (non-Advaitin) interpretations of scripture, thereby failing to realize their identity with Brahman. We shall begin with Śaṅkara's presentation of the design argument in his *Brahmasūtra Bhāṣya* and subsequently turn to his cosmological argument, as we see these employed against the Sāṃkhya. We shall then briefly summarize his parallel refutations of the non-theistic Vaiśeṣika.[4]

The design argument against the Sāṃkhya

In his commentary on the second aphorism of the *Brahma-Sūtras*, Śaṅkara introduces the design argument in elucidating Bādarāyaṇa's cryptic allusion to Brahman as "that from which origin and so forth" arise (*BSB* 1.1.2; Thibaut 1890: 15–16). Śaṅkara clarifies that the elliptical phrase, "origin and so forth," refers to the creation, maintenance, and destruction of the universe from Brahman.[5] He then explains that the world serves as a realm for diverse agents and enjoyers to experience the fruits of action according to definite places, times, and causes. This ever-cycling universe, he concludes, "is produced by the omniscient, omnipotent Brahman, and is characterized by its arrangement (*racanā*) transcending comprehension." Śaṅkara's use of the term *racanā*, arrangement or design, presents a teleological understanding of the natural world that encompasses both its wondrous physical organization and its interlocked karmic or moral order.

Without a superintending, omniscient, and omnipotent Lord, Śaṅkara adds, the production of the cosmic cycle would not be possible. He discounts all non-intelligent causes, such as the insentient Primordial Matter (*pradhāna*) of the Sāṃkhya, the atoms (*aṇu*) of the Vaiśeṣika, the void (*abhāva*) of the Buddhists, and the world's inherent nature (*svabhāva*) of the Materialists, as well as any being subject to *saṃsāra*. None of these is able to take into account the places and times crucial for karmic compensation, nor to employ the causes necessary to produce the universe.[6]

Śaṅkara for the moment offers little further rational argumentation in support of the design argument and his teleological claims. In this part of his commentary he focuses on establishing the Advaita view as the single, correct meaning of the texts, namely that Brahman is the one efficient and material cause of the world, and the goal of knowledge. In the process, Śaṅkara disputes alternative interpretations of the texts such as those of the Materialists, Buddhists, and Sāṃkhyas. Having argued against his opponents on the basis of scripture in the first part of his commentary, with only brief allusions to notions of design, he later turns to rational refutations, with less reliance on scriptural references.[7]

Rational argumentation against the false teachings of other schools is not simply a philosophical exercise for Śaṅkara, since misunderstanding of Brahman and its relation to the world leads to continued bondage. The Sāṃkhya, in particular, is troubling for Śaṅkara, since its advocacy of the idea that cause and effect are not different (*kārya-kāraṇānanya*)[8] resembles the Advaita view that Brahman and the world are not different, despite what he considers the school's pernicious spirit–matter dualism (*BSB* 1.4.28; Thibaut 1890: 288–9). Thus he finds it incumbent upon expositors of the true meaning of the Upaniṣads to combat the Sāṃkhya doctrine and others like it. He begins his logical refutation of opposing views, then, with the Sāṃkhya.[9]

Early in the second division of the *Brahma-Sūtras* Bādarāyaṇa states in his usual elliptical manner, "And the inference fails because of the existence of design (*racanā*)" (*Brahma-Sūtras* 2.2.1). Bādarāyaṇa alludes here to the Sāṃkhya doctrine that Primordial Matter, Pradhāna, is the cause of the world, which he rejects as it cannot explain the design manifest in the universe. As Śaṅkara explains, the Sāṃ khya relies on a faulty analogical inference that the world as an effect must have as its source Primordial Matter, since they share the same qualities (*guṇas*). Just as earthen pots and dishes have the common qualities of their material source, clay, so the world, endowed with the qualities of pleasure (*sukha*), pain (*duḥkha*), and dullness or delusion (*moha*), shares these with and derives them from its source, the Pradhāna.[10] Śaṅkara then concludes his summary of the Sāṃkhya: "like clay, the non-conscious Pradhāna, in order to facilitate the purpose of the conscious soul, just by its own nature (*svabhāva*) begins transforming itself in various ways" (*BSB* 2.2.1; Thibaut 1890: 364).

The whole Sāṃkhya argument rests on the sound idea that cause and effect are not different, but according to Śaṅkara, the Sāṃkhya misapply it. He specifically calls attention to the incompleteness of the Sāṃkhya presentation of the clay analogy, which leaves out any reference to the potter. His critique is worth quoting at length, as it provides one of the earliest and clearest statements of a Hindu version of the design argument:

> If this issue [regarding the cause of the world] is to be resolved by force of analogy alone, then consider that in ordinary life no non-intelligent entity is observed to produce modifications suitable for satisfying the purposes of some particular person, by itself, without being superintended by an intelligent agent. In ordinary life what we do see is that houses, palaces, couches, seats, pleasure-gardens and the like, which are useful for obtaining pleasure and avoiding pain at appropriate times, are constructed by intelligent craftsmen. In like manner, observe that this entire universe, externally consisting of the earth and other elements, is suitable for experiencing the fruits of various acts. With regard to the self, starting with its physical body whose various forms are fixed by birth and whose limbs have a regular arrangement, observe that such body is an abode for the soul to experience the fruits of diverse acts. Since even the most competent craftsmen cannot comprehend [the world's construction], how could the non-intelligent Material Nature (*pradhāna*) devise it? In the case of such things as a lump of earth or a stone, no (power of contrivance) is seen, but the design (*racanā*) of special forms out of such things as clay is seen when they are superintended by potters and the like. In the same way, Material Nature transforms itself only when connected with a superintending, external intelligence. And in ascertaining the essential nature of the root cause of the world, it is not appropriate to note only the material causality of such things as clay while ignoring external agents like potters. It thus follows that there is some guiding ruler. Nor is this view at all inconsistent with, but rather it conforms to scripture which designates the cause as intelligent. Therefore, since the design of the world is otherwise inexplicable, its cause is not to be inferred as non-intelligent.
>
> (*BSB* 2.2.1; Thibaut 1890: 364–6)

Worth noting in Śaṅkara's account above is the pervasive concern with karmic fruits. The universe as a whole is the stage upon which sentient beings reap the consequences of their former acts. Organic bodies are designed with regular arrangement of limbs in order for embodied souls to undergo the pleasures and pains found in the world. Śaṅkara pays little attention to the ordered and complex mechanical aspects of the cosmos or of the anatomical body that so intrigue Western natural theologians. He emphasizes not adaptation of organisms to the environment for survival, but adaptation of sentient beings to experience pain and pleasure, in order to benefit from karmic opportunities for spiritual progress.

Śaṅkara was aware of a number of objections that could be brought against his design argument. He anticipated these critiques in several passages prior to refuting the Sāṃkhya's notion of a self-transforming Primordial Matter. These include the problems of an artisan lacking tools and a body, of a creator acting without purpose, of a creator who appears partial and cruel, and of a creator who is not just sadistic but masochistic.

With regard to artisans lacking tools, Śaṅkara argues that Brahman differs from such human agents as potters, who need wheels and clay in order to produce their goods. He illustrates his claim by noting the self-transformation of milk turning into curds and water into ice, thereby avoiding for the moment dealing directly with the problem of the artisan analogy (*BSB* 2.1.24; Thibaut 1890: 346–7). Turning from non-sentient entities (like milk) to intelligent beings, Śaṅkara merely insists that, while normal human craftsmen require instruments, this is not true of Brahman and other beings endowed with supernatural powers, like the gods and holy seers, who produce mansions and chariots by mere desire (*BSB* 2.1.25; Thibaut 1890: 347–9).

As for the Supreme Being creating without having a body, Śaṅkara defers to scriptural authority. Scripture confirms that while Brahman has no eyes, ears, or mind (*Bṛhadāraṇyaka Upaniṣad* 3.8.8), it also indicates that Brahman can see without eyes, grasp without hands, and the like (*Śvetāśvatara Upaniṣad* 3.18). Thus, unlike gods and other supernaturally powerful beings who need bodies to create, Brahman requires no organs or limbs to produce the universe, although his ability to do so transcends reasoning (*BSB* 2.1.31; Thibaut 1890: 355–6). Śaṅkara's deference to scriptural affirmations of God's mysterious and inherent powers when he comes to difficult challenges, needless to say, significantly weakens the force of his design argument, utilizing it simply as an expedient means when arguing against certain opponents.

More problematic for Śaṅkara is the matter of a creator acting without purpose, for Brahman, being complete and perfect, can have no desire to attain or accomplish anything. But in that case, why does Brahman bother with manifesting as Īśvara and creating the world? Śaṅkara has recourse to two interrelated ideas: play (*līlā*) and inherent power (*svabhāva*). He explains:

> Just as in common earthly life a king or his minister, with all desires fulfilled and without any goal or purpose, may engage in sportive activities amidst pleasure grounds (*krīḍā-vihāra*), and just as exhalation and inhalation transpire without reference to any external goal or purpose but by innate propensity alone, in the same way the Lord, without reference to any other, external goal, by innate propensity alone, may engage in sportive activity.
>
> (*BSB* 2.1.33; Thibaut 1890: 356–7)[11]

We may recall Śaṅkara's inclusion of a pleasure-garden (*vihāra-bhūmi*) among examples of consciously constructed creations in his design argument. The term *vihāra* refers to recreation, strolling for pleasure, as well as to a pleasure-garden, thus evoking not only the idea of karmic enjoyment, but also of Brahman's purposeless cosmic sport. And we have already met with the "inherent-propensity" defense of God's purposeless creative activity in the *Śvetāśvatara Upaniṣad* (6.8), which Śaṅkara cites in defending Brahman's self-transformative power (*BSB* 2.1.24; Thibaut 1890: 346–7).[12]

As in the West, the problem of reconciling a benevolent creator with the suffering in the world has exercised the best of theological and philosophical minds in India. In Śaṅkara's case, we may ask if Brahman's cosmic play is simply a cruel and bloody sport, given the violence and destruction affecting all sentient beings sooner or later. Śaṅkara tackles the question following Bādarāyaṇa's elliptical statement that "injustice (*vaiṣamya*) and cruelty (*nairghṛṇya*) cannot be [attributed to the Lord] because of his regard" (*Brahma-Sūtras* 2.1.34). Regard for what? As Śaṅkara explains, it is regard for the merits and demerits of individual beings. Karma thus accounts for the joyous lives of gods, the miserable lives of animals, and the intermediate lives of humans, just as rain causes all seeds to sprout, but the diverse potential of seeds of various kinds, not the rain-god, causes the differences in the plants. To be sure, if at the beginning of *saṃsāra* all individuals were equal, the Lord would be guilty of unfair discrimination. But *saṃsāra* has no beginning. The karmic cycle likewise is without beginning, resembling the endless vegetative cycle of seed and sprout (*BSB* 2.1.35–6; Thibaut 1890: 359–61).

The idea of the Lord having to take into account merit and demerit suggests a significant restriction on his freedom. Śaṅkara simply affirms that we know the Lord is bound to observe karmic fruits on the basis of scriptural authority. He cites among various confirming passages the following statement from the *Kauśītaki Upaniṣad* (3.8): "The Lord causes a person he wishes to lead upwards to act righteously, and one he wishes to fall backwards to act unrighteously." This raises a puzzling question, for it seems to place at least some responsibility for karmic destiny back on the Lord.

However convincingly a karmic theodicy may justify the lack of cruelty on the part of the Lord in dispensing unequal fruits to diverse souls, there remains a deeper problem: the apparent masochistic proclivity of Brahman to inflict self-injury, given Śaṅkara's insistence on the non-difference between Brahman (or supreme Self) and embodied souls.[13] For as the *Chāndogya Upaniṣad* proclaims, Brahman entered into his creation by means of the individual soul (*jīva*) (6.3.2), and also "That Ātman [Self, i.e., Brahman], that you are (*tat tvam asi*)" (6.8.7). Surely, a free and independent Brahman, as Śaṅkara admits, would

> create only what is beneficial for itself, delighting the mind, and not what is detrimental such as the myriad assortments of senseless ills like birth, death, old age, and disease. For no person, not coerced by another, would make a prison for himself and then enter it. Nor would he, being perfectly stainless, acknowledge on his own this wholly unclean body as himself. Also, in every way he would abandon by his desire whatever he has created that produces misery, appropriating for himself only the pleasant.
>
> (*BSB* 2.1.21; Thibaut 1890: 343–4)

And, remembering that he himself has spread out the illusion of the world like a magician, would not the Self reabsorb that illusion whenever he wished? Yet we know that the embodied Self cannot readily absorb even its own body.

To resolve the problem, Śaṅkara invokes the idea of the lower and higher truths. When the misapprehension of the manifest world is dissipated and the truth of non-difference is realized, then the whole of *saṃsāra*, including all transmigrating beings, vanishes. He concludes: "Under these circumstances, what shall we make of creation? Or what shall we make of such faults as not doing what is beneficial?" (*BSB* 2.1.22; Thibaut 1890: 345). For Brahman, then, there is ultimately nothing beneficial to be achieved and nothing harmful to be avoided. Suffering is reduced to a magical show, and like the cosmos, is apparent only. The idea of cosmic design, then, is equally a delusion. But it is a helpful stepping stone to the higher truth, especially in avoiding the snares of false doctrines like the Sāṃkhya. So let us turn back to the lower level of empirical reality for a moment and consider the second major argument that Śaṅkara wields against the Sāṃkhya, the cosmological.

The cosmological argument against the Sāṃkhya

The design argument is teleological, concerning itself with the ultimate end or purpose (telos) of the universe, its final cause. The complementary cosmological argument turns to the initial cause of the universe—which in the Indian context does not presuppose an absolute beginning of the cosmos but only ever repeating re-creations.[14] The basic assumption of the cosmological argument is that a living, conscious being is necessary to initiate motion. In the West, this assumption is clearly evident in Plato's assertion that anything that moves of itself is alive, and that a living entity, the soul, "by her own motions stirs all things in sky, earth, or sea" (Plato 1961: 1452). The same assumption inspires Śaṅkara's response to the Sāṃkhya.

The complementary nature of the teleological and cosmological arguments is evident in Śaṅkara's continuing refutation of the Sāṃkhya:

> Putting aside this matter of arrangement, consider the initiating activity by which this arrangement is accomplished. [The Sāṃkhya hold that] the three qualities of Primordial Matter,[15] departing from equilibrium and entering into an imbalanced state, commence activity intending to produce particular effects. But the non-intelligent Primordial Matter cannot by itself initiate activity, for this is not seen in such things as clay, or in chariots and the like. For we do not see such non-intelligent things as clay or chariots by themselves, when not directed by potters, horses and so forth, commencing activity intending to produce particular effects.
>
> (*BSB* 2.2.2; Thibaut 1890: 367)

Śaṅkara then turns to a number of objections that the Sāṃkhya might raise. For our purposes, four are especially important. First, consciousness (*caitanya*) responsible for initiating activity is only seen where bodies are present and not where bodies are absent, so consciousness is an attribute of the body, just as the Materialists maintain. Thus initiating activity belongs to the body, not to an independent conscious principle. Śaṅkara in reply does not deny that activity is seen in insentient entities, but this does not mean, he argues, that an independent consciousness does not exist, just as fire independently exists, even if only seen when fuel is present (*BSB* 2.2.2; Thibaut 1890: 368). Later in his commentary (*BSB* 3.3.53–4; Thibaut 1904: 269–72), he addresses the same problem in more detail, noting the Materialist claim that consciousness arises from the material elements like intoxicating power from fermenting liquids. The body, the Materialists insist, is the conscious Self, for consciousness is seen only within bodies. Thus, all the defining characteristics of the Self, such as life, movement, and memory, as well as consciousness, being seen only in and not outside bodies, must not be different from the body. Śaṅkara counters that such characteristics do not belong essentially to the body, for they are absent in a dead body. Further, we cannot establish that they do not exist after the death of the body, for they may pass into another body. And just as lamps may assist consciousness to perceive objects, so the Self may use the body as a means of perception, although in dreaming not even the body is needed, as it lies motionless.

The second objection is that an entity of pure consciousness devoid of activity, as Śaṅkara claims Brahman to be, cannot initiate activity. Śaṅkara responds, utilizing the Sāṃkhya analogy of a non-moving magnet activating iron filings in explaining how the inactive Puruṣa causes the disequilibrium in Prakṛti, by mere proximity. In just such a manner, the unmoving Lord can move the universe.

Third, there are some material entities that are self-transformative, such as milk and water. Śaṅkara summarizes the Sāṃkhya challenge:

> Just as insentient milk by its inherent power (*svabhāva*) flows to nourish a calf, and insentient water by its inherent nature flows to serve humankind, so does the insentient Primordial Matter by its inherent power initiate activity for the sake of liberating Puruṣa.[16]
>
> (*BSB* 2.2.3; Thibaut 1890: 369)

Śaṅkara counters that both milk and water are activated by conscious beings: the milk by the desires of the cow and calf together, and the water, ultimately, by the will of the Lord—as scripture declares. We have seen earlier that Śaṅkara uses the examples of milk and water to justify the idea that Brahman, without the use of external instruments, can transform itself into the universe. The earlier argument, he now explains, relies merely on popular illustrations and does not represent the higher truth known from scripture (*BSB* 2.2.3; Thibaut 1890: 369–70). In any case, even if one accepted that Primordial Matter and its abiding qualities are inherently capable of self-transforming activity, it is not possible to account for the Pradhāna sometimes evolving and sometimes not evolving the universe. He concludes: "But the alternating initiation and cessation of activity on the part of the Lord is not problematic, since he is endowed with omniscience, omnipotence, and has the great creative power of Māyā" (*BSB* 2.2.4; Thibaut 1890: 370-1). And if one remains unconvinced, Śaṅkara is willing to appeal once again to the design argument: however self-active Primordial Matter may be, without the faculty of consciousness it cannot account for the arrangement of the universe (*BSB* 2.2.9; Thibaut 1890: 375).

Śaṅkara's refutation of the Vaiśeṣika

Śaṅkara's critique of the Vaiśeṣika focuses on the non-theistic teachings of Kaṇāda, legendary author of the *Vaiśeṣika-Sūtras* (ca. first centuries BCE). Śaṅkara's key points against Kaṇāda and his followers can be quickly summarized as they parallel his arguments against the Sāṃkhya. The eternal but insentient atoms (*paramāṇus*) posited by the Vaiśeṣika are just as incapable of deciding to re-initiate purposive, cosmic transformation following a period of dissolution as the Sāṃkhya's eternal but insentient Pradhāna with its qualities. Further, the eternal souls of the Vaiśeṣika during dissolution have no bodies and no capacity for will, and thus they cannot effect the joining together of atoms that must occur at the beginning of a new creation. If, as the Vaiśeṣika argue, the atoms are impelled into action by the "unseen force" (*adṛṣṭa*) that consists of the merits and demerits of sentient beings, this too creates irresolvable problems, due to the non-conscious nature of *adṛṣṭa* (*BSB* 2.2.11–14; Thibaut 1890: 381–91).[17]

Śaṅkara was aware that both the Sāṃkhya and the Vaiśeṣika schools had articulated theistic versions of their cosmologies, but he still had problems with them because they denied material causality on the part of God (*BSB* 2.2.37–8; Thibaut 1890: 434–7). That is, they argued for the eternal and independent existence of matter, however subtle or non-manifest its primordial form or forms, such as atoms, might be. To refute this position, Śaṅkara turns back to scripture, for as he and at least some of his opponents agree, "scripture alone is a valid means for providing knowledge of objects beyond the senses" (*BSB* 2.3.1; Thibaut 1904: 4). Śaṅkara argues that the four elemental kinds of atoms—earth, water, fire, and air—are traceable back to the fifth, non-atomic, and all-pervasive element, ether. And then, on the basis of such Upaniṣadic statements as "From the Self ether arises" (*Taittirīya Upaniṣad* 2.1.1), he concludes that all the elements of the Vaiśeṣika, whether in their gross, perceptible forms or their atomic forms, as well as the all-pervasive ether, ultimately are traceable back to Self, that is, to Brahman—a clear assertion of descending evolutionism. And at the time of dissolution, the gross, perceptible substances dissolve, not just back into different sorts of eternal atoms as the Atomists claim, but rather further dissolve, successively, back into each other (earth into water, water into fire, fire into air, and air into ether), and ether finally is absorbed back into Brahman, from which ether re-emerges at the time of a new creation (*BSB* 2.3.2–14; Thibaut 1904: 4–26). Thus, the whole of the universe is non-different from Brahman.

Conclusion

Śaṅkara's arguments in the second section of his *Brahmasūtra Bhāṣya*, while supposedly based on logical argument rather than texts, are "thoroughly informed by the upaniṣads" as interpreted within the Advaita world view (Clooney 1993: 111). The ideal of Brahman as efficient and material cause is solidly based in the Upaniṣads and is not a claim that can be established by rational argument alone. Later Advaitins, in accordance with Śaṅkara's basic scriptural perspective, fully acknowledge that rational arguments attempting to prove God's existence cannot give certain knowledge and thus are not true inferences. They can only establish the probability of God's existence, at best, and in any case, sole reliance on reason can mislead one into rejecting scripture, as happened with the Buddhists (Murty 1959: 142–3). And as David Hume points out with regard to the cosmological argument: "The beginning of motion in matter itself is as conceivable *a priori* as its communication from mind and intelligence" (1964: 426). Hume then provides a number of examples of self-activating matter: "Motion, in many instances, from gravity, from elasticity, from electricity, begins in matter, without any known voluntary agent; and to suppose always, in these cases, an unknown voluntary agent, is mere hypothesis" (ibid.).

We may also note that the Māyā doctrine, as creative energy and illusory force, is also rooted in the Upaniṣads rather than reason. Śaṅkara repeatedly utilizes aspects of the Māyā theory to dismiss resistant and troublesome logical arguments against his interpretation of Brahman as non-different from the world. Accordingly, his commentary is meant for insiders who seek the truth within Advaita, "not to persuade or convert outsiders" (Clooney 1993: 110). At the same time, Advaita "formalizes its position as propositional claims intended to refer to realities outside the Text" (ibid.: 117).

What then is the contemporary reader of the *Brahmasūtra Bhāṣya* to make of Śaṅkara's arguments, with their claims that not only aim to explicate the higher truth of Brahman, but that also impinge upon the lower truth of the empirical world? The issue, I suggest, is especially acute for a sympathetic reader not only of the Vedāntic texts but also the texts of modern science. A number of themes in Śaṅkara's Advaita, in particular, have relevance to contemporary Neo-Vedānta responses to and interpretations of Darwinian evolution, beginning with Swami Vivekananda in the late nineteenth century. Relevant themes include the notion that nothing comes from nothing, the unity of the universe, the karmic teleology of the manifest or empirical realm combined with the purposeless, playful character of the Ultimate, and the dismissal of the design argument as an inferior approach to the Ultimate, in favor of an apparent descending evolutionism. Such motifs in the modern period were developed to allow for an acceptance of organic evolution, but a very limited acceptance constrained by the requirements of Neo-Advaitin spiritual evolution.

Let us now turn to the theistic Atomism of the Nyāya-Vaiśeṣika, and to the natural theology of Udayana in particular, which "represents the acme of perfection attained by the Nyāya-Vaiśeṣikas in their rational search for God" and at the same time provides an overview of "Indian rational or natural theology in general" (Chemparathy 1972: 8).

4 Theism and atomistic design in the Nyāya-Vaiśeṣika of Udayana

Within the classical Hindu Dharma traditions, the joint school of the Nyāya-Vaiśeṣika offers the most elaborate rational arguments for the existence of God.[1] Yet the founding texts for the two systems, the *Vaiśeṣika-Sūtras* (ca. first centuries BCE) attributed to the legendary figure, Kaṇāda, and the *Nyāya-Sūtras* (ca. first centuries CE) attributed to Gotama, were non-theistic.[2] Moreover, both the Nyāya and the Vaiśeṣika systems seem to have developed originally from materialistic speculations about the natural world, only later assimilating Upaniṣadic ideals of spiritual liberation (P.T. Raju 1985: 251).[3]

The inclusion of Upaniṣadic ideas is already found in the *Vaiśeṣika-* and *Nyāya-Sūtras*, which acknowledge and argue for the existence of souls, rebirth, karma, and liberation. For instance, in Gotama's *Nyāya-Sūtras* (3.1.19), the soul is said to exist "because the joy, fear, and sorrow manifest in new-born infants can only result from memory of what was previously experienced in former lives." The *Nyāya-Sūtras* (3.1.22) further specify that "the desire for a mother's milk is acquired from repeated feedings in prior lives."[4]

Subsequent Nyāya-Vaiśeṣika thinkers continued to develop arguments for the existence of souls and of karma, in part to combat the skepticism of the Materialists. But they also came to elaborate such arguments in defense of God's existence. One of the perennial objections to theistic creationism is that God is an unnecessary hypothesis, not needed to set things in motion since the universe itself may well be eternally self-activating. And during the earliest period of Atomism, as Ninian Smart notes, "such concepts as gravity (to explain the falling of bodies), unseen force (to explain the circulation of fluids in plants and so on), etc., indicated that no intrinsic difficulty was felt in assigning motion to inanimate objects quite independently of volitional activity" (1964: 156). Nonetheless, as noted in the last chapter, Śaṅkara criticized Kaṇāda's non-theistic Atomism for its inability to explain the origins of physical and moral order in the universe without reference to a superintending Lord. Śaṅkara specifically dismissed the invisible force (*adṛṣṭa*) of karma as an explanatory principle, since it is insentient and thus not up to the task of assigning to souls their appropriate embodied conditions.

Early theistic Nyāya-Vaiśeṣika philosophers like Uddyotakara (ca. sixth century CE) anticipated much of Śaṅkara's critique of non-theistic perspectives, including Kaṇāda's Atomism and the Sāṃkhyas' notion of self-evolving Primordial Matter.

We may recall the Sāṃkhyas' appeal to the example of insentient milk flowing from a cow to nourish her calf, and Śaṅkara's rejoinder that it is not the insentient milk by itself but the desires of the cow and calf together that cause the milk to flow. In a memorable variant of Śaṅkara's answer, Uddyotakara, after noting that milk does indeed flow from cows, exclaims: "But not from dead cows" (1985: 947).[5] Applying the dead-cow logic to "dead" or unconscious atoms, as well as to unconscious Nature and the insentient karmic force (*adṛṣṭa*), Uddyotakara argues: "Just as axes become active only when overseen by a wood-cutter, since they lack consciousness, in the same manner Primordial Matter, atoms, and karma, lacking consciousness, become active [only when superintended by a conscious agent]" (ibid.: 945).[6] A conscious, willing being like the Lord is thus required to activate atoms and set the cosmos in motion.[7]

We also noted earlier Śaṅkara's critique of the Sāṃkhyan notion of an ever active Pradhāna constantly creating, for that would prevent the alternating creation and dissolution of the universe. He thus posited a Lord who at times wills to create, and at other times wills to desist from such activity. But this simply transfers the problem of alternating activity and inactivity from Matter to God, for God's nature, including his inherent activity, is also, supposedly, eternal and unchanging. Śaṅkara's counter invoked the Lord's omnipotence, omniscience, and creative power of Māyā. Uddyotakara similarly notes the quandary of God's inherently active nature (*svābhāvika*) and the consequent apparent impossibility of his ever being inactive (ibid.: 949).[8] Uddyotakara appeals to God's discriminating intelligence (*buddhi*), but without reference to any mystical power of Māyā. Rather, for Uddyotakara, God must wait for the convergence of necessary causes and conditions before engaging his creative energies. Among these causes is the ripening of karmic fruits, whose distribution he oversees.[9]

Within Nyāya-Vaiśeṣika circles dissenting views on the nature of God gradually emerged, but Uddyotakara's ideas are representative and quite influential on later writers. For Uddyotakara the major characteristics of God include: (1) his omnipresence; (2) his everlastingness, as he passes unscathed through cosmic dissolutions; (3) his omniscience and permanent consciousness, in contrast to ordinary souls who are conscious only when conjoined with *manas* (mental instruments) and sense organs; (4) his unique competency to initiate atomic motion and combination at the beginning of a cosmic cycle; (5) his unique ability to manipulate matter without a body; and (6) his dependence on the merit and demerit of souls in creating bodies for diverse souls (cf. Potter 1977: 100–1).

By the time of Udayana (tenth–eleventh century CE), the reality of a transcendent moral order overseen by an omniscient God was regarded as largely self-evident among Nyāya-Vaiśeṣika thinkers. But the continuing skepticism of those like the Buddhists and Sāṃkhyas persisted, continuing to inspire rational justifications for belief in God of increasing sophistication. Udayana's *Nyāya-Kusumāñjali* (*Flower Offerings of Logical Arguments*) is considered to be "by general acclaim the definitive treatment of the question of how to prove God's existence" (ibid.: 7).

Arguments for the existence of God in the *Nyāya-Kusumāñjali*

Many of Udayana's proofs for the existence of God combine causal and teleological arguments, interfused with moral or karmic arguments. Two general varieties of these arguments are especially relevant with regard to their implications for modern evolutionary theory.[10] One focuses on the order and structure of the cosmos, from atoms to the universe as a whole, the other on society and culture. The former is what Karl H. Potter refers to as the "cosmoteleological argument," in which Udayana reasons "from the world as effect to God as the purposive agent which is its cause" (1977: 101). The latter, what I call the "socioteleological argument," argues from the existence of human culture, including language, customs, arts and crafts, religion, and specifically the Vedas, back to God as the primordial tutor.

Cosmoteleological arguments

Udayana's cosmoteleological arguments encompass atomistic, terrestrial, organic, and cosmic phenomena. Let us begin with his atomistic considerations, for in these the interweaving of the cosmological and teleological is most evident. We see this interweaving clearly in his summary introduction of the atomistic argument, which he calls "conjoining" (*āyojana*). Conjoining involves both the imparting of motion to atoms at the beginning of a cosmic cycle and the bringing together of the activated atoms in the right combinations. Udayana asserts: "If [atoms were] self-moving, they would not be insentient. The unseen force (*karma*) does not rule out other known causes. Effects require causes, and special effects undoubtedly have special causes" (Udayana 1996: 401).

Regarding the cosmological aspect of the conjoining argument, dealing with the first cause or first impeller of motion, Udayana does not totally deny the causative efficacy of karma, but sees it as needing direction from a superintending and intelligent agent. We noted earlier that Uddyotakara, with his dead-cow logic, denied the possibility of atoms or karma being self-activating, as both are insentient. Udayana repeats the latter's basic argument: "Atoms and the like commence motion only when directed by consciousness, for they are unconscious like an axe and so on" (ibid.).

The teleological aspect in Udayana's summary statement is indicated in his reference to special effects requiring special causes. An intelligent and purposive director of atoms is required to combine them in the appropriate way to produce the special effects we see in the world. Earlier in the *Kusumāñjali*, Udayana points out that the different kinds of invisible atoms (earth, water, fire, and air) with their eternal, individual qualities like color, produce specific kinds of gross substances (the visible gross elements of earth, water, fire, and air), with their appropriate qualities. Udayana assumes the standard Nyāya-Vaiśeṣika atomistic cosmogony, in which two atoms of the same type must combine to form an atomic dyad, and three such dyads must combine to form a triad, constituting the first gross or sensible elements. These triads can then be combined to form the various large bodies found in the universe. If in the initial stages unlike atoms or dyads are brought together, they would be unable to manifest definite qualities like color and taste, and thus would fail to serve the purpose of providing pleasurable and painful experiences to souls (Bhattacharyya 1961: 125–6). Accordingly, a knowledgeable director who can see atomic structures is indispensable for bringing about the required combinations to produce the diverse effects in the world in accord with the merits and demerits of all beings. Only an omniscient God can accomplish this. Otherwise, as Udayana concludes, the correspondence of qualities between subtle atomic elements and gross elements would be purely accidental (*ākasmika*) (1996: 370).[11]

Turning from atoms to the world, Udayana considers the idea of the Earth as an intended effect (*kārya*). The Sanskrit term *kārya* in this context means not only an effect, but also the idea of a purpose to be effected, something accomplished by a doer or creator, *kartṛ*. As Udayana succinctly summarizes the argument: "The Earth and such things are preceded by a creator (*kartṛ*), for they are intended effects" (1996: 369). The Earth is an intended effect for it is composed of parts, and thus must have a composer. During cosmic dissolutions, according to the Nyāya-Vaiśeṣika, no embodied beings survive, only the disembodied Lord with his knowledge of the fundamental structures of the universe. The creation of the world with its mountains and trees, which are intended effects, must then depend upon the eternal and omniscient Lord (Chemparathy 1972: 88).

Buddhist and Jain critics have frequently pointed out that the world with its mountains and oceans, whose parts exhibit no obvious signs of intentional design, differ from the clearly contrived nature of temples, water wells, or pots. The Nyāya-Vaiśeṣika counter that, although we see no immediate, embodied artisan in the case of mountains and oceans as we do with pots, God creates some things directly and invisibly, like the world, and others indirectly and visibly, like pots, using humans as co-creators.[12] As Udayana declares in another of his works, the *Kiraṇāvalī*:

> What becomes [constructed] in the absence of a [visible] embodied being is dependent upon a prior [invisible] intelligent causal agent, as it has the nature of possessing a cause. Whatever possesses a cause, like a chariot, is dependent upon a prior intelligent causal agent. Similar in character is this world; therefore it is dependent upon a prior intelligent causal agent.[13]

Furthermore, from the Nyāya-Vaiśeṣika perspective, mountains and oceans produce pleasure and pain, and thus their role in the distribution of karmic fruits requires that they are under intelligent supervision.

In any case, while mountains and oceans may not exhibit obvious signs of contrivance, organisms do in Udayana's estimation. In explaining the origin of species, he once again relies on an argument from karmic necessity, combined with the idea that the whole cosmos came into being more or less all at once at the beginning of the current cycle. The sexual reproduction of species we now observe going on around us is not without a beginning, he argues. The successive generations of all plants and animals go back to an initial, non-sexual production. This non-sexual, initial production, Udayana insists, applies even to the four social classes: "The origin of the social classes is like that of scorpions, plants and the like. Scorpions today have ancestors, but initially they arose spontaneously from cow-dung, and plants have ancestors, but at the beginning arose from a single seed" (Udayana 1996: 181). Relying on his conception of the causal role of karma at the beginning of an eon, Udayana concludes: "So humans, beasts, cows, and Brahmins all have ancestors, but their initial cause is the distinct combination of the material elements according to karma" (ibid.). Implicit here, of course, is the role of God in overseeing the karmic combinations.

Just as creation requires a supernatural intelligence, in Udayana's view, so also do the maintenance and destruction of the cosmic cycle.[14] Regarding maintenance, or support, Udayana argues that the existence of God can be inferred

> because of support (*dhṛti*): for the whole universe, from the earth to [all the worlds contained in] the Egg of Brahmā, although heavy, does not fall, being supported by the volitional effort of some upholding agent, either directly or by intermediate means.
>
> (1996: 403)

To illustrate direct and indirect means of support, he cites the example of a bird flying, its body supported directly by the bird, while something carried in its claws is indirectly supported. Such arguments when applied to the earth reflect an intuitive folk physics that unobstructed objects in space fall downward towards some unspecified but intuitively imagined lowest ground, unless held up by a volitional agent such as God. Richard Dawkins cites a contemporary example of this folk physics, quoting from a "recovering theist" regarding her first experiment with the idea that there is no God:

> I'm embarrassed to report that I initially felt dizzy. I actually had the thought, "Well, how does the Earth stay up in the sky? You mean, we're just hurtling through space? That's so vulnerable!" I wanted to run out and catch the Earth as it fell out of space into my hands.
>
> (2006: 324)

Udayana regards the dissolution of the universe, including the dismantling of the atomic dyads, as also requiring conscious will and effort. The existence of God can be inferred, he says, "because of destruction (*saṃharaṇa*): for the whole universe from the Egg of Brahmā to the atomic dyads, being liable to destruction, is destroyed by an agent endowed with volitional effort" (1996: 403). Just as a piece of cloth does not unravel by itself, Udayana explains, so also the universe needs a conscious being to unweave it.

Udayana's cosmoteleological arguments indicating that God has created for some purpose seem to compromise God's eternal self-sufficiency. Like Nyāya-Vaiśeṣika in general, Udayana downplays Śaṅkara's solution of this problem that invokes the Lord's sportive play, for even sport suggests that God is seeking relief from some sort of misery. Udayana does, on rare occasions, refer to the Lord's cosmic activities as play, but as George Chemparathy notes, Udayana seems only to mean that God creates spontaneously and with ease (1972: 162). In any case, Udayana's basic answer is that God creates not for himself but only out of compassion for others. This accords with the general Nyāya-Vaiśeṣika view that God creates the world so that all creatures can grow spiritually by experiencing the fruits of positive and negative karma. The latter, of course, accounts for the suffering in the world (ibid.: 158–9).

Among other common objections to Udayana's cosmoteleological arguments is the problem of a disembodied creator. Like Uddyotakara, Udayana maintains that God is able to manipulate material bodies without himself being embodied.[15] But this claim undermines the fundamental analogy on which the cosmoteleological arguments rest. All intelligent agents with whom we are familiar through ordinary experience are physically embodied and act upon or utilize materials through their bodily limbs. But for Udayana, if God has a body, this would imply that he has organs of sense-perception as well as organs of action (hands, feet). Such a supposition would militate against his eternal and uncaused omniscience since his knowledge would then depend in part on the information received through his senses. Also, possessing a body would subject the Lord to experiences of pain and pleasure.

In defending the idea of disembodied or immaterial causation of physical events, Udayana offers largely abstract, rational arguments in his *Kusumāñjali*. In his *Ātmatattvaviveka*, however, he presents more empirical justifications for the disembodied creatorship of God.[16] Udayana refers, for instance, to the phenomenon of frogs found firmly encased within solid stones and avers that it is impossible for even an invisible being, if embodied, to create these frogs without breaking the stones. Only a disembodied agent, namely God, can enter within the solid rock to carry out this rather odd creative whim (Udayana 1995: 408–9). In the same passage Udayana cites the ability of snake-charmers to remove poison from bitten persons through mental concentration, perhaps with God's assistance, but without recourse to any physical means. Agents with limited capacities, like most humans, require a body to be causally effective, but that does not rule out a bodiless agent if that agent possesses, as God does, eternal cognition, desire to act, and effort.

The socioteleological argument

The classical Hindu perspective generally assumed that all culture and civilization were fully set up at the beginning of a cosmic eon, inaugurating a Golden Age, and then gradually declined. From within such a viewpoint, one may well wonder how all the arts of civilization were initiated, since humankind must have learned all these skills more or less in their entirety right at the start, not over several generations. It is within this traditional cultural context that Udayana formulated his socioteleological argument for the existence of God. Humankind in its infancy must have been taught all the arts of civilization, and this knowledge must have been imparted by some superhuman and omniscient tutor. As Udayana summarizes, God can be inferred

> on account of the arts, for the skill of artisans such as weavers in manufacturing cloth, the clear use of language on the part of articulate speakers, the writing of letters of the alphabet in order on the part of young children, since all these must go back to a [knowledgeable] being who is himself independent and untaught.
>
> (1996: 409)

Epitomizing the knowledge taught to humans at the start of a cosmic cycle is the Veda. Like language and other skills, Vedic knowledge is forgotten by human beings during cosmic dissolution, and thus needs to be re-taught by God in each new creation.

Not all Hindus, however, accepted the prevailing view of an eternally alternating cycle of universal creation and total dissolution. The Mīmāṃsakas, the non-theistic school of Vedic exegesis, asserted that the world in its completeness was eternal, with perhaps occasional or local periods of moderate decline. There was thus no first day of creation, but only an endless succession of days. They viewed the Vedas as eternal, not composed or revealed by God, but passed down from eternity by a beginningless chain of teachers and students (see Dravid's commentary in Udayana 1996: 179). The denial of cosmic dissolution is a clear challenge to Udayana's notion of supernatural tutorial necessity.

Udayana's response to the Mīmāṃsakas initially focuses on establishing the reasonable possibility that there was a beginning to the present cosmic cycle; in other words, that there was a first day. Just as a rainy day may follow a non-rainy day, as at the beginning of the monsoon season, so may a first day follow a period of cosmic dissolution (Udayana 1996: 180). Likewise, the regular succession of generations of plants and animals goes back to an initial, non-sexual production, an argument we noted earlier. Since the sequence of days and generations of species have abrupt beginnings at the start of a cycle, it is reasonable to conclude that human knowledge—now handed down through the generations—also has a sudden beginning.

Such a conclusion is not only reasonable, but unavoidable, Udayana claims, given the state of society and of Vedic learning in particular. In his eyes, society at the time was in a deplorable state of corruption, evidenced by an allegedly huge decline in morality and spiritual discipline—a decline commencing right after the start of the present cycle. Specifically, Udayana notes degeneration in the mode of procreation through the ages, deterioration in education, in spiritual knowledge, in human capacity, in the study of the Vedas, in the performance of religious rituals, and in intellect, resulting in the loss of transmission of the sacred traditions. Regarding the first, procreation, Udayana notes that formerly offspring were mind-born; then they were produced by sexual union for the sake of progeny; later they were simply the unintended consequence of copulation to satisfy lust; and now, in the Kali Yuga, copulation occurs with almost total disregard for time and place, following the custom of beasts (ibid.: 189). Udayana further catalogues the decline in consecratory observances for pregnant women, in Brahminical eating practices, in Brahminical observation of purity rules, and increasing Brahminical engagement in commercial enterprises, including the charging of interest.

In summing up, Udayana utilizes the classic metaphor of the dharma-cow, who in the first of the four ages, the ideal Krita or Golden Age, walks on four legs (austerity, knowledge, sacrifice, and charity), but who loses the use of a leg in each of the succeeding ages. In the current Kali Yuga, the cow has only the leg of charity to stand on, and even this leg is faltering as its foot is covered with the sores of ill-gotten moneys, stained with the filth of unbelief, and stung with the barbs of greed and anger (ibid.).[17] Such ongoing degeneration will inevitably result in the radical loss of all culture and Vedic learning by the end of the current cycle. Since no knowledgeable, omniscient being other than God survives the end, only he is capable of re-establishing the teachings at the beginning of a new cosmic regeneration.

The Mīmāṃsakas, to be sure, admit that society on Earth may be in general decline, and some Vedic books may even be lost in India. They maintain, however, that those books may turn up in another country, and in any case, even when the Earth is in a state of severe decline, Vedic knowledge continues unabated elsewhere in the universe. The Mīmāṃsakas propose that Vedic learning can be renewed on Earth by teachers coming here from other planets to instruct men in religious duties. Udayana rejects such sophistry, arguing that it is difficult enough to travel from one earthly continent to another, and that journeying to or from other worlds is impossible. He dismisses any supernatural means for cosmic travel, such as making oneself the size of an atom or as large as the universe, asserting that these are fanciful guesses. There is no empirical evidence that such travel is possible (and thus no evidence that we have been visited by extraterrestrial aliens). The simplest solution to the problem of Vedic regeneration, for Udayana, is that any "visit" from another world by a being of supernatural power is merely the manifestation of the omniscient and omnipotent God on Earth to instruct human beings at the beginning of a renewed cycle (ibid.: 194; cf. Chemparathy 1972: 151).

The Sāṃkhyas, unlike the Mīmāṃsakas, accept the notion of periodic cosmic dissolutions and re-creations. But given their atheistic system, they also object to the idea of God as cosmic tutor, though on different grounds. They explain the re-establishment of Vedic lore after a period of dissolution by reference to the intuitive powers of sages like Kapila (legendary founder of the Sāṃkhya), who were able to "perceive the contents of the eternally existing Veda," and who, moved by compassion, taught it to the new generation of human beings (Chemparathy 1972: 150–1). Udayana accepted that Kapila and others might have extraordinary intuitive powers and supernatural vision attained by ascetic or ritual practices, but not the ability to perceive completely the truth of the Veda.[18] Accordingly, Kapila-like sages could never inspire the sort of unquestioning and absolute trust in the Vedas that Udayana observes among people in the world. Thus God is still required as author and teacher of the Veda.

Conclusion

The theistic arguments of Udayana and the Nyāya-Vaiśeṣika, however cogent they may appear, in the end presuppose belief in God. Two particularly unconvincing arguments for skeptics, according to the renowned Indian philosopher K. Sadchidananda Murty, are those directed against the Cārvāka view "that matter can of itself evolve into the world," and the notion that there may be multiple creators (1959: 208).[19] Udayana himself recognized that a person who does not accept the reality of God will not be persuaded by inferential arguments for God's existence. These arguments, for Udayana, constitute merely "a kind of meditation on truth which is already known" (ibid.: 208–9). Such meditation affirms what a believer may have already experienced of God, assists in defending belief against the attacks of atheists, and constitutes a form of worship following scriptural study by which a person may come to full knowledge of God. Interestingly, as Murty notes, Udayana views all humans, even supposed atheists, as accepting God in some form, however vague and unacknowledged. And even the atheists, by God's compassion, will eventually come to know him directly (Udayana 1996: 3).

Udayana clearly recognizes the limitations of rational arguments for God's existence. He argues that inferential arguments, as well as scriptural revelation and our general experience of God, are supplemented and confirmed by immediate, supernatural, or yogic perception. While yogic vision may not be sufficient to perceive all Vedic truth, it is adequate at least to see God. As he states: "The Lord revealed in scripture and known through inference is also directly perceived by some, just as pots are perceived" (ibid.: 489). Such direct perception, arising from yogic discipline and virtue, is true and certain in Udayana's view, and more compelling than the mediated knowledge coming from scripture or inference. Yogic vision, available only to a spiritual elite, transcends and is in tension with the morally neutral, sensory based, empirical methodology integral to modern science. It has become a major and contentious epistemological issue in the contemporary Hinduism and science discourse.

Even aside from its epistemological and metaphysical views, Nyāya-Vaiśeṣika claims about the empirical world are hard to defend in the light of modern scientific discoveries. For instance, it is difficult to see any way of reconciling the notion of eternal and indestructible atoms with the so-called Big Bang cosmology, or with the ever-increasing zoo of sub-atomic particles produced in today's high-energy atom-smashing particle accelerators, or with basic laws of chemistry that explain the formation of chemical compounds. The early twentieth-century Sanskritist Arthur Berriedale Keith noted some problematic claims; for instance, the *Vaiśeṣika-Sūtra*'s insistence that only like atoms could combine, thereby denying the possibility "of a combination of heterogeneous atoms to form the body" (1921: 204, n 2). The Indian scholar of Nyāya-Vaiśeṣika, Gopikamohan Bhattacharyya, calls attention to several related problems, especially in the theistic interpretation of atomic interactions:

> It should be observed in this connection that modern scientific discoveries do not lend any support to an idea of God as the guiding principle of atoms ...

> Matter is held to be an autonomous body and does not require the aid of any super-being to account for the phenomenal world.
>
> (1961: 139)

I sympathize with those, like the Sinologist Nathan Sivin, who reject smug and arrogant comparisons between current scientific understandings of the world with the "primitive," proto-scientific explanations of pre-modern cultures. But I find troubling contemporary Hindu apologists who blithely assert the complete harmony of ancient and modern science, and who often pronounce ancient science as the superior of the two regarding their respective understandings of how the natural world works. For instance, Aruna Goel has argued that the insights into the natural world of the Nyāya-Vaiśeṣika system have been confirmed rather than discounted by modern scientific knowledge. Such ancient insights were possible, she explains in general agreement with Udayana, due to the extraordinary perceptions of yogis who were able to perceive atoms directly (Goel 1984: 115). She further asserts that the atomic system of the Nyāya-Vaiśeṣika is

> an improvement on the materialistic theory that all things, including life, mind and consciousness, are transformulations and mechanical products of material atoms. The Vaiśeṣikas harmonize the atomic theory with the moral and spiritual aspects of life and theistic faith in God as the creator of the world.
>
> (ibid.: 185)

She concludes that

> it may be inferred that the basic assumptions of the joint-system [the Nyāya-Vaiśeṣika] present the reality in a more meaningful way than does the modern science … the joint-system points out that there is an unlimited energy hidden in the elements, of which we know only a little.
>
> (ibid.: 186)

She then calls for "a great effort to unearth this knowledge [of the hidden energy] and bring it to the notice of modern scientists and technologists" (ibid.). Such enthusiasm for integrating classical theistic Atomism with modern chemistry and physics reflects a concern to validate traditional Hindu Dharmic ideals by invoking the authority of modern science. Seeking such validation poses one of the great challenges for the tradition in coming to grips with both the theoretical conceptions and the methodological naturalism of modern science.[20]

Whatever the tension between classical Hindu and modern atomic theories, the conflict between Nyāya-Vaiśeṣika and Darwinian accounts of the origin of species and history of life is perhaps even greater. Five major Nyāya-Vaiśeṣika ideas conflict with modern evolutionary views. First is the belief that all species currently reproducing sexually were initially produced at the start of a world cycle through karmically determined atomic combinations overseen by God. Second is the notion that human culture in its fullness was revealed by God to humans immediately following the karmic assembling of their bodies, with civilization declining ever since. Third is the view of consciousness as an intermittent manifestation occurring when an eternal soul is conjoined with an eternal but independent mind-base (*manas*) and a temporary physical body. This theory denies both the Advaita notion of consciousness as innate to the soul and the modern evolutionary notion of the gradual emergence of consciousness from increasingly complex neural networks in the brain. Fourth is the karmic explanation for innate psychological propensities such as an infant's suckling at the breast immediately after birth. And fifth is the assumption that any insentient object or principle must ever and constantly manifest its innate character and activity, highly problematic when applied to a process like evolution.

Regarding the explanatory role of karma or the unseen force (*adṛṣṭa*) in the Nyāya-Vaiśeṣika, Erich Frauwallner points to some of the negative implications for science. The introduction of this religious and moral notion into the system undermined "the fundamental attempt to explain the phenomenal world from natural causes" (1973: vol. 2, p. 64). The result was a kind of "karma-of-the-gaps" approach to the natural world. In Frauwallner's words, "Where no reason could be discovered for a phenomenon, resort was had to the convenient explanation through the Invisible (*adṛṣṭa*) which was no explanation at all and one became accustomed to the renouncement of a penetrating explanation of things" (ibid.).[21] I would add that inserting God into the system to explicate the workings of karma does not provide any more of an explanation, but only shrouds unexplained phenomena in a further veil of mystery.

The problem with attributing unexplained phenomena to the mysterious working of some supernatural agent or force is that naturalistic explanations may someday be found, forcing a theological retreat. For instance, recall Udayana's citing the phenomenon of stone-encased frogs in support of the idea of disembodied causation. Today we have a number of plausible naturalistic explanations, including the burrowing of frogs into mud during the rainy season, the mud later drying out to rock-solid hardness entrapping the creatures, which can survive many months if they have any slight access to air (Darling, n.d.).

Traditionally, of course, objections to the Nyāya-Vaiśeṣika were based not so much on empirical grounds as on logical and theological considerations. We have noted several objections to theistic Atomism raised by such opponents as the Buddhists and Mīmāṃsakas. Murty observes that these opponents anticipated by centuries "all the criticisms of the teleological and cosmological arguments, which were made by Hume and Kant" (1959: 207). Many of these objections are nicely summarized by the Vedāntic theologian Rāmānuja, the subject of our next chapter.

Before turning to Rāmānuja, a brief word on the legacy of the Nyāya-Vaiśeṣika in the modern period is in order. Their Atomist cosmology provided Hindus in colonial and post-colonial times with a sophisticated and rational model of the world that, in part or whole, could serve as an alternative to the modern scientific, evolutionary perspective. We see this perhaps most distinctly in the theistic world view of Dayananda Saraswati and his Arya Samaj. But even the Neo-Advaitin Vivekananda accepts certain of the Nyāya-Vaiśeṣika arguments for the pre-existence of the soul in prior lives.

5 Design discounted in the Viśiṣṭādvaita Vedānta of Rāmānuja

All Vedāntins affirm Brahman as the Supreme Reality, infinite consciousness, and ultimate cause of the universe. They all respect the fundamental Vedāntic texts, the Upaniṣads, *Brahma-Sūtras*, and *Bhagavad-Gītā*. They insist that these scriptures are essential for obtaining knowledge of Brahman, denying that humans on their own have the capacity to attain such knowledge through perception, intuition, or inference.[1] Interpretations of these texts, however, vary considerably among Vedāntins, particularly with regard to the relation of Brahman to the world and individual souls.[2]

Advaita Vedāntins like Śaṅkara proclaim that Brahman, the world, and individual souls are ultimately identical or non-dual, with Brahman being the only full and complete reality. Other Vedāntins like Madhva (1238–1317) offer a dualistic view, Dvaita-Vedānta, affirming the eternal and radical distinctness of Brahman or God, souls, and matter. The Vedāntic theologian Rāmānuja (traditional dates: 1017–1137) provides an intermediate view known as Viśiṣṭādvaita, an "identity-in-difference" or "non-duality of qualified or differentiated beings" (Lipner 1986: 120, 142).[3] According to this perspective God, souls, and matter constitute a unity in which each is real and eternally existing, but inseparably linked and interdependent. Dvaita and Viśiṣṭādvaita are both realistic and pluralistic, like the Nyāya-Vaiśeṣika, rejecting the monistic illusionism of Śaṅkara's Advaita.

For Rāmānuja, Brahman in its ultimate form is not Śaṅkara's quality-less (*nirguṇa*) Absolute, but rather is the Supreme Person (Puruṣottama) or Lord (Īśvara) endowed with all auspicious qualities (*saguṇa*).[4] At times, as during cosmic dissolution prior to evolution of the universe, or when souls attain liberation, matter and souls, while not distinct from God, still continue to exist as themselves in a very subtle state. When Brahman is not evolved, being one without a (distinct) second, he is the cause of all that will unfold. When evolved and manifold, he is also the effect (Rāmānuja 1985–1991 [*Śrībhāṣyam*, henceforth abbreviated *ŚB*]: 1.4.23; Thibaut 1966: 399). Rāmānuja envisions the Lord as the supreme Self of the world, enlivening and directing the cosmos and all its individual souls that together constitute his body.[5] He is their inner self and inner controller. This soul–body metaphor, significantly, suggests an organic rather than mechanistic model of the cosmos, with evolutionary implications to be examined below.

Rāmānuja agrees with Śaṅkara in affirming that Brahman is both the efficient and material cause of the world.[6] He frequently cites the *Chāndogya Upaniṣad* (6.2.3) statement, "May I be many," to affirm Brahman's efficient causality.[7] His insistence that Brahman is the material cause of the universe is made explicit in his refutation of the Sāṃkhyan doctrine that Primordial Matter (Pradhāna) is wholly distinct from Primal Spirit (Puruṣa). To support his view, he invokes the famous analogy of the clay, gold, and iron in the *Chāndogya Upaniṣad* (6.1.4–6): just as by one lump of clay, one nugget of gold, or one pair of iron nail-scissors, everything made of clay, gold, or iron, respectively, is known, so also by knowing Brahman the whole world is known. As Rāmānuja points out, this can only be the case if Brahman is the material cause of the world and not just its efficient cause, for we never know about a pot from only knowing the potter, who is only an efficient cause (*ŚB* 1.4.23; Thibaut 1966: 398).

These Upaniṣadic analogies might suggest that, by knowing particular effects, their material causes can be known, since they are of the same substance. This would suggest that knowledge of the world could lead to knowledge of its material cause, Brahman. But Rāmānuja insists, like Vedāntins generally, that Brahman is to be known only from scripture. Knowledge of Brahman allows us to know the world, as the *Chāndogya Upaniṣad* clearly indicates, but observing the world, or its arrangement and order, does not lead us to knowledge of Brahman. Such a perspective would seem to rule out the design argument, but in certain contexts for Rāmānuja it can play a role in removing erroneous conceptions about the nature and origin of the universe.

Tactical use of the design argument

As fellow Vedāntins, Śaṅkara and Rāmānuja share a number of general notions regarding the nature and moral purpose of the cosmos, while still differing on significant points. Such similarities and differences are nicely reflected in their respective commentaries on the *Brahma-Sūtras*, and in particular in their treatment and utilization of the design argument. We may recall that Śaṅkara, in commenting on the second aphorism of the *Brahma-Sūtras* (1.1.2), refers to the incomprehensible arrangement of the world (*acintya-racanā*), which serves as a theater of karmic fulfillment for sentient beings according to definite places, times, and causes. He concludes that the ultimate cause from which this world has arisen is the omniscient, omnipotent Brahman. Commenting on the same aphorism, Rāmānuja similarly notes the incomprehensible, diverse, and wonderful arrangement of the world (*acintya-vividha-vicitra-racanā*), which allows all sentient souls to experience the fruits of their actions according to definite times and places. But he then identifies the cause of the world as "the Lord of All (*sarveśvara*), antagonistic to all evil, whose purposes are good and true, endowed with limitless auspicious qualities like wisdom and bliss, omniscient, omnipotent, and supremely compassionate, the Supreme Person (*para puṃs*)." He concludes, "That [Supreme Person] is Brahman," thus punctuating that Brahman is not the transpersonal Absolute of Śaṅkara (*ŚB* 1.1.2; Thibaut 1966: 156).

Both Rāmānuja and Śaṅkara clearly accept that the world, whatever its ultimate ontological status, is marvelously arranged. And both acknowledge that this marvelous arrangement is due to a conscious and intelligent Lord, whatever his ultimate status may be. Their insistence that perception, inferences from observed phenomena, and logical argument are incompetent means for knowing Brahman lead them in the end to dismiss or disparage the design argument. But neither is averse to an occasional, tactical use of the argument when dealing with atheistic opponents.

Rāmānuja's and Śaṅkara's most extensive elaborations of the design argument occur in their commentaries on *Brahma-Sūtras* 2.2.1–10, in refuting the Sāṃkhyan notion that the non-intelligent Primordial Matter (Pradhāna) by itself is the cause of the world. Rāmānuja closely follows the Advaitin's refutation at several points. Like Śaṅkara, Rāmānuja notes the misleading similarity of the Sāṃkhya to Vedānta regarding the notion that effects pre-exist in their cause (the *satkārya-vāda*). He notes in the same manner as Śaṅkara the crucial Sāṃkhyan analogy: just as Primordial Matter alone, consisting of the three *guṇas* of pleasure, pain, and dullness, gives rise to the world consisting of those same qualities, so also clay and gold give rise to jars and gold ornaments. Rāmānuja rejects this analogy in the following familiar manner:

> The Pradhāna you Sāṃkhyas affirm is not competent to produce the arrangement (*racanā*) of this variegated world, for it is non-intelligent and not superintended by an agent understanding its essential nature. And so it is in similar situations, just as wood and other materials by themselves are incompetent to construct a chariot or palace and so forth. From observation we know that unconscious materials like wood that are not superintended by an agent who knows their nature fail to bring about effects, and also from observation we know that materials produce effects when supervised by a knowledgeable agent. Thus the Pradhāna unsupervised by a knowledgeable agent cannot be the cause [of the world].
>
> (*ŚB* 2.2.1; Thibaut 1966: 484)[8]

In *ŚB* 2.2.3, Rāmānuja, again like Śaṅkara, invokes the "ever-active-essence" argument against the Sāṃkhya notion that the Pradhāna is innately creative: if this were the case and Primordial Matter were not guided by an intelligent Lord, then creation would always exist and there could be no cosmic dissolution. Thus a Lord whose purposes are ever realized is necessary to explain the alternating creation and dissolution of the universe. Rāmānuja takes note of the customary objections: if the Lord is perfect, he cannot have the desire to create or destroy the universe, and if despite this he does create the universe with all its inequities, then he is without mercy (*nirdaya*). Rāmānuja responds in the standard manner: even a perfect being may, for the sake of play (*līlārtha*), commence creative activity, and the inequalities in the world are simply due to the good and evil deeds of beings.[9] As for whether karma, in conjunction with Pradhāna, is sufficient to create the inequalities in the universe without requiring a superintending Lord, Rāmānuja replies that such a view exhibits a complete ignorance of scripture. Scripture defines karma, good and evil deeds, as actions pleasing or not pleasing to the Supreme Person and which result in pleasure or pain through the Lord's favor or punishment.

In his comments on the next several aphorisms (*ŚB* 2.2.4–9), Rāmānuja repeats many of Śaṅkara's arguments against the Sāṃkhya. These include the idea that the Pradhāna itself can have no purpose or motivation to create, being non-intelligent. And if the soul (*puruṣa*) is ever unchanging, pure, and inactive, then the soul must be eternally liberated and can never experience bondage (or release). Rāmānuja concludes, however, by maintaining that the above arguments militate not only against the Sāṃkhya, but also against those (like Śaṅkara) who uphold an unchanging Brahman that, by cognizing ignorance (*avidyā*), supposedly experiences an unreal bondage and an equally unreal liberation. For how can an unchanging Brahman come to be deluded? Rāmānuja states his preference at this point for the Sāṃkhya view, which despite its rejection of God at least admits a plurality of souls that accounts for the various births and deaths of different beings (*ŚB* 2.2.9; Thibaut 1966: 492–3).

While Rāmānuja and Śaṅkara utilize the design argument to counter various non-theistic positions, both ultimately dismiss the argument as inconclusive, but for quite different reasons. For Śaṅkara, all such rational proofs of God's existence belong to the lower and illusory realm of mundane reality and ordinary knowledge. When confronted with logical and theological difficulties in defending the design argument, he resorts to the Māyā doctrine. Rāmānuja rejects the idea that there are two levels of knowledge, equally with the notion of the world as illusion. He thus dismisses the design argument on quite different grounds: at best it is a distraction for devotees, and at worst it is very bad theology.

Critique of the design argument

Rāmānuja provides a detailed rebuttal of the design argument in commenting on the third aphorism of the *Brahma-Sūtras* (1.1.3).[10] We may recall that, in commenting upon the previous aphorism, he briefly alluded to the incomprehensible arrangement (*acintya-racanā*) of the world that is produced by Brahman, identified with the Supreme Person. Thus, his refutation of the design argument is not a rejection of God as the creator of the marvelously constructed world, but only an assertion that God's creatorship can be known only from scripture, so graciously revealed by Him, and not by inference.

In *ŚB* 1.1.3, Rāmānuja first and quickly dismisses perception as a valid means of knowing Brahman, since Brahman is not an object of sense. It is not knowable even through the subtle perception of the yogic, intuitive kind, as this latter merely reproduces things previously perceived in the normal way. The remainder of Rāmānuja's lengthy comment on *ŚB* 1.1.3 is taken up with proving the inadequacy of inference, and in particular the failure of the argument from design in proving the existence of God.[11]

Rāmānuja's critique begins with a summary statement of the design argument, presenting the view of his opponent who, though unnamed, often represents the standpoint of theistic Nyāya-Vaiśeṣika. According to the opponent's argument, the inference to God or a Supreme Being (Puruṣa) as the intelligent designer of the universe is based on the characteristic of the world being an effect. For we invariably find that effects are produced by competent agents knowledgeable about: (1) the material cause (*upādāna*), (2) the instrumental means (*upakaraṇa*), (3) the recipient to be benefited (*saṃpradāna*), and (4) the purpose (*prayojana*). This is exemplified by the familiar case of effects like jars.

Rāmānuja assumes, like his opponent, that many important features of the universe manifest signs of design that are due to intelligent supervision. But this does not mean, for Rāmānuja, that inferences from these designed features of the universe can establish the existence of one supreme designer. His critique focuses on two interrelated aspects of the analogy underlying the design argument: that similar effects have similar causes, and vice versa. Effects such as mountains and organic bodies are supposed to closely resemble human artifacts; and regarding causes, the supreme creator-designer must resemble a human artisan. Like David Hume several centuries later, Rāmānuja recognizes that the force of the overall argument depends on the appropriateness of both sets of similarities. As Hume's skeptical protagonist Philo declares in the *Dialogues*: "Now it is certain, that the liker the effects are, which are seen, and the liker the causes, which are inferred, the stronger is the argument. Every departure on either side diminishes the probability, and renders the experiment [theological inference] less conclusive" (Hume 1964: 411).

Regarding such grand and inanimate features of the natural world as mountains, oceans, and the earth as a whole, Rāmānuja argues that experience does not reveal to us their material and other causes, even if we accept that they are effects. Nor can we reproduce them. They are not like jars and water pots, whose material and other causal factors we know and which we can reproduce. It accords with experience to attribute the production of jars and water pots to an intelligent agent, but not the production of large-scale natural features of the universe. Any such inference is vitiated by the radical difference between them and human artifacts.

Unlike inanimate entities like mountains and oceans, the bodies of animals and persons are comparatively diminutive in size, and they breathe. Thus, even if it could be shown that animate bodies are produced by an intelligent being, this would still not show that inanimate natural objects are. But inference also fails to show that animate bodies, while consisting of integrated parts and limbs, are consciously contrived. It may be argued that the world with its different parts is guided by an intelligent agent, just as a healthy body is controlled by its indwelling soul. But the soul does not produce or maintain the body. The integration of body parts, Rāmānuja claims, requires no intelligent principle to conjoin them or keep them integrated. Rather, it is due to the non-sentient, unseen force of karma not only of the person himself but also of all other conscious beings who enjoy that soul's body, such as his wife. Rāmānuja thus ignores the protests of Śaṅkara and Udayana that the unseen force, being insentient, cannot produce the well-contrived bodies of animate beings. It seems that for Rāmānuja, as far as reason and observation go, karma may act on its own, like a law of nature.

Turning to the problematic nature of the analogy between human agents and one supreme creator God, Rāmānuja raises four basic issues that we have encountered before. The first is whether inference alone can establish that an intelligent cause of the universe is one and not many. The second concerns the competency and necessary qualifications of a world creator. Third is whether an intelligent cause must be embodied. And the fourth concerns negative implications regarding God's nature and character.

Regarding the problem of possible multiple creators, we may recall that Udayana largely ignored the issue, relying on the arguments of earlier Nyāya-Vaiśeṣika theologians. His predecessors claimed that the unity of the causal agent was inferable from the unity of the design and the harmony of parts manifested in the world.[12] But can such unity of design be explained without recourse to a solitary designer? Rāmānuja's example of the healthy body referred to above is meant to show that the body's integrated design can be accounted for not only by a non-intelligent cause, karma, but also by multiple causes, the several karmic forces of the person's associates acting in concert.

As regards the cosmological argument for the supposed necessity of a single prime mover, as argued by Śaṅkara and Udayana, Rāmānuja is thoroughly skeptical. The setting in motion of any inanimate object may be dependent upon an animate being, but it may equally be dependent upon many such beings, as is obvious in the case of moving heavy objects like chariots, large rocks, and trees.

Another argument Rāmānuja offers for the possibility of multiple creators is that we have no evidence that the earth and the oceans were created—or will be destroyed—all at one time. The universe is not like a single jar that is created more or less in a single burst of activity. Rather, the universe consists of multiple effects. So there may be multiple creators and destroyers, just as we see in the everyday world with its multiple products: things created at different times often have different producers. Rāmānuja concludes: "Both perception and inference go against the line of argument attempting to show that there is just one causal agent" (*ŚB* 1.1.3; Thibaut 1966: 170–1).

As for the qualifications of a world creator, Rāmānuja and his opponents agree that a competent creator must have sufficient knowledge of the material, instrumental, and final causes of the intended effect, as well as the ability to produce the effect in question. The specific qualifications of the creator must be inferred from the nature of the object produced, and the more complex the product, the greater must be the knowledge and proficiency of its creative agent. As Rāmānuja's opponent, a defender of the design argument, declares:

> After inspecting manufactured products like jars and water-pots, a person can infer the agent's ability and knowledge needed to construct those things. On seeing for the first time a royal palace of wonderful arrangement (*saṃniveśa*) and ascertaining from the special arrangement of its parts that it is an effect, that person will certainly then infer the diverse abilities and knowledge of the causal agent. Thus, when such things as organic bodies and the universe are established as being effects, then some special being (*puruṣa*), having both

> direct knowledge of all those things and proficiency in constructing them, may indeed be inferred.
>
> (*ŚB* 1.1.3; Thibaut 1966: 166)

Rāmānuja summarizes the opponent's conclusion, contrasting the qualifications of the "special being" just mentioned and identified as the Lord, with those of ordinary souls:

> It is clear that ordinary souls with their limited powers and knowledge, and being under the control of their good and evil acts, lack competence for constructing such entities as organic bodies and the universe with their remarkable arrangement (*saṃniveśa*) of various parts. One may therefore infer that a special being, the Lord who is endowed with superhuman, incomprehensible, unlimited powers and knowledge, and who possesses the skill necessary to construct the entire universe, although he is bodiless, by mere volition has fully effected the wonderful arrangement (*racanā*) of the visible universe infinite in extent.
>
> (*ŚB* 1.1.3; Thibaut 1966: 169)

We shall return to the matter of the Lord's bodiless, purely volitional creative activity in a moment, but first we must look at Rāmānuja's critique of the opponent's discounting of the capacity of "ordinary souls."

Rāmānuja makes two points. First, competent creators do not have to be omniscient. All they need is immediate knowledge (*sākṣātkāra*) and perception (*pratyakṣa*) of the relevant material and instrumental causes, but not complete knowledge of the powers inherent in those causes. What is required is merely sufficient knowledge, just as potters commence making pots having direct knowledge only of their implements, the wheel and molding sticks and the like, without knowing all the inherent powers in those items. Thus, nothing prevents agents lacking omniscience from producing pots, or constructing the world.

Second, regarding the possible objection that souls lack the ability to perceive very subtle, obscure, and remote entities as would be required of a world creator, Rāmānuja claims:

> The conclusion that there is just one supreme causal agent surpassing all ordinary souls cannot be justified on the grounds that the latter lack the capacity for constructing this wonderful world, which is produced by some agent. For we can see the surprising powers of specially meritorious souls who, through their superior karma, are able to produce extraordinary effects—thus the idea that there is just one supreme person of infinite superior karma [capable of world-creation] is faulty.
>
> (*ŚB* 1.1.3; Thibaut 1966: 170)

Having established that reason by itself cannot demonstrate the oneness of the cosmic creator, nor that God is uniquely qualified to construct the world, Rāmānuja turns to the old problem of whether or not a creative agent must be embodied. If the opponent wishes to argue that the Supreme Lord can create by mere volition alone, Rāmānuja counters that we do not actually observe such disembodied creators: "We see that potters and the like by means of their bodies wield their sticks, wheels, and so forth. So one cannot conclude that a disembodied Lord can wield the material and instrumental causes for producing the world" (*ŚB* 1.1.3; Thibaut 1966: 172–3).

If the opponent further replies that there are instances of producing material effects by volition alone, as in the driving out of demons, ghosts, and venom from bodies by mere will power, then Rāmānuja questions how this is possible. The opponent here apparently holds the Nyāya-Vaiśeṣika position that volition depends not on the body but upon the internal mental organ or *manas*. This eternal *manas* can exist separately from the body, combining at times with the soul to produce a conscious being. But even if the *manas* is eternal and independent, Rāmānuja argues, there is no evidence that it can produce any effects unless and until it is associated with a body.

Of course, it might be argued that God has a body. Rāmānuja considers two possibilities in this case: that God is permanently embodied, or only intermittently so. In the former case, then something made of parts, as bodies are, can be eternal. But then one may as well accept that the world with all its various parts is eternal. In the latter case, how can one account for an occasionally existing body of the Lord? The possible answer that the Lord himself, by his will alone, is the cause of his own material body is problematic, for as already shown, a disembodied being cannot by mere volition produce anything. Nor can he attain a body by assuming the body of someone else, for then we would have an endless regression of such body-borrowing without ever explaining how the body was created in the first place.

Rāmānuja's final and most important theological critique of the similarity of the supreme creator-designer to a human artisan concerns the negative implications regarding the character of God. Briefly stated, Rāmānuja argues that the stronger one makes the analogy between human creators and God, then the more disparaging of God the design argument as a whole becomes. It is quite irreverent to suppose that the Lord shares the limited capacities and odious qualities that characterize humans. Only scripture can provide true knowledge of God, and scripture declares that God is "utterly the opposite of all that is offensive, and radically different from all entities [like human artisans] known by the other means of knowledge" (*ŚB* 1.1.3; Thibaut 1966: 173).

One specific example of such possible disparagement, met with before, concerns the motivation of the Lord. The causal agents we actually observe, that is individual souls, are subject to karma and desire, and possess purposes based on those desires. If the analogy for proving God's existence is to be convincing, then the Lord must also be subject to karma, desires, and possessing purposes. But this would be derogatory of the Lord, who has no aims. While Rāmānuja elsewhere makes clear that God possesses no purposes and thus creates only in sport (*līlā*),[13] the important point in the present context is that reason or inference based on experience and analogical argument cannot establish the existence of a non-purposive causal agent but can only imply that God is subject to desires and motives. Such attempts result in bad theology. Hume's skeptical Philo fully agrees: "I was from the beginning scandalized, I must own, with this resemblance, which is asserted, between the Deity and human creatures; and must conceive it to imply such a degradation of the Supreme Being as no sound Theist could endure" (Hume 1964: 395).

Evolutionary aspects of Rāmānuja's Viśiṣṭādvaita

Rāmānuja's rejection of inference as a means of attaining knowledge of God clearly underlies his denigration of the design argument, leading him to emphasize its analogical failings. Perhaps the greatest of these failings, implicit throughout his theology, is that the universe as a whole resembles more an evolving organic body than a fabricated artifact. Hume explicitly makes this point:

> Now if we survey the universe, so far as it falls under our knowledge, it bears a great resemblance to an animal or organized body, and seems actuated with a like principle of life and motion … The world, therefore, I infer, is an animal, and the Deity is the SOUL of the world, actuating it, and activated by it.
> (1964: 416)

Hume would likely have endorsed Rāmānuja's soul-body model of the universe had he known of it. Noting that the animal analogue of the cosmos was the prevalent one among the ancient philosophers known to him, Hume continues:

> For though sometimes the ancient philosophers reason from final causes, as if they thought the world the workmanship of God; yet it appears rather their favourite notion to consider it as his body, whose organization renders it subservient to him. And it must be confessed, that as the universe resembles more a human body than it does the works of human art and contrivance …, the inference seems juster in favour of the ancient than the modern theory.
> (ibid.)

Hume was not entirely ignorant of Hindu organic models of creation. For instance, he notes: "The Brahmins assert, that the world arose from an infinite spider, who spun this whole complicated mass from his bowels, and annihilates afterwards the whole or any part of it, by absorbing it again" (ibid.: 425). As ludicrous as such a model may have seemed to Europeans, Hume concludes: "Why an orderly system may not be spun from the belly as well as from the brain, it will be difficult … to give a satisfactory reason" (ibid.).

We have noted earlier Śaṅkara's use of the famous analogies of the spider and the potter. Rāmānuja also makes use of the potter analogy in his argument against the Sāṃkhya, but in general, his soul-body model of the universe seems more aligned with the spider model:

> According to Rāmānuja and his followers the individual souls and the material world constitute the body of the Supreme Self (God) ... There is also in place a suggestion that since God's will acts directly upon the atoms, the atoms themselves constitute the body of God. We must remember however that such an account will come closer to the "spider-model" than to the "potter-model"!
>
> (A. Chakrabarti 1989: 33–4)

The spider-model and Rāmānuja's body-of-God model clearly resonate with organic notions of reality:

> God is the immanent principle of Prakṛti as well as of individual souls. The exact significance of representing them as his body is that they are sustained by him, are altogether subject to his control and entirely subserve his purposes ... In other words, God together with the souls and matter is an organic whole ...
>
> (Hiriyanna 1949: 181)

Such an organic model lends itself to evolutionary—but hardly Darwinian—ideas. Rāmānuja's cosmological thinking is evolutionary, "because creation takes place not out of nothing but out of pre-existent matter" (Kumar 1991: 496). This pre-existent matter for Rāmānuja, of course, is not the independent atoms of Nyāya-Vaiśeṣika, but rather the body of God in subtle form. Rāmānuja thus shies away from the Nyāya-Vaiśeṣika creationist perspective in favor of the Sāṃkhyan notion of *pariṇāma* or evolutionary transformation. Sāṃkhyan evolutionary theory presupposes the *satkārya* doctrine that effects are latent in the cause, a view accepted by most Vedāntins (Carman 1974: 115).[14] The subtle form of matter and individual souls existing eternally in Brahman evolve into the differentiated forms that constitute his manifest body and are eventually reabsorbed by it.

Occasionally Rāmānuja relates the evolution and dissolution of the universe to the expansion and contraction of the consciousness of souls, even though such consciousness in its essential nature is eternal and unchanging. When Brahman transforms from the causal state into the effected state, "matter becomes differentiated into forms fit to be experienced, and souls undergo an expansion (*vikāsa*) of consciousness or awareness (*jñāna*) suitable for experiencing the various fruits of their karma" (*ŚB* 2.3.18; Thibaut 1966: 542). Yet elsewhere Rāmānuja argues that scriptural references to "the birth and death of the soul (*jīva*) indicate merely the union or separation of the soul with a body, resulting respectively in the contraction (*saṃkoca*) and expansion (*vikāsa*) of consciousness" (*ŚB* 2.3.18; Thibaut 1966: 543). And with regards to liberation, he states that "when karmic bonds are removed, the various qualities of the soul like bliss and consciousness that were contracted by karma become manifest and expand when the supreme light is attained" (*ŚB* 4.4.3; Thibaut 1966: 758).

Interestingly, in the Sāṃkhya the terms *expansion* and *contraction* are used to refer to the alternating evolution of the cosmos from Primordial Matter and its return or involution back into it (Killingley 1990: 173), a usage that may be indirectly reflected in Rāmānuja's thought. In Rāmānuja's case, however, the emphasis is on a spiritual rather than physical evolutionary process. Indeed, he shows considerable indifference to the details of cosmogony, simply accepting the general Vedāntic scheme with its incorporation of Sāṃkhyan elements. The particulars of the Sāṃkhyan evolutionary model are of little interest to Rāmānuja, as he attributes all cosmic causality directly to the "will and intelligent reflection" of the Supreme Brahman or God and his "initial resolve to become many … repeated at each stage of the creative process" (Carman 1974: 117).

We may note here, in anticipation of developments in the colonial period, that the Neo-Advaitin Vivekananda found in Rāmānuja's idea of expansion and contraction of the souls an anticipation of the modern ideas of evolution and atavism, affirming thereby both the scientific nature of tradition and the scientific credibility of Vivekananda's own theory of evolution-involution. This scheme of evolution-involution, as we shall see in Parts II and III, has come to dominate much of contemporary Hindu responses to Darwinism.

Conclusion

Before we turn to the colonial period, it is worth summarizing certain salient points in the classical Hindu Dharmic perspectives that have played important roles in Hindu responses to modern science and Darwinian evolution. Despite frequent claims for the harmony between traditional Dharmic knowledge and modern science, we may note a number of tensions between the orthodox views and modern evolutionary theory. My purpose here is not to denigrate the often sophisticated and multifaceted theological and philosophical world views of the classical thinkers, but rather to note the many challenges facing those who would attempt to reconcile traditional Dharmic views with a modern scientific world view.

Regarding attitudes towards the natural world, all the classical orthodox schools we have considered view the cosmos in teleological terms, as subserving the karmic and spiritual needs of the soul—a view in marked contrast to the Cārvākas. This teleological perspective is maintained in the Sāṃkhya despite the absence of a creator god. It is also upheld in the theistic schools even in the face of God's absolute perfection and absence of motivation, for while God lacks any purposeful intentions, he may engage in sport. And in Advaita, the world we experience may belong to a lower level of reality, but within that level, it appears as designed by the Lord for the karmic benefit of sentient beings. All thus agree that the world is a karmic stage for the unfolding of the drama of spiritual evolution. Such a view runs counter to the concerns and methods of modern science, whose non-teleological approach to the cosmos is an underlying cause of current Hindu apprehension about Darwinian theory.

While certain traditional philosophical notions may seem to have some resonance with modern scientific theories, the resonance often quickly dissipates upon closer examination. For instance the *satkārya* view found in Sāṃkhya and

Vedānta, that this world as effect is latent in or has its source in an ultimate unity, seems to many modern Hindu interpreters to readily harmonize with modern evolution. But in the Vedāntic version, the ultimate unity is consciousness, from which matter devolves, in contrast to the modern scientific view that consciousness is an emergent property of matter.

Traditional Hindu concerns with karmic evolution complement a general indifference to cosmology. Ancient and venerable notions about the physical workings of the universe passed down from inspired seers, such as the eternal cycling of the cosmic ages, are largely accepted without much question. Human history, accordingly, is believed to be astronomically ancient and characterized within any given cycle not by progress but inevitable decline. These views, it should be noted, are in radical conflict with modern scientific understandings of the cosmos. In any case, interest in nature among the orthodox Dharmic thinkers is mostly confined to citing simple examples from the natural world and everyday life to illustrate logical points confirming theological or spiritual teachings. The classical writers make no appeals to scientific discoveries about the natural world to affirm their views, nor do their teleological perspectives of the cosmos inspire any interest in investigating nature. The appeal to science to confirm ancient spiritual insights becomes one of the defining characteristics of Neo-Hindu views in the modern period.

Finally, one of the most intractable difficulties in attempting to harmonize the Dharmic traditions with modern science concerns the matter of epistemology, of how we discover knowledge about the universe in which we live. Is all knowledge already known to the ancient sages, perceived through yogic perception or revealed in scripture, and thus merely to be rediscovered by later generations? Or is knowledge part of an open-ended quest for genuinely new discoveries, accumulating over centuries, and lacking—at least in principle—absolute certitude?

We have seen that, in the classical schools, reason and perception (at least of the ordinary, non-yogic kind) are radically subordinated to the authority of scripture, or to direct experience of God. Reason is used as a handmaid to revelation. The two Vedāntins, Śaṅkara and Rāmānuja, both utilize the rational design argument in denouncing the views of atheistic schools like the Sāṃkhya, but such arguments are intended merely to allay the possible doubts of wavering believers rather than as ultimately compelling and independent proofs of God's existence. We have also noted that Udayana, the greatest of the Nyāya-Vaiśeṣika rational theists, ultimately relies on God's grace and compassion, not reason nor even scripture, as the final resolution for the doubts of skeptics. Needless to say, modern science in the realm of empirical investigation has enshrined the skeptical principle of questioning all traditional authority and so-called received wisdom, since these have so often proved unreliable in ascertaining the ways in which the world works. As the physicist Pervez Hoodbhoy has remarked regarding science and its quest for knowledge, "the ultimate arbiter of truth is appeal to experiment and observation" (1991: 11).

In turning to the modern period, we will find that the descending evolutionism of Śaṅkara, conjoined with Sāṃkhyan notions of ascending evolutionism, has become the dominant model for Hindus in dealing with modern evolutionary thought. This model allows for the acceptance of a limited Darwinian perspective, but only as subordinated to ideals of karmic or spiritual evolution. Theistic creationism, while not as influential overall in the modern period, still has its avid supporters. Updated versions of the design argument continue to appear, and perhaps paradoxically, the design argument—despite the traditional critiques—is re-emerging within non-theistic contexts in the form of consciousness-guided evolutionism.

Part II

The colonial period

6 Design and Darwin in the colonial period

In 1869 a young Hindu graduate of Calcutta Medical College delivered a lecture, "On the Physiological Basis of the Mind," at the Canning Institute in Howrah, near Kolkata. He informed his audience that the mind was undoubtedly the natural product of a gradual, progressive evolution of life. He argued that if the evolutionary origin of the mind is denied then

> We must suppose that at some period of intra-uterine life, the mind or the soul is by the Creator breathed into the forming organism …, or the soul having been once breathed into the first man, is continued into successive generations by the sperm-cell and the germ-cell. The greatest stumbling block to this belief is offered by the Darwinian hypothesis of progressive development … a hypothesis which is daily receiving confirmation, and is soon likely to take the rank as an established fact of science. The only other alternative is to revert to the crude doctrine of transmigration of souls.
>
> (Sircar 2003a: 37)

In these stark terms Mahendralal Sircar (1833–1904), one of the first Hindus to comment on the "Darwinian hypothesis," asserted the radical opposition of Darwin's theory to both traditional Christian and Hindu accountings for the presence of the soul or mind in the physical body.[1] He went on specifically to ridicule the idea of rebirth, which he claimed would require us to think of "spirits floating and flying about every where," seeking a suitable organism to inhabit "either from choice, or doomed … by an unalterable fate, or by an irrevocable curse of some greater being" (ibid.).

Despite his rejection of traditional religious explanations of the mind and his enthusiastic endorsement of modern science, Sircar was no atheist, rigorously promoting a theistic evolutionary perspective.[2] God in Sircar's view had brought order out of the chaos of crude matter, and mind is simply "matter in its most exalted state" (ibid.: 38). Sircar viewed the evolutionary process as thoroughly teleological in nature, reaching "nearly its full effulgence in man, in whom is found the final and the most profound stamp of the Divinity" (ibid.).

Sircar found confirmation of his theistic views in the theories of the German neuroanatomist and phrenologist Franz Joseph Gall (1758–1828), who proposed that God, in creating humankind, had provided a majestic forehead—contrasted with the depressed forehead of other animals—thereby giving all humans the ability to discern "the Supreme Intelligence." Sircar concluded:

> Who will say after this, that cerebral organology leads to atheism and irreligion? On the contrary, has it not established the existence of a First Cause on the most irrefragable of proofs, the proof afforded by the initiations of original and fundamental faculties? And by virtue of this, has it not established religion on foundation[s] that are to remain unshaken for ever?
>
> (ibid.: 39)

It is ironic that the British colonial masters of Sircar's India used Gall's phrenological theories to rationalize imperial policies towards their "inferior" colonial subjects (Sabbatini 1997). As two American promoters of Gall's phrenology argued:

> The Hindoo skull is remarkably narrow in the region of the organs which have reference to the maintenance of existence, and it is said of that people, that their indifference to life is so great, that if fatigued on a journey or march, they ask no greater boon than to lie down and sleep, even at the risk of being devoured by wild beasts, or overtaken and slain by the enemy.
>
> (Drayton and McNeill 1882: 73)

This characterization implied that Hindus were utterly incapable of intellectually rigorous endeavors, including those in the demanding fields of scientific research. Sircar spent much of his life attempting to disprove such views, arguing specifically that Hindus were an energetic race and quite competent for undertaking cutting-edge science if given the opportunity, often denied in colonial India.[3]

In his defense of Hindu competency, Sircar accepted European notions that there was some sort of essential character of the Indian people, subscribing to the idea that there existed a "Hindu mind." At the same time, he was compelled to acknowledge that this Hindu mind in its present state was lethargic and undisciplined. In an article entitled "On the Desirability of Cultivation of the Sciences by the Natives of India," published shortly after his Howrah lecture, Sircar portrayed the "Hindu religion" as "a heterogeneous medley of theology, philosophy, science and what not—in other words, … a chaotic mass of crude and undigested and unfounded opinions on all subjects" (2003b: 41).[4] The Hindu mind, he continued, had once been quite spirited, but under the influence of its surroundings and centuries of dogmatic promulgation and enforcement of these crude and unfounded opinions, the "Hindu mind … has lost much of its original Aryan vigor and energy," becoming averse to patient observation and investigation of the physical world, preferring its own inspired speculations (ibid.). He entreated the British rulers of India to help restore India, insisting that they had "a most solemn trust, the regeneration of a vast dependency, of the once glorious Hindu nation … downtrodden for centuries by foreign yoke and a most de-energizing religion" (ibid.).

He praised Britain for doing her duty in raising India "in the scale of nations," despite Britain's shortcomings characteristic of a colonial power.

The only way for improving the Hindu mind in its present state of lethargy, Sircar believed, was "by the cultivation of the Physical Sciences" (ibid.: 42). The Hindu religion—like all religions, according to Sircar—had divorced God from his works. The present sad state of the Hindu mind "could only be remedied by the training which results from the investigation of natural phenomena" (ibid.). The despotism of the Hindu religion could "only be shaken to its inundations [sic] by the irresistible force of recoil which the mind will acquire, when fed upon the substantial nourishment of pure truths as presented direct by the handiwork of God" (ibid.). To accomplish this, Hindus must not only be taught about these truths, but must also be allowed to discover these truths for themselves. He concluded, "It is thus only that the Hindu mind will be in a position to shake off its inherent indolence and apathy and to appreciate the laws of nature or the working of the Divine Mind" (ibid.: 43). Science, for Sircar, was a moral enterprise leading to the regeneration of the Indian nation. And India, by its own contributions to science, would repay its debt to Europe by helping to humanize and re-spiritualize Western materialistic science (P. Chakrabarti 2004: 151, 168–70). This theme was soon to resound in the writings and speeches of such great Neo-Vedāntic teachers as Swami Vivekananda.

For Sircar, natural theology was employed to bolster his appeal for support of Indians to actually engage in scientific research, and also to affirm the complete harmony of religion and science. But despite his insistence on this harmony, the underlying teleological assumptions of natural theology led him to qualify his endorsement of the "Darwinian hypothesis" that he had lauded in his Howrah lecture as "soon likely to take the rank of an established fact of science" (Sircar 2003a: 37). Like many of his Hindu successors grappling with Darwinian evolution, he accepted a progressive evolutionary view of the cosmos, but with regard to organic evolution he questioned the key mechanism of the Darwinian theory, natural selection, with its anti-teleological implications (cf. P. Chakrabarti 2004: 170, 175).

The investigation of nature by Hindu scientists in the latter part of the nineteenth century was increasingly inspired by Indian nationalist sentiment and used to confirm, not the theistic conceptions of God familiar to Christian scientific researchers, but the ideals of Vedānta, especially the Advaita Vedānta of Śaṅkara. Such a research program is well exemplified in the work of the physicist Jagadish Chandra Bose (1858-1937), whose claim to have discovered emotional sensitivity in plants and a vital pulse or divine spark even in "inanimate" materials like metals served to corroborate for many Hindus at the time—and down to the present day—the all-pervading nature of the consciousness of Brahman.[5] Bose was inspired by the Vedāntic teachings of Vivekananda and his disciple Sister Nivedita, as well as by the notions of cosmic unity espoused by the Brahmo Samaj, a society of tremendous importance in its impact on Hindu responses to modern science. Interestingly, Henri Bergson, a critic of Darwin's natural selection, found in Bose's research illuminating insights into the continuities and beauties of life (Nandy 1995: 68).

Bose's desire to become a scientist was first sparked while attending St. Xavier's college in Kolkata from 1874 to 1880. He was encouraged by Sircar's colleague, the Jesuit Belgian missionary Eugene Lafont (1837–1908), professor of physics at the school. Sircar and Lafont were co-founders in 1876 of the Indian Association for the Cultivation of the Sciences (IACS), the first major institution to promote native Indian learning of modern science using both lectures and demonstrations (Lourdusamy 2004: 5).[6] A complementary purpose of the IACS was "to rescue from oblivion whatever is connected with India, ancient or modern," thus accommodating Sircar's interest in India's glorious Aryan past (Sircar 2003c: 46). The success of Sircar and Lafont's project, incidentally, is evidenced by the fact that the first Asian Nobel Prize in the sciences went to C.V. Raman in 1930 for his work in physics carried out at the IACS.

The introduction of Darwinism into India, as Sircar's remarks at Howrah suggest, could easily have led to polarization of Hindu intellectuals into modernizers supportive of freethinking, atheism, and evolution, over against traditionalists opposed to modern science and its corrosive effects on traditional ideals. The potential for controversy is confirmed by Lafont in an 1869 letter to the *Indo-European Correspondence*, a Catholic periodical. Lafont, like Sircar, saw modern science in its mature form as confirming the existence of God and the trustworthiness of religion. In his letter, he dismisses the freethinkers and atheists, on the one hand, and the science-deniers on the other, observing that both sides resort to "scoffs and jeers" rather than logical argumentation in denouncing their opponents. He notes that the issue resulted in a "great fuss" in the editorials in the local newspapers in Kolkata, and cites a number of instances, the first of which concerns the ancestry of humans. He reports that an atheist supporter of evolution, arguing that humans "are all the offspring of apes," offered to raise money for an expedition to search for "tailed men." Lafont concludes his summary of this example by noting that the theory of ape ancestry and tailed men was debated in the local Kolkata media, "supported by editorials and correspondence columns" and yet also "combated in earnest by respectable men" (Lafont 2003: 17–18).

The initial controversy in the late 1860s in Kolkata over ape ancestry seems to have quickly abated. It was perhaps overwhelmed by issues of social reform and growing tensions between the British and Indians. When Hindu religious thinkers and reformers began to grapple seriously with Darwinism in the mid-to-late 1870s, an array of theological and philosophical responses emerged ranging from outright rejection to cautious assimilation. But polarization between liberal modernizers supporting science and traditionalists rejecting it did not materialize. Many of the Hindu religious thinkers involved, including those who denounced Darwinism, saw modern science in general, and modern technology in particular, as a means of advancing Indian civilization and as quite compatible with the Hindu tradition—although modern science was increasingly viewed as requiring an infusion of Indian spirituality.[7]

I have focused on Sircar above not only because he was one of the earliest Hindu commentators on Darwin, but also because his views highlight a number of issues critical to the Hindu reception of Darwinism, including theological speculations about cosmology and notions of design, the idea of a former glorious Hindu civilization, the seemingly degenerate state of contemporary India, the role of the British government in India's future, and the relation of science and religion. The complex of responses to modern evolutionary theory emerging in the last quarter of the nineteenth century reflects various social, cultural, political, and religious factors at play in the decades before the arrival of Darwinism in India. We shall focus on three of these.

Three major factors in the Hindu reception of Darwinism

The first major factor consisted of the diverse cosmogonic and cosmological views found in the classical Hindu Dharmic traditions. These classical views, considered in Part I, provided Hindu intellectuals with rich and varied sources to draw upon when confronted with the challenge of Darwinian evolution. The three theories of production (*ārambha*, or theistic creationism), transformation (*pariṇāma*, or ascending evolutionism), and manifestation (*vivarta*, or descending apparent evolutionism), associated with the schools of Nyāya-Vaiśeṣika, Sāṃkhya, and Advaita Vedānta, respectively, played significant roles in this encounter. Noteworthy are the various design arguments found in the Advaita and Nyāya-Vaiśeṣika philosophies, and the critical refutations of the argument exemplified in the writings of Rāmānuja.

As in Europe and America during the last 150 years, so in India, notions of design have frequently been in tension with Darwinian perspectives, resulting at times in outright rejection of modern evolutionary theories. In the Hindu case, however, classical evolutionist ideas, especially those associated with Śaṅkara's apparent evolutionism, created a third alternative between evolution and creationism that, initially at least, seemed more open to Darwin's ideas. Yet the classical Dharmic notions of evolution, whether material evolution (*pariṇāma*) or apparent evolution/manifestation (*vivarta*), assumed that consciousness, variously rendered as spirit, soul, or Self, was not a product of evolution but either independent of it, as in Sāṃkhya, or the source of it, as in Advaita. Thus classical Dharmic evolutionary theories are in critical tension with, if not outright opposition to, modern Darwinian evolutionary theory, a point implicit in Sircar's dismissal of "the crude doctrine of transmigration of souls."

The second fundamental factor affecting Dharmic responses to Darwinism was that modern evolutionary theory came to Hindus from outside their own cultural-religious sphere and was embedded within a host of alien ideas. For Christian natural philosophers in the sixteenth through mid-nineteenth centuries, the investigation of nature was a means of coming to know the works of the Master Craftsman. Hindus had no such philosophical-theological framework. For Śaṅkara and Udayana, natural theology was employed to counter materialistic and atheistic world views, not to inspire empirical investigation of the world. Śaṅkara's version, especially, in which all cosmic design is an illusory manifestation, discouraged interest in such inquiry. Sircar, anticipated in part by Rammohan Roy and A.K. Dutt, was one of the first Hindus to justify scientific research in terms of studying the handiwork of God.

While the ancient and medieval Indians made significant discoveries in the realms of astronomy, mathematics, and medicine, such advances were not necessarily inspired by orthodox Brahminism and often were in tension with it (Nanda 2003: 77–9).[8] And though technological improvements in agriculture, textiles, metallurgy, military technology, medicine, and other fields continued well into the period of European influx (Baber 1998: 53–95), such developments were largely in the hands of the lower classes, skilled artisans and craftsmen, often under state patronage. In the millennium preceding the arrival of the Europeans in the sixteenth century, it appears that, for the elite Hindu intellectual classes, the naturalistic and skeptical aspects of ancient Indian civilization became increasingly submerged beneath metaphysical claims and mystical yearnings for other-worldly liberation.

Western-educated Hindu elites of the nineteenth century like Sircar were profoundly influenced by the flood of European literature coming into India. This literature, including historical works, philosophical essays, theological treatises, and socio-political tracts, often reflected Western liberal ideals of tolerance, equality, and faith in reason, science, and human progress. Such ideas presented major challenges to traditional Dharmic world views that generally assumed the sacredness of social hierarchy as well as the non-progressive nature of the world, a world caught in a never-ending series of cycles, each cycle itself marked by gradual but inevitable degeneration from an ancient Golden Age.

The third significant factor was the colonial context within which Hindus first encountered modern science, including technological innovations like the printing press, steam engines, railroads, and the telegraph. Thus many of the Hindu intelligentsia viewed British rule as providential, promising social and material benefits to the country. Sircar was hardly the first to beseech the British administration to provide modern scientific education to Indians as a means of uplifting society, a plea already set forth in Rammohan Roy's famous 1823 petition to Lord Amherst.[9] But as the century wore on there was growing ambivalence towards science, for Indians increasingly perceived British rule as oppressive, riddled with racial discrimination, and driven by greed, all facilitated and enhanced by science and modern technology. Sircar himself worried that the West was "threatening to be, more and more, the civilisation of iron and blood" (quoted in Lourdusamy 2004: 13).

Commercial interests had dominated the earliest contacts of the British with India in the seventeenth and eighteenth centuries, and were of paramount concern in the first few decades of rule by the East India Company, beginning in 1757 after the Battle of Plassey. Company officials, apprehensive that interference in local customs and religious beliefs could jeopardize their position in India, as well as their profits, banned missionary activity in native lands under Company rule. But increasing unease back home with the exploitative nature of the British presence in India, combined with religious concern for the souls of Indian "heathens" and the argument that political-economic aims and moral-religious goals were not incompatible, led to the acceptance of missionary activity in British India beginning in 1813. Needless to say, evangelical proselytizing, especially in its more aggressive modes, was felt by many Hindus to be as oppressive as economic plundering or political subjugation.

An integral aspect of the colonial context was the variety of images that Europeans held of India. These images or interpretive perspectives were important not only for how the British conceived their own role in governing India, but also for how Hindus came to understand themselves and their culture vis-à-vis the West. It was almost inevitable that aspects of the Western approaches to Indian civilization would be turned around by the Hindus to critique Western civilization. The multifaceted interplay between the various perspectives on the part of both the British and the Indians played a major role in molding Hindu perceptions of Darwinian evolution. It is thus worth examining these interpretive approaches in detail.

Four Western interpretations of Indian/Hindu culture

Four distinctive, often conflicting but sometimes complementary, interpretations of Indian culture and religion are especially relevant for our inquiry. Amartya Sen nicely lays out three of the Western approaches I will deal with here, which he designates as the curatorial, the magisterial, and the exoticist (2005: 139–60). An additional category will prove especially helpful for our purposes: the missionary.

The curatorial approach

This approach to Indian civilization sought to describe, classify, and discover the diverse and fascinating or distinctive aspects of India and Indian culture. Ancient texts were uncovered, their languages studied, and their contents translated into modern European languages with the help of Indian pandits. This curatorial approach was motivated in large part by intellectual curiosity, infused with Romanticist empathy towards its subject. It concluded that India had once possessed a great civilization that had undergone a long process of decline up to the present day. This assessment even suggested that ancient Indian scientific ideas underlay the natural philosophy of such European intellectual giants as Bacon, Locke, and Newton, although advocates of the curatorial perspective like William Jones (1746–1794) qualified such claims by insisting that "the 'Asiatics' were 'mere children' in comparison to the scientific Europeans" (Baber 1998: 18; cf. Sugirtharajah 2003: 31–3). This approach also proposed that the original Indian religion was monotheistic but had become corrupted on the popular level by superstition and idolatry (Trautmann 2004: 65).

The curatorial approach has often been termed "Orientalist," but this latter name has become highly problematic since the publication of Edward Said's *Orientalism*. Part of the difficulty is that many of the so-called British Orientalists, like William Jones, worked for the British colonial rulers, so that motivations of intellectual curiosity could easily blend with considerations of imperial policy. At the same time, in the hands of Indian nationalists, the Orientalist motif of an ancient Indian Golden Age could be used to confirm the intellectual and political capacity of native Indians for self-rule.[10]

The magisterial approach

This approach, gradually eclipsing the curatorial, became dominant in the 1830s. It assumed an attitude of radical European and British superiority, viewing Indian culture as containing nothing of value for the progress of civilization. Supporters of the magisterial approach disparaged Orientalists like Jones as romantic and sentimental dupes of their Brahminical tutors. The negative appraisal of Indian civilization inspired a concomitant sense of responsibility on the part of the British for protecting and improving the lives of their supposedly benighted native subjects. This perspective, conjoined with a Utilitarian outlook, favored English as the language of instruction for educating the natives in "useful knowledge." Needless to say, such a viewpoint readily served to justify British rule in India.

Foremost among the proponents of the magisterial approach was the influential British colonial historian James Mill (1773–1836), whose views significantly impacted the cultural context in which Hindus first encountered modern science. In his *History of British India*, first published in 1817, Mill attempted to assess the state of Indian cultural advancement at the beginning of the nineteenth century.[11] As a Utilitarian, Mill wished to determine the extent to which India was in need of useful knowledge to achieve social and material advancement. While acknowledging that "Hindus had made considerable progress beyond the first and lowest stage of human society" (1820: vol. 1, p. 285), he insisted that Indian society had stagnated at an early point. He claimed that from the time of Alexander's invasion (in 326 BCE) until the arrival of modern Europeans in the sixteenth century, the state of Hindu society and knowledge had remained "exactly the same" (ibid.: vol. 1, p. 146). He rejected ancient Hindu myths and legends as "a maze of unnatural fictions" (ibid.: vol. 1, p. 142) and denied as recent Brahminical fabrications all claims of Indian scientific achievement and intellectual originality (ibid.: vol. 2, pp. 44–109).[12]

Several factors, for Mill, needed to be considered in judging "the true state of the Hindus in the scale of civilization," including their progress in literature, arts, science, law, government, manners, and religion (ibid.: vol. 2, p. 135). With regard to the stage of Hindu religion, Mill turned his attention to notions of divine power and cosmological speculations. Of special interest to him were the various creation stories in the Vedas, Purāṇas, and "Gentoo code" (*Laws of Manu*), available to him in recent English translations—he knew no Indian language and had never been to India. In perusing the translations, Mill despaired of finding anything in Hindu cosmological thinking suggestive of a refined and civilized state. While noting variations in the creation accounts, he saw them all as sharing several basic ideas epitomized in Manu's cosmogony, which he characterized as the speculative guess of a rude but overconfident mind, producing "all vagueness and darkness, incoherence, inconsistency, and confusion" (ibid.: vol. 1, p. 286). Manu's account, in Mill's estimation, utterly lacked the humility and simplicity of Moses' portrayal in Genesis of God's power and wisdom in creating the universe.

What particularly disturbed Mill in the Hindu cosmogonies was the profusion of vegetative and sexual motifs, and the lack of any notion of the rational ordering of the universe by a superintending divine being. Especially fantastic, and downright licentious in Mill's rather Puritanical mind, was Manu's notion that the creator Brahmā had split into male and female for the purposes of reproduction, "As if 'The Mighty Power' could not produce [other creatures] by his male virtue alone" (ibid.: vol. 1, p. 289). Mill cautions us that if a description of creation

> includes not even a portion of that design and contrivance which appear in the ordinary works of man; if it carries the common analogies of production, in animal and vegetable life, to the production of the universe, we cannot be mistaken in ascribing it to a people, whose ideas of the Divine Being were grovelling.
>
> (ibid.: vol. 1, p. 287)

Finding no evidence of any Hindu familiarity with the idea of divine design and contrivance of the universe—with the possible exception of what he deemed a likely recent borrowing from a foreign source (ibid.: vol. 1, p. 330n)—Mill set forth his relentless and unforgiving conclusion:

> No [literate] people, how rude and ignorant soever, … have ever drawn a more gross and disgusting picture of the universe than what is presented in the writings of the Hindus. In the conception of it no coherence, wisdom, or beauty, ever appears: all is disorder, caprice, passion, contest, portents, prodigies, violence, and deformity. It is perfectly evident that the Hindus never contemplated the universe as a connected and perfect system, governed by general laws, and directed to benevolent ends …
>
> (ibid.: vol. 1, pp. 329–31)

Mill's surpassingly negative assessment of Indian culture in general and of Hindu religion in particular, aside from its obviously subjective preference for rationality over passion, was mistaken on nearly all counts regarding the scientific and intellectual achievements of the Indians.[13] Regarding his denial of any Hindu rational theology, the design and other logical arguments for the existence of God propounded by Śaṅkara, Udayana, and other Hindus clearly belie Mill's disparaging critique. His contemptuous appraisal, of course, was well suited to justifying Britain's fledgling colonial enterprise: a country as mired in darkness and depravity as he judged India to be surely needed the benign and enlightening rule of the British Raj.

The clearest rebuttal to Mill's verdict is Udayana's *Kusumāñjali*, a work apparently unknown to the British in Mill's day. But even when Udayana's writings came to be known by Westerners in the middle of the nineteenth century, the magisterial assessment remained. E.B. Cowell, for instance, who translated and published sections of the *Kusumāñjali* in 1864, after noting that its logic and metaphysics were "unborrowed from any other land," belittled Udayana's efforts in this backhanded compliment:

> [I]t cannot be uninteresting to trace the brave efforts of a Hindu thinker, far away from the circle of Christianity, who, perplexed by the doubts or open disbelief taught in many of the systems current in his day, endeavoured, however vainly, to build for his countrymen the first truth of Theology on a firm logical foundation.
>
> (Cowell 1983: v)[14]

Cowell's appraisal, a rather meager advance upon Mill's, manifests a strong theological bias that reveals the frequent convergence of the magisterial with missionary perspectives.

The missionary approach

The prejudices expressed by Mill and Cowell partly reflected the attitudes of the Catholic and Protestant missionaries of the seventeenth through nineteenth centuries. Missionary perspectives often complemented the magisterial concern for the well-being and improvement of the natives, stressing the moral and spiritual dimensions. But the missionary approach, while asserting that Indians were lost without conversion to Christianity, was not monolithic. Two contending strands can be discerned. First, the "dark" view—akin to Mill's—espoused the notion that Indians, dark-skinned "heathens," had long been mired in a state of blind idolatry.[15] Second, the "light" view—having some affinity with the curatorial perspective—promoted the idea that the lighter-skinned Brahmin classes, with the light of natural reason, had originally promulgated a primitive monotheism, childlike but non-idolatrous, almost biblical in its approach to the divine.

One missionary whose views were similar to those of Mill was fellow countryman and Baptist Evangelical, William Ward (1769–1823), who arrived in India in 1793.[16] Unlike Mill, however, who never traveled to India nor knew any Indian language, Ward lived many years in India working with Hindu pandits in translating and interpreting Sanskrit philosophical works, including those of the six so-called orthodox schools. The results of Ward's researches were reported in his *History, Literature, and Mythology of the Hindoos* (first published in India in 1811, and in London in 1817). Far better informed than Mill, Ward acknowledged that the Hindus had indeed developed rational arguments for the existence of God. He noted in one instance how Hindu thinkers had dismissed the Buddhist view that the world had "sprung into existence of itself" by insisting that God's existence could be inferred "from his works" (1985: vol. 2, p. 117). Ward also provided a translation of an intriguing summary of Indian religious schools, written by a Hindu, entitled "Account of All the Hindoo Sects." The account includes a dialogue between a rational theist (a Nyāyika) and an atheist representing Buddhist, Jain, and Cārvāka points of view. The theist's basic argument is that "From the works of creation we are constrained to infer that God exists. If you say there is no God, from whence arose creation?" (ibid.: vol. 4, p. 303).[17]

In his own summary of Hindu schools, Ward noted the three basic cosmogonic perspectives of the classical traditions discussed in Part I:

> On creation, three opinions, derived from the dŭrshŭnŭs [*darśanas*], are current: the one is that of the atomic philosophy, another that of matter possessing in itself the power of assuming all manner of forms, and the other, that spirit operates upon matter, and produces the universe in all its various appearances.
>
> (ibid.: vol. 2, p. 119)

Ward identified the three with the Nyāya-Vaiśeṣika, the Sāṃkhya, and the Vedānta, respectively. In the missionary's mind, these diverse views simply demonstrated the inability of human reason alone, without the aid of revelation, to arrive at certain truth regarding the ultimate mysteries of the universe. Thus, despite his greater familiarity with the Hindu religion, Ward's overall assessment was no less gloomy than Mill's:

> There is scarcely any thing in Hindooism, when truly known, in which a learned man can delight, or of which a benevolent man can approve; and I am fully persuaded, that there will soon be but one opinion on the subject, and that this opinion will be, that the Hindoo system is less ancient than the Egyptian, and that it is the most PUERILE, IMPURE, AND BLOODY OF ANY SYSTEM OF IDOLATRY THAT WAS EVER ESTABLISHED ON EARTH.
>
> (ibid.: vol. 3, p. ciii)

The affinity of this bleak missionary perspective with the magisterial is made eminently clear in Ward's preface to his work. He begins:

> It must have been to accomplish some very important moral changes in the Eastern world, that so vast an empire as is comprised in British India … should have been placed under the dominion of one of the smallest portions of the civilized world, and that at the other extremity of the globe.
>
> (ibid.: vol. 1, preface xvii)

Ward makes explicit that it is Divine Providence that has given this task to Britain, the only country capable of accomplishing "the intellectual and moral improvement of India" (ibid.). Near the end of the preface, having regaled his readers with detailed accounts of the moral depravity and intellectual stagnation of the Indians, he points to the divine call of the British not only to help India, but all of Asia:

> Never were such miseries to be removed—never was such a mighty good put within the power of one nation—the raising a population of sixty millions to a rational and happy existence, and through them the illumination and civilization of all Asia!
>
> (ibid.: vol. 1, preface liv)

Part of the missionary strategy for the moral improvement of the Hindu mind and eventual conversion was the introduction of modern scientific ideas, since these were thought likely to undermine traditional myths and religious rituals.

As indicated earlier, the missionary approach to India was not one of consensus. Some of the Jesuits made exceptions to the dark generalizations countenanced above, at least with regard to elite literati groups, especially the Brahmins, whom they came to regard as endowed with "the light of natural reason" (Županov 2005: 17). Roberto Nobili (1577–1656), for instance, in researching the Upaniṣads and other Brahminical literature, found in the concept of Brahman the expression of a pristine monotheism unsullied by any materialistic, idolatrous notions of God, that had been discovered through natural reason (Halbfass 1988: 40).[18] While the "dark" missionary assessment calling for the radical conversion of the Indian heathen from all things Hindu was the dominant missionary discourse early in the century, the failure of such an approach led eventually to a renewed interest in the notion of a primitive Hindu monotheism. The rehabilitation of this more tolerant or "light" interpretation of the indigenous tradition developed into a major missionary strategy of embracing the notion that the Hindu religion already contained genuine spiritual insights and yearnings that would be fulfilled in accepting Christ.

The exoticist approach

A perspective drastically opposed to the missionary idea of fulfillment in Christ was that of the exoticists. Building on the curatorial-orientalist notion of an ancient Golden Age and elevating ancient Hindu civilization to the pinnacle of gentleness and wisdom, this approach focused on the extra-rational and mystical. During the seventeenth and eighteenth centuries, European Romantics frequently used the exoticist interpretation of India to critique Western rationality and its mechanistic interpretation of the world (Sen 2005: 152, 155). For our purposes the most significant expression of the exoticist approach is that of the Theosophists, founded by Helena Petrovna Blavatsky (1831–1891) and Colonel Henry S. Alcott (1832–1907) in New York in 1875. The two visited India in the late 1870s and early 1880s, met with various Hindu leaders, including Dayananda Saraswati, and propounded the superiority of Hinduism over Western civilization, not only with regards to spiritual matters, but also scientific.[19] As Blavatsky wrote in her 1877 *Isis Unveiled*:

> Beside the discoverers of geometry and algebra, the constructors of human speech, the parents of philosophy, the primal expounders of religion, the adepts in psychological and physical science, how even the greatest of our biologists and theologians seem dwarfed! Name to us any modern discovery, and we venture to say, that Indian history need not long be searched before the prototype will be found of record. Here we are with the transit of science half accomplished, and all our ideas in process of readjustment to the theories of force-correlation, natural selection, atomic polarity, and evolution. And here, to mock our conceit, our apprehensions, and our despair, we may read what Manu said, perhaps 10,000 years before the birth of Christ …
>
> (Pt. One, p. 620)

Blavatsky went on to explain that the ancient Indian sage Manu had propounded the theory of evolution of life on Earth (Brown 2007a: 438). She also claimed that the traditional Hindu theological doctrine of the ten avatars or incarnations of the God Viṣṇu in various animal and human forms was really an allegorical presentation of Darwinian evolution. Hinduism and eastern religions in general, she concluded, with their rejection of a radical dualism between God and the world, were fully in accord with modern science, in contrast to the outdated dogma of Christianity and its insistence upon an extra-cosmic creator God. This exoticist approach had tremendous appeal to the Hindu intelligentsia suffering under the humiliation of British imperialism.

Assimilations and reinterpretations of Western approaches

The various Western approaches to India had profound consequences not only for the British understanding of their role in India and their own self-understanding, but also for the self-perception of the Indians themselves. But as Sen cautions us, the Indian reaction to the Western approaches "need not take the form of simple acceptance—it sometimes includes strategic responses to the variety of Western perceptions of India that suit the interests of internal imagining" (2005: 155). These internal imaginings were critical to the Hindu reinterpretation of their own religious traditions and to the articulation of responses to Darwinism based on those reinterpretations. Running throughout the Hindu assimilations of the Western approaches are the following persistent themes: (1) a Vedic Golden Age encompassing spiritual insights *and* scientific discoveries; (2) a subsequent period of degeneration; (3) the possibility of social, cultural, and scientific revitalization and progress; (4) the comparative study of Indian and Western civilizations with respect to their particular roles and responsibilities towards each other; and (5) epistemological considerations regarding the role of scripture, perception, reason, intuition, and suprarational consciousness in ascertaining truth, both empirical and spiritual.

The Hindu assimilation and reinterpretation of Western perspectives during the colonial period from 1800 to Independence in 1947, with regard to Darwinism, may be conveniently divided into three general phases: a pre-Darwinian period extending into the 1860s; a phase of initial encounter with Western, often non-Darwinian evolutionary ideas from the 1870s to the end of the century; and a final phase of re-envisioning traditional Vedāntic ideas of evolution in light of new European philosophical critiques of Darwinian theory, a phase bringing us to the end of the colonial period.

The first, pre-Darwinian period was centered in Bengal, in the city of Kolkata, established in the late seventeenth century as a trading port that became in 1772 the seat of the first capital of British India. The Hindu intellectual elites of Bengal at the beginning of the nineteenth century were comprised of a newly emergent urban landlord class who became enormously wealthy under the East India Company rule. They sought out Western education, including proficiency in English, for themselves and their sons, in order to gain employment in the British administration

(Baber 1998: 195). As Tapan Raychaudhuri notes, "the Bengali intelligentsia was the first Asian social group of any size whose mental world was transformed through its interactions with the West" (1988: ix). Among the intelligentsia, some looked to the Vedānta for inspiration, dismissing Purāṇic and Tantric Hinduism as the superstitious and idolatrous religion so scorned by Westerners. Others revered the Purāṇ ic-Tantric deities like Śiva, Śakti, and Kṛṣṇa, celebrated the associated religious rituals, and were deeply suspicious of "Westernizing" approaches that threatened traditional beliefs and practices. Yet neither of these two factions was unaffected by the ideas and ideals of the other side, so any sharp demarcation between modernist liberals and anti-Western conservatives is problematic.

The first few decades of this pre-Darwinian phase, from 1800 until about 1830, were generally marked by a curatorial-orientalist empathy and admiration on the part of the British for Hindu, Vedic culture—Mill's and Ward's views notwithstanding. The admiration on the part of the British was complemented by a Hindu appreciation and emulation of the West. This time was followed by a period of British "Anglicist," anti-Hindu sentiment, accompanied by increasing Hindu resentment, defensiveness, and cultural self-assertion. The major intellectual issues during the pre-Darwinian period as a whole focused on the role of reason, revelation, and emotion or intuition in ascertaining truth; the strengths and liabilities of the design argument; the "essential" nature of Hinduism as polytheistic, monotheistic, or monistic; universalism versus cultural affirmation; and the possibility of different kinds of science, lower and higher. We shall look closely at two major figures of the Bengali intelligentsia involved in these issues: Rammohan Roy, founder of the Brahmo Samaj, and Debendranath Tagore, restorer of the Brahmo Samaj after Rammohan's death. The Brahmo Samaj was one of the foremost instruments for the modernization of India, playing "a crucial role in the genesis and development of every major religious, social, and political movement in India from 1820 to 1930" (Kopf 1979: xiii).

The second period, from the 1860s to the end of the nineteenth century, began when much of the country was preoccupied with political issues stemming from the aftermath of the Great Indian Rebellion of 1857. This traumatic event had occurred just two years prior to publication of *On the Origin of Species* and may explain why Darwinism did not attract much immediate attention. Regarding the relative lack of controversy at the time, David Gosling notes that a number of illustrated articles on the evolution of humans began to appear from 1873 on in the *Tattvabodhini Patrika*, the Brahmo monthly periodical founded by Debendranath Tagore in 1843. And yet there is no suggestion in these articles "of a debate concerning the implications of biological theories for philosophy and religious belief" (Gosling 1976: 15).[20] Gosling concludes that Hindus were less concerned about Darwinism than their Victorian counterparts in England because of the prevalent Hindu idea of reincarnation that already suggested "a common ancestry for human beings and animals" (ibid.). But as Sircar's 1869 lecture on the physiological or evolutionary origin of the mind indicates, the problematic implications of Darwinian theory for traditional beliefs, especially reincarnation, were present from the start.

In any case, from the late 1870s on, we find the development of a spectrum of responses to Western evolutionary ideas. These early responses were most often based either on very slight familiarity with Darwin's theory itself, or on primarily Lamarckian, Spencerian, and Theosophical interpretations of evolution. At one end of the spectrum stands Swami Dayananda Saraswati, with his outright rejection of Darwinian evolution. We shall examine his reemployment of the atomistic design argument of the Nyāya-Vaiśeṣika in his rejection of Christianity, along with his brief but uninformed condemnation of Darwinism.

At the opposite end we find the great Hindu nationalist Bankim Chandra Chatterji, who in an 1875 essay, "Mill, Darwin and Hinduism," not only rejected the design argument but also accepted the Darwinian theory including its mechanism of natural selection that troubled so many theologians, Christian and Hindu. He saw in the destructive role of natural selection a power complementary to the creative and preserving powers overseeing the evolutionary development of species, and thus Darwinism as a whole for Bankim constituted a parallel to the Hindu trinity of Brahmā, Viṣṇu, and Śiva (the creator, preserver, and destroyer of the universe, respectively). Bankim was careful not to claim that the Hindu trinity originated in scientific speculation, nor that science offers any evidence for the existence of the three gods, only asserting that "it must be acknowledged that this Hindu worship of the three gods is more natural and more in accord with science than the Christian religion supported by the scientific European people" (1986: 70). While Bankim's use of science to defend—however mildly—Hindu doctrines against Christianity, his specific ideas regarding science and evolution were so rooted in Positivism that they have had relatively little impact on subsequent Hindu religious responses to Darwinism.[21]

Among the religious thinkers relatively open to Darwinian notions, we shall consider in detail the pivotal figures of Keshab Chandra Sen and Swami Vivekananda. Keshab, the controversial Brahmo leader from 1866 until his death in 1884, along with Bankim, was among the first Indians to attempt to reconcile modern evolutionary theories with traditional Hindu ideas, but far more assertively than Bankim. Keshab's evolutionary views of religious history and his openness to Christianity inspired others like Vivekananda, who was, briefly, a member of Keshab's Brahmo Samaj.

Another Keshab-inspired figure, whom we can only briefly mention here, is Mahadev Govind Ranade. A leading member of the Prarthana Samaj in Mumbai that was modeled on the Brahmo Samaj, Ranade was favorably disposed towards modern science and saw evolution as providing new insight into the genesis of life. But like Śaṅkara and Udayana of old, he felt the need for an initial prime mover to set inert matter into motion, and to account for how molecules are "attracted and repelled in a particular way through all time and space" (Ranade 1902: 9).[22] This prime mover and intelligent guiding principle for Ranade was neither the impersonal Brahman of Advaita nor the extra-cosmic designer of Deism, but the immanent Lord of Rāmānuja, whom he saw reflected both in Paul's declaration that it is God in whom "we live and move, and have our being," and in the *Taittirīya Upaniṣad* statement (3.1.1) that Brahman is that from which beings arise, by which they live, and to which they return (Ranade 1902: 1, 3).

Neither Ranade nor the Prarthana Samaj were to be nearly as influential as their Bengali counterparts, and especially Vivekananda, regarding Hindu approaches to modern science and evolutionary ideas. Vivekananda, while in many ways indebted to Keshab's universalism and general harmonization of religion and science, was far more critical of Christian beliefs and practices, gained a much greater international reputation, and in the end vastly outshone his erstwhile Brahmo mentor.

Vivekananda's elaborate if non-systematic synthesis of Lamarckian, Spencerian, Theosophical, and Vedāntic notions of evolution has deeply influenced almost all subsequent attempts at harmonizing Dharmic and Darwinian ideas. Vivekananda, like Keshab, typifies those Hindus attempting to assimilate Darwinian notions who engage not so much with Darwinism per se, but rather with various Western socio-cultural and philosophical interpretations of "evolution." These interpretations were frequently construed as lending support to traditional Hindu formulations of evolutionism, as well as providing ammunition against Christian missionaries and their creationist doctrines.

The third phase, from the turn of the century until Independence, saw an increasing engagement of Hindu Vedāntic thinkers with the new ideals of "creative" and "emergent evolution" developing in the West. These innovative interpretations of evolution presented a new style of natural theology, reinserting a teleological dimension into organic evolution by means of a detailed critique of alleged weaknesses in a purely naturalistic, Darwinian account. Particularly striking in these new Hindu assimilations of Western thinking, epitomized by Aurobindo Ghose, is an increasing celebration of the evolving reality of this world, in contrast to the more world-deflating, if not wholly world-denying, perspective of traditional Advaita Vedānta that is seen in Vivekananda.

Before leaving this overview of colonial developments, it is worth noting that there were many Advaitins in the first half of the twentieth century who remained indifferent to the political situation and who continued to espouse an illusory view of the world with its apparent changes. Such Advaitins made no attempt to accommodate Darwinian evolution within an idealistic, monistic framework. Perhaps the most famous of these is the South Indian saint Ramana Maharshi (1879–1950), who dismissed questions of evolution, both physical and spiritual, as irrelevant to the spiritual quest, since in reality, from his Advaitin perspective, there is no change of the Ātman, no birth, no death, and certainly no evolution. He advised disciples not to worry about the past or future (and thus not to worry about evolution), but to live in the present.[23] When asked whether he believed in evolution, he replied: "Evolution must be from one state to another. When no differences are admitted, how can evolution arise?"[24]

It is time now to turn to the first great figure of the modern Hindu revival, the early nineteenth-century Bengali social and religious reformer Rammohan Roy.

7 The Vedāntic Deism of Rammohan Roy

Early in the nineteenth century in Bengal, a remarkable tract on religious toleration was published entitled the *Tuḥfat al-Muwaḥḥidīn* or *Gift to Deists* (1803 or 1804).[1] Composed by the pioneering Hindu social and religious reformer Rammohan Roy (1772?–1833), it draws a sharp distinction between what humans believe by "natural tendency" and what they learn from "habit and training" (*The English Works of Rammohuṇ Roy* 1978 [hereafter abbreviated as *EW*]: 943). He bemoans the fact that adherents of each religion promote their own particular tenets while condemning those of others. Nonetheless, he is confident that an impartial inquiry into the doctrines of different nations will separate truth from untruth, leading to repudiation of dogmas and practices that foster prejudice and ill-will among disparate communities.

Rammohan specifically hopes that reasonable persons of whatever faith community "will turn to the One Being who is the fountain of the harmonious organization of the universe" (*EW*: 947). This optimism is bolstered by his belief that all humankind is endowed with the natural ability to infer the truth of God from his creation. As he summarizes:

> … each individual …, simply by keen insight into, and deep observation of, the mysteries of nature such as different modes of life fixed for different kinds of animals and vegetables and propagation of their species and the rules of the movements of the planets and stars and endowment of innate affection in animals towards their offspring …, has an innate faculty in him by which he can infer that there exists a Being Who (with His wisdom) governs the whole universe …
>
> (*EW*: 948)

Rammohan goes on to lament that some believers, not realizing the One Being who is the creator, foolishly "give Divine attributes to large created beings and make them objects of worship" (*EW*: 948). Rammohan thus employs notions of divine design not only to argue for a universal religious rationalism, but also to deplore the polytheistic beliefs and idolatrous practices of his fellow Hindus.

It is revealing that this brief allusion to the design argument, among the first to be voiced in British India, was written not in English but in Persian with an Arabic introduction, the languages of the Mughal rulers whose grand Indian empire of the sixteenth and seventeenth centuries was in serious decline by the end of the eighteenth.[2] Rammohan was not addressing an English audience, and any influence of European rational theology upon him was at best indirect as he was just learning English at the time. In his youth he had felt great antipathy towards the British in India, but a growing acquaintance with Europeans caused him to give up his prejudice against them and persuaded him to believe that British rule was beneficial to India (*EW*: 224). His study of European literature and culture seems to have begun in earnest in 1805, after publication of his *Gift to Deists* (Subrata Dasgupta 2007: 59).[3]

A likely inspiration for Rammohan's tract was the mid-seventeenth-century Persian work, the *Dabistān-i Mazāhib* or *School of Religious Doctrines*, in which the beliefs and practices of several religions, including Zoroastrianism, Hinduism, Judaism, Christianity, and Islam, are summarized.[4] The *Dabistān* criticizes the different religions for their miracle-mongering, dubious morals, exclusivist tendencies, and advocacy of superstition and idolatry. It ridicules as superstitious notions and practices the Hindu belief in avatars and the worship of images like Śiva's male organ, concluding "that the irrational [inanimate and insentient objects] cannot be the creator of the rational" (*Dabistān* 1973: 365). To overcome the false creeds of the mutually contradictory prophets and revelations of the various sects, one must rely upon reason, which "renders it evident that the world has a creator, all-mighty and all-wise, who has diffused upon the [world] … various benefits" including a rational system of moral precepts (ibid.).

These themes of universal rationalism, rejection of miracles and supernatural revelation, passing reference to the providential design argument as evidence for an omnipotent and beneficent creator, and reason-based morality, are all evident in Rammohan's *Gift to Deists*. These same themes reverberate throughout the discussions and controversies of the next several decades among various Hindu intellectual circles in Bengal, and between them and the British administrators and missionaries. Rammohan was among the first of the Bengali intelligentsia to engage in these debates and certainly became the most famous and influential. Such discussions provided a broad cultural framework for the reception of Darwinian ideas later in the century.

After publication of the *Gift to Deists*, Rammohan became interested not only in English literature, but also in the sacred texts of Vedānta, inspired in all likelihood by British Orientalists like H. T. Colebrooke who espoused the notion of an ancient and pristine age of Hinduism. Colebrooke's "Essay of the Vedas," published in 1805, "contained the essence of Rammohun's later arguments centering around the discrepancies between ancient textual requirements and contemporary practices" (Kopf 1969: 198). In addition, Ramram Basu, one of the pandits at Fort William College, whom Rammohan may well have met, had already published in 1801–1802 tracts that expounded "Rammohun's idea of a theistic *Brahmo*, the cornerstone of his reformation movement" (ibid.). The notion of a "theistic *Brahmo*"—the idea that the Brahman of the Upaniṣads is a personal God and not an impersonal absolute—was the ideological foundation of the society Rammohan established near the end of his life, the Brahmo Sabha, or Brahmo Samaj as it later became known.

In 1815 around the age of forty-three, Rammohan settled permanently in Kolkata, living in a house inherited from his father, a once prosperous zamindar (hereditary landlord and government tax collector). By skillful management of his inherited properties and through his own business dealings with the East India Company, Rammohan had become quite wealthy and yearned for the cosmopolitan life of the city. While deeply impressed by the teachings of Vedānta, he was not attracted to its ascetic ideals.

In the four years following his move to Kolkata, Rammohan published his major works on the classical Vedānta, writing in Bengali as well as in English. These included a translation of the *Vedānta Sūtras* and summaries thereof, as well as translations of five Upaniṣads. As he had done in *Gift to Deists*, he continued to excoriate the superstitions and idolatrous practices of all sects, but he now emphasized the core truth to be found in all religions. In particular, he found in the Upaniṣads and *Vedānta Sūtras* a pure monotheism radically opposed to the idolatrous polytheism of a later age. This Vedāntic monotheism, bordering on Deism, he deemed the equal of Christian teaching at its best and far superior to the Trinitarian travesty proffered by the Protestant missionaries whom he met, and with whom he entered into bitter controversy in the early 1820s.

Throughout the 1820s Rammohan devoted himself to a number of social causes, including especially the rights of women and most famously the abolition of sati. The British administration in Bengal under Lord William Bentinck banned sati in 1829, a ban soon challenged in the courts. To ensure that it was not rescinded Rammohan traveled to England in 1830 to lobby the House of Commons. He never returned to India, dying in England in 1833.

In his writings against idolatrous practices, in his apologetics defending a pure Hindu monotheism, and in his criticisms of Trinitarian doctrine, Rammohan frequently drew upon the design argument. Let us now examine in some detail his various uses of design, before turning to the epistemological views underlying his Vedāntic Deism.

The design argument as evidence against idolatry

In his Bengali writings, clearly intended for his fellow Bengali Hindus, Rammohan laments that the true monotheistic teaching of the Upaniṣads, a teaching confirmed by the *Vedānta Sūtras*, is no longer well known.[5] As a result, he claims, various misunderstandings about the Upaniṣads have arisen, promoted by self-serving pandits. Among these misunderstandings is the notion that one cannot worship a God beyond speech and mind, beyond form and qualities, and thus recourse to concrete, visible beings—idols—is necessary. Rammohan responds that, while God's ultimate nature is indeed unknowable, attention to the orderliness of the universe may still reveal to us his creative and protective character:

> Yet we can succeed in our attempt [to understand something of God] if, on seeing the manifold construction and regulation of the universe, we ascertain that he is its creator and regulator. A little attention will suffice to make it

> clear that the creator of this impenetrable and variously constructed universe must be more pervasive and powerful than it. By what reasoning can we be persuaded that some object which is a part of it, or can be pervaded by it, is its creator?
>
> (Translated by Killingley 1982: 9)

Rammohan makes a similar point in English in his 1817 tract, "A Defence of Hindoo Theism." The "Defence" was a response to a letter appearing late the previous year in a Madras newspaper attacking Rammohan's theological views expressed in his early Vedāntic writings. In the "Defence" Rammohan summarizes the various arguments of the letter writer, whom he calls "an advocate for idolatry." Among the advocate's arguments is the difficulty of knowing "the Invisible and Almighty Spirit" (*EW*: 94). Rammohan again acknowledges that "perfect knowledge" of God's nature "is certainly difficult, or rather impossible." But he reiterates that it is not so difficult for someone endowed with common sense and not bound by religious prejudice "to read the existence of the Almighty Being in his works of nature" (*EW*: 94). He adds that Muslims, Protestant Christians, and the followers of Kabir and Nanak have no need of images in worshiping God.

Rammohan admits that the scriptures, including the Vedas as well as the Purāṇas, occasionally recommend the worship of concrete images to the ignorant, as a means of steadying their minds (Killingley 1977: 293). He notes that the distinction between inferior, ignorant people and those already capable of true worship is based in part on their respective competence to comprehend the idea of divine design in the universe. With reference to the elephant-headed god Ganeśa, he explains:

> By considering that Brahman, who is beyond the reach of words and the mind, has nothing to do with an elephant's head, and by other considerations, these [inferior] people will form a desire to know the truth of Brahman … But it is not the intention of the śāstras [scriptures] to give teachings about elephant-headed beings to those who have wisdom, and who are capable of seeing the varied rules and design of the world, and remaining intent on its Regulator.
>
> (Translated by Killingley 1977: 293)

This reflects an interesting modification of Śaṅkara's two levels of truth: those engaged in idol worship are engaged with a lower knowledge; the higher knowledge recognizes that what is illusory (a manifestation of Māyā) is not the world itself or the notion of Īśvara, but rather the mistaken perception of God as visible in concrete forms (cf. Killingley 1977: 294–5).

The design argument as vindication of "the Hindoo religion"

In his English summary of the *Vedānta Sūtras*, entitled *Abridgement of the Vedant*, Rammohan sets out "to prove to my European friends, that the superstitious practices which deform the Hindoo religion have nothing to do with the pure spirit of its dictates!" (*EW*: 4). He specifically rejects the views of some European scholars who in a spirit of false generosity attribute wholly symbolic meaning to the various Hindu polytheistic and material representations of divinity.[6] He then provides an interpretation of the *Vedānta Sūtras* loosely based on Śaṅkara's *Vedānta Sūtra Bhāṣya*,[7] incorporating several aspects of Śaṅkara's design argument into his exposition.[8]

Rammohan begins his exposition by summarizing the first two aphorisms or *sūtras* (1.1.1–2) of the *Vedānta Sūtras*. The first *sūtra*, he claims, points to the necessity for humankind to acquire knowledge of the Supreme Being, who is beyond human vision and comprehension. Accordingly, the second *sūtra* explains that the Supreme Being can be known through his works. Rammohan paraphrases the *sūtra*: "He by whom the birth, existence, and annihilation of the world is regulated, is the Supreme Being,"[9] and then, in a manner reminiscent of Śaṅkara, elaborates:

> We see the multifarious, wonderful universe, as well as the birth, existence, and annihilation of its different parts; hence, we naturally infer the existence of a Being who regulates the whole, and call him Supreme: in the same manner as from the sight of a pot we conclude the existence of its artificer.
> (*EW*: 7–8)

We may recall that the example of the potter and pot is utilized by Śaṅkara in his design argument refuting the Sāṃkhya (*BSB* 2.2.1).[10] Rammohan soon turns his attention to the Sāṃkhya, attacking its notion of Nature as an efficient cause of the universe:

> … the Vedanta thus declares: "Nature is not the Creator of the world not being represented so by the Veda," for it expressly says, "God has by his sight created the Universe." Nature is an insensible Being, she is, therefore, void of sight or intention, and consequently unable to create the regular world.
> (*EW*: 9)

And just as Śaṅkara rejects the claims of the atheistic Atomists (*BSB* 2.2.11–14), so Rammohan argues:

> *Atoms* are not supposed to be the cause of the world … Because an atom is an insensible particle, and from the above [Vedic] authority it is proved, that no Being void of understanding can be the author of a system so skillfully arranged.
> (*EW*: 9)

And finally, like Śaṅkara, Rammohan insists that the highest reality is both the efficient and the material cause of the universe:

> God is the efficient cause of the universe, as a potter is of earthen pots; and he is also the material cause of it, the same as the earth is the material cause

> of the different earthen pots, or as a rope, at an inadvertent view taken for a snake, is the material cause of the conceived existence of the rope.
>
> (*EW*: 12)

What is of special interest here is Rammohan's allusion to the rope-as-snake delusion, a standard Advaita metaphor signifying that the universe is only an apparent manifestation, not real in itself. Yet an analysis of Rammohan's Vedāntic works suggests that he was gradually moving away from the illusionism of traditional Advaita, despite his frequent claims that he was following Śaṅkara.[11] In his interpretation of Vedānta, he transformed Śaṅkara's teaching into a less ascetical, more humanitarian, ethical system (cf. Killingley 1977: 245–9).

Much of this transformation may have been due to Rammohan's growing familiarity with European thought, as well as with his own lifestyle as a prosperous businessman. His shift away from illusionism was almost a logical necessity of his arguments for an original Hindu monotheism, rather than a monistic, impersonalist view of Brahman. He seems to have sensed that emphasizing the illusory nature of the world significantly undermines the force of the design argument, thus weakening one of the major supports for his ethical monotheism. This development is reflected in his increasing neglect of Śaṅkara's specific understanding of the two levels of reality, offering his own interpretation of them, noted earlier. In any case, Rammohan's rejection of illusionism was to be reiterated with varying force by later Hindu interpreters of the Vedāntic tradition, especially in the twentieth century.

Running throughout Rammohan's defense of an original Hindu monotheism and reflecting the universalist outlook of his *Gift to Deists* is the notion that the rational worship of the One True God is not confined to the Vedāntic tradition. In fact, he believed that the design argument "is principally inculcated by Christianity" (*EW*: 483).[12] His endorsement of Christianity, however, applied only to its Unitarian version, and he caustically denounced Trinitarianism as a woeful corruption smacking of polytheism. In the ensuing debates with the Baptist missionaries, his basic tactic was to show that the missionary critiques of several Hindu religious ideas applied with equal if not greater force to the teachings of Trinitarian Christianity.

The design argument as evidence against Trinitarianism

In 1821 a letter from a Baptist clergyman was published in the missionary weekly newspaper, the Bengali-language *Samarchar Darpan*, in which the author raised a number of questions regarding various Hindu śāstras. Specifically, he targeted the teachings of four of the classical *darśanas*: Vedānta, Nyāya, Mīmāṃsaka, and Sāṃkhya. He also attacked the mythological fancies of the Purāṇas and Tantras. The author of the letter invited an answer, and Rammohan soon sent a reply in Bengali, which was not published. Rammohan therefore took it upon himself to publish a response in *The Brahmunical Magazine, or The Missionary and the Brahmun, Being a Vindication of the Hindoo Religion against the Attacks of the*

Christian Missionaries, in Bengali with an English translation. Eventually four numbers of *The Brahmunical Magazine* appeared between 1821 and 1823, as the rancorous debate continued over the next two years.

In introducing his first response, Rammohan sketches the broad history of the British and the missionaries in Bengal. He notes that the British had been in exclusive possession of Bengal for nearly fifty years, during the first thirty of which they had refrained from interfering with the religious lives of their subjects, but this had changed. Using the printing press, street preaching, and material enticements, Christian missionaries were now seeking to convert Muslims and Hindus by denouncing those religions and extolling the virtues of their own. The arrogance that comes with political domination, he charges, seems inevitably to induce derision of the religion of the conquered, although the religion of the conquerors may itself be quite ridiculous. In the case of the English in India, such behavior on the part of men renowned for their humanity and justice is most unjust and unreasonable. This charge of arrogance and cultural blindness lies behind Rammohan's main argument against the *Samarchar Darpan* letter writer: on one hand, the sacred teachings of the Vedas explained in the śāstras are far more rational than the doctrines professed by the missionaries, and on the other, the doctrines and mythologies of the Purāṇas and Tantras, even if unreasonable, are no more irrational than those of the Christian proselytizers (*EW*: 147).

More specifically, Rammohan makes three points relevant to the design argument. First, the missionary claim that God makes souls out of nothing is intrinsically illogical and unreasonable. Underlying Rammohan's statement is the classical Hindu *satkārya* notion that something cannot arise from nothing. Since God is beyond our senses, we must rely on reason and experience. But if the irrational belief in creation out of nothing is accepted, then all rational proof of God's existence—including implicitly the design argument—is called into question, thereby encouraging atheism (*EW*: 155).

Second, a related missionary argument faulting the Nyāya doctrine that souls, time, space, and atoms, as well as God, are eternal is based on false assumptions. Abandoning any Advaitic notions of monism and accepting the basic realist ontology of the Nyāya-Vaiśeṣika, Rammohan claims that the smallest material particles, or atoms, are indestructible. Accordingly:

> The immaterial God cannot be supposed the material cause of those particles, nor can Nothing be supposed to be the cause of them: therefore these particles must be eternal, and are only brought into different forms, at different times and places, by the will of God.
>
> (*EW*: 156)

Udayana could not have said it more succinctly.

Third, the Trinitarian doctrine professed by the missionaries is inane and theologically disastrous. The missionary claim that the three members of the Trinity are equal in giving grace to humankind, pardoning sin, and guiding men in righteousness, and that these are functions that "omniscience, omnipotence, infinite love and mercy can alone perform," receives a caustic reply from Rammohan. Invoking the traditional design-argument notion that there is only one designer of the universe, he asserts that Trinitarian thinking is thoroughly polytheistic and little better than belief in an astronomical number of gods. He asks whether one infinite and omnipotent person is able "to arrange the universal system and preserve its harmony." If one suffices, then additional persons are "superfluous and absurd" (*EW*: 174).

During the period of the Trinitarian controversy Rammohan was studying the Bible with two Baptist missionaries in a collaborative effort to produce a Bengali translation of the Gospels. He eventually persuaded one of them, William Adam, to abandon the Trinitarian dogma as a later and corrupt interpretation of the pure, monotheistic teachings of the biblical text and of Christ himself.[13] Adam founded a Unitarian church in Kolkata, which Rammohan attended while avoiding the "established" or Trinitarian churches. In justifying his practice, Rammohan argues that the established churches preach logical absurdities and glorify violence. Defending such illogicalities as a virgin birth or the necessity of a bloody sacrifice to appease God's wrath in terms of some divine mystery simply cannot appeal to a rational person. On the other hand, Rammohan declares,

> Unitarian prayer, worship, and preaching constantly put me in mind of the beneficial design kept in view by the wise and benevolent Author of all, in organizing the members of the animal body, such as bones, veins, vessels, limbs, &c., and in preparing the manifold necessaries of life for our maintenance, as proofs of his gratuitous blessing and free grace … [In addition] the Unitarian mode of worship exhibits how that infinite and Supreme author has designedly stationed the heavenly bodies, in systematic order, capable of producing and nourishing all the animal and vegetable objects under his divine control …
>
> (*EW*: 201)

Despite the similarities between Rammohan's earliest presentation of the design argument in his *Gift to Deists* and his later use of design to defend Unitarianism, there occurred a significant shift in his underlying epistemological views. Beginning with his Vedāntic works, we see his increasing doubts about the sufficiency of reason for ascertaining truth.

Reason, revelation, and investigation of nature

In the various introductions to Rammohan's Upaniṣadic translations, we can discern a backing away from the complete reliance on reason manifest in his *Gift to Deists*, as he cautiously acknowledges a qualified but key role for scripture. He laments that inquiry into the religious doctrines of ancient nations often reveals them to be mutually contradictory and thus of little help in discovering theological truth. But if we then turn to reason, "we soon find how incompetent it is, alone, to conduct us to the object of our pursuit" (*EW*: 37).[14] In fact, he claims, reason readily leads to skepticism destructive of religious faith, and thus the ascertainment of truth may require recourse to both reason and tradition, and specifically to scripture. As Rammohan now insists in the preface to his Bengali translation of the *Maṇḍūkya Upaniṣad*: "Anyone who as a result of past or present virtue has an inclination to know the nature of Brahman, must daily hear the texts of the Vedānta and think of their meaning" (translated by Killingley 1977: 288).

The hearing of Upaniṣadic texts and reflecting thereon are essential steps in the traditional practice of *jñāna-yoga*, the path of knowledge favored by Śaṅkara. Rammohan also begins to emphasize the need for inner devotion. As Stephen N. Hay summarizes, we see in the introductions to his Upaniṣadic translations a clear "shift from reliance on reason alone to a combination of tradition, reason, and devotional faith" (1996: 753). Nonetheless, for Rammohan "scriptural revelation did not have absolute value; its purpose was to direct our thoughts to the general revelation of God and morality in nature and the human mind" (Killingley 1993: 147).

While in the West the idea of God's revelation in nature frequently inspired and rationalized the scientific investigation of the physical universe, in India the initial interest in science and especially technology in the colonial period was largely utilitarian rather than theological or moral. Rammohan, like many of the Westernized intelligentsia of Bengal in his day, was duly impressed by European science and technological accomplishments, seeing these as necessary for India's advancement. In his 1823 petition to Lord Amherst, Rammohan called upon the British administration to institute the instruction of Indians by European gentlemen in the fields of "Mathematics, Natural Philosophy, Chemistry, Anatomy, and other useful sciences, which the natives of Europe have carried to a degree of perfection that has raised them above the inhabitants of other parts of the world" (*EW*: 472). At the same time, he insisted, that while India may be indebted to the British for introducing "useful mechanical arts," the world owes a debt of gratitude to India, where the "dawn of knowledge" in science, literature, and religion first arose (*EW*: 906)—a theme reflecting the Orientalist idea of an ancient Vedic Golden Age.

While the study of the natural sciences would aid India's progress, Rammohan also came to see it as playing a role in matters of faith. He apparently agreed with the missionaries "that a training in the 'useful sciences' would make people proof against fraud of all kinds, and thereby enable them to find the truth in religious matters for themselves" (Killingley 1977: 353–4). Such a view is clearly seen in Rammohan's last statement on the design argument, written in England in 1832 as part of the introduction to a collection of his English translations of Vedic and Vedāntic texts. He proclaims that the intent of the Vedāntic writings is to warn against idolatrous conceptions of the deity, and thus they recommend that humankind

> direct all researches towards the surrounding objects, viewed either collectively or individually, bearing in mind their regular, wise and wonderful combinations and arrangements, since such researches cannot fail … to lead

> an unbiased mind to a notion of a Supreme Existence, who so sublimely designs and disposes of them, as is everywhere traced through the universe.
>
> (*EW*: 227)

Research into the natural environment thus leads us towards greater belief in one divine and sublime Designer. Here, the Vedas are subordinated to nature as "the true source of revelation; the Veda serves only as a guide" (Killingley 1993: 89).

When Rammohan failed to return from England, he left his fellow Hindus back in India to carry on his agenda of social, religious, and educational reforms. His ideals were to inspire many of his Bengali colleagues to work out on their own the philosophical and theological implications of his Vedāntic monotheism in the face of continuing missionary activity and the advancement of European science, accompanied by the introduction into the sub-continent of ever new and more inescapable modern technologies like the telegraph and railroad.

Conclusion

The purposes for which Rammohan employs the design argument are molded to considerable extent by the colonial context. Certain terminology he employs, referring for instance to the Supreme Designer as "Author of all," and some of his specific design analogies, are clearly borrowed from eighteenth-century European Deism. But it should be evident that traditional Hindu sources, primarily Vedāntic but also those of theistic Atomism, inspire his rational theology. One clear example of Rammohan's dependence upon Śaṅkara's Advaita augmented by an infusion of Enlightenment natural theology is found in a short essay published in 1829, entitled *Religious Instructions founded on Sacred Authorities*. Responding to the query "To whom is worship due?" Rammohan responds that it is due

> To the AUTHOR and Governor of the universe, which is incomprehensibly formed, and filled with an endless variety of men and things; in which, as shown by the zodiac, in a manner far more wonderful than the machinery of a watch, the sun, the moon, the planets and the stars perform their rapid courses; and which is fraught with animate and inanimate matter of various kinds, locomotive and immoveable, of which there is not one particle but has its function to perform.
>
> (*EW*: 135)

In support of this answer, he quotes from a number of Vedāntic texts, including the *Vedānta Sūtras* (1.1.2) and Śaṅkara's commentary thereon which refers to the incomprehensible design (*acintya racanā*) of the universe (*EW*: 139). At the same time, the references to "the AUTHOR and Governor of the universe" and "the machinery of a watch" are thoroughly European.

Whatever the sources of Rammohan's natural theology, his rationalist tendencies and utilitarian views of religion as a means of social improvement, along with his critique of creeds and idolatry, often provoked both Christians and Hindus to denounce him as a Deist. The missionaries saw Deism as one of the most formidable obstacles to their success in India. The Baptist William Carey already complained in 1807 that "India swarms with deists; and deists are in my opinion, the most intolerant of mankind. Their great desire is to exterminate true religion from the earth" (quoted in Kopf 1969: 139; cf. Halbfass 1988: 56). Rammohan soon became a focus of such concerns. As early as 1816, the *Government Gazette* in Kolkata, commenting on Rammohan's just-published *Abridgement of the Vedant*, observed: "Those conversant in Hindoo mythology, will be surprised to find that a BRAHMIN [Rammohan] should attempt to trace the origin of *his* religion to pure Deism" (quoted in Killingley 1993: 116).[15] In 1823 the Anglican bishop of Kolkata remonstrated, almost certainly with reference to Rammohan and his followers: "Our chief hindrances are some deistical Brahmins, who have left their own religion, and desire to form a sect of their own" (quoted in Killingley 1993: 113). All this was a bitter disappointment to the missionaries, for at one time they had hoped that Rammohan would become a Christian. His formal founding of the Brahmo Samaj in 1830 was a clear and final signal that this was not going to happen (N.S. Bose 1976: 42). The English newspaper, the *John Bull* of Kolkata, reporting on the initial meeting of the Brahmo Sabha in August of 1828, lamented that Rammohan's group of liberal Hindus had "from Unitarianism very naturally slid into pure Deism" (quoted in Collet 1914: 134).[16]

Orthodox Hindus also expressed their disapproval of Rammohan's Vedāntic rationalism by labeling him a Deist. Thus, in Bombay the modernizing pioneer Bal Gangadhar Jambhekar lamented: "With the love of general literature, he [Rammohan] indulged also in reflection on religious subjects, which induced him at last to totally forsake the faith in which he was born and to embrace Deism" (quoted in Halbfass 1988: 202). Interestingly, the Marathi version of Jambhekar's article translated the term *deism* as *vedānta-mata* (creed of Vedānta) (ibid.).

Rammohan's Vedāntic Deism, if we may call his Vedāntic monotheism such, was an attempt to integrate European rationalist, deistic ideas with the religious and philosophical insights of his own Hindu tradition. As Halbfass remarks, "No matter what Rammohun may have adopted [from the West], he brought it into his own particular hermeneutic situation of appealing to and reflecting upon different traditions, of appropriating the alien, and of asserting himself against that alien" (1988: 207).

Building upon Orientalist notions of a Vedic Golden Age, Rammohan's ideas fostered a new self-understanding on the part of Hindus with regards both to their own partially forgotten past and to the modernizing culture of the British imperialists. His efforts to expound his deistic Vedānta, well exemplified in his attempt to blend Western versions of the design argument with that of Śaṅkara, became a potent agent of modernity within the Indian context, just as Deism had been in the West.[17] Śaṅkara's Vedānta, which Rammohan did so much to revitalize—even though he modified or rejected its illusionism and monistic idealism—served to inspire one of the major responses to Darwinism at the end of the nineteenth century, that of the Neo-Vedāntins.

While Rammohan may have wavered in his views on the relative role of reason and revelation, he never seems to have undergone any sort of existentialist crisis regarding the reality of God. With the influx of Western rationalistic and atheistic ideas in the early nineteenth century, English-educated Bengalis were sometimes painfully confronted with the notion that there may be no God. One Hindu who experienced this sort of crisis was Debendranath Tagore, who played a critical role in reviving Rammohan's Brahmo Samaj that had languished after Rammohan left for England. Debendranath's crisis of faith was precipitated in part by growing doubts about the sufficiency of reason and the design argument to establish the existence of a personal and providential God, rather than the indifferent God of Deism. Debendranath thus came to emphasize the need for an intuitive approach to God, and in the process developed further Rammohan's idea of two levels of knowing into two types of science, empirical and spiritual—a notion that was to be of vital importance in the Hindu interpretation of Darwinism for the generation following Debendranath. Debendranath is the subject of our next chapter.

8 Design and the intuitive theism of Debendranath Tagore

For three nights in 1835 the eighteen-year-old Debendranath Tagore (1817–1905) stayed close to his beloved grandmother as she lay dying on the banks of the Ganges River near one of Kolkata's burning grounds. On the last night, sitting on a coarse mat by the shed where she had been laid, he faintly heard the holy name of God (Hari) being sung to her. A full moon had risen. Suddenly he experienced a profound sense of the unreality of all material things and an aversion to the luxuries of his home life, producing ecstatic joy. This rapturous delight, Debendranath tells us in the opening pages of his autobiography, was vouchsafed to him at this emotionally climactic moment by God himself. He concludes: "Who says there is no God? This [experience of joy] is proof enough" (1994: 3). But it did not remain proof enough for long.

After the funeral Debendranath's joy soon evaporated, leaving him despondent and desperately yearning to re-experience that joy. This longing, in turn, led to an awakening of love for God, of a desire to know God, but he felt encumbered by the luxury and pleasures that surrounded him at home. He found some solace in the following months studying Sanskrit and reading English philosophy. Yet a sense of the meaninglessness of life remained as he pondered the relentless ways of nature that consumed all, "this Nature-fiend" as he described the power that had taken his grandmother away (ibid.: 6). He rejected the atheist's contentment with mere material nature, unwilling to accept what seemed to him a despairing and empty view of the universe. Instead, Debendranath began a quest to obtain direct awareness of God.

In the midst of his struggles, all at once a flash of insight swept over him: knowledge of the body and of the outer world leads to awareness of the spirit within. And likewise, knowledge of the physical universe provides the key to awareness of God:

> I now realized that with the knowledge of the outer world we come to know our inner self. After this, the more I thought over it, the more did I recognize the sway of wisdom operating throughout the whole world. For us the sun and moon rise and set at regular intervals, for us the wind and rain are set in motion in the proper seasons. All these combine to fulfill the one design of preserving our life. Whose design is this? It cannot be the design of matter, it

> must be the design of mind. Therefore this universe is propelled by the power of an intelligent being.
>
> (ibid.: 7)

Debendranath goes on to note God's providential benevolence in endowing humans with the instincts necessary for survival:

> I saw that the child, as soon as born, drinks at its mother's breast. Who taught it to do this? He alone, who gave it life. Again, who put love into the mother's heart? Who but He that put milk into her breast. He is that God who knows all our wants, whose rule the universe obeys.
>
> (ibid.)

With this teleological understanding of nature, Debendranath's "Nature-fiend" seems to have been put to rest, at least for a while. Incidentally, we may recall that the *Nyāya-Sūtras* (3.1.22) provides a different, non-theistic explanation for the infant's instinctive suckling: its unconscious memory of nursing in a prior life—this karmic explanation being favored by non-theistic Neo-Advaitins like Vivekananda, as we shall see.[1]

At this time Debendranath also experienced a powerful sense of God's all-pervading, invisible presence in the universe. These linked insights into God's compassionate wisdom and God's infinite being transcending all forms undermined any lingering attachment to his family's customary modes of worshiping God in concrete images. Debendranath concluded that God

> is neither the *kali* of Kalighat, nor the family *Shalgram*.[2] Thus was laid the axe at the root of idolatry. In studying the mechanism of creation, we find evidences of the wisdom of the Creator. On looking at the starry sky, we feel that He is infinite.
>
> (ibid.: 8)

This critique of his family's religious practices evokes Rammohan Roy's invocation of design to disparage idol-worship. In fact, Debendranath at this point confesses that, once he realized God was without form, "I remembered Rammohun Roy—I came to my senses" (ibid.: 9). In Debendranath's case, his appeal to design served not only as a critique of idolatry, but also as a means of finding spiritual comfort and purpose in the universe. In his grief following his grandmother's death, he found great consolation in Rammohan's teachings and soon accepted the Brahmo faith, although he did not become a formal member of the Brahmo Sabha until 1843.

Debendranath had become well acquainted with Rammohan early in life. Debendranath's family, like Rammohan's, was part of the wealthy land-owning elite emerging in Kolkata at the beginning of the nineteenth century. One of Debendranath's playmates was a son of Rammohan, whose home he often visited. At age nine he enrolled in the Anglo Hindu School that Rammohan had helped to establish in 1822 for the study of European science, philosophy, and literature. From age fourteen to eighteen (ca. 1831–1835), Debendranath attended the Hindu College that promoted an education in European subjects for the Bengali elite to facilitate their entry into the British colonial administration.[3]

The survival of the Brahmo Sabha after Rammohan's departure for England was largely due to the financial support of Debendranath's father, Dwarkanath Tagore, a co-founder and wealthy trustee of the Sabha. But the Society's prospects were in doubt and its prestige in serious decline, until Dwarkanath's eldest son, Debendranath, joined the Sabha in 1843 and took over the leadership. The Brahmo Samaj, as Debendranath renamed it, came to play a critical role in inspiring a variety of Hindu responses to modern science, including responses to Darwinian evolution.

Rammohan's teachings led the young Debendranath to investigate the ancient Hindu religious treatises or *śāstras* on his own. He found a mixed bag, since many of the old *śāstras* "were full of idolatry," a depressing discovery (ibid.: 10). In the midst of this despair, as Debendranath tells the story, a page of Sanskrit text fluttered by him, and out of curiosity examined it. Unable to understand it, he consulted the pandit-minister of the Brahmo Sabha and learned that the text was from the first verse of the *Īśa Upaniṣad*. The *Īśa* verse affirmed that the whole world is encompassed by God, confirming Debendranath's growing feeling of God's all-pervasive nature.

He went on to explore other Upaniṣads in depth. His reflections on them were developed and elaborated upon in his *Brahmo Dharma Grantha* (*Book of the Brahmo Religion*), published in 1848 as a guidebook for the growing membership of the Brahmo Samaj. In this work Debendranath clarified his views on the design argument, the roles of reason and intuition, and set forth a critical interpretive strategy of higher and lower sciences in dealing with the challenge of modern science.

To understand the work itself, we need to look briefly at Debendranath's role in reviving the Brahmo Samaj and at the tumultuous events affecting the Society during the 1840s. Inspired by his Upaniṣadic studies, Debendranath established in 1839 the Tattvabodhini Sabha (Truth-Awakening Society) dedicated to spreading the ideals of Rammohan's Brahmo Sabha. It was also intended to counter the advances of the Christian missionaries, especially the Presbyterian missionary Alexander Duff, in attracting Bengali youth (Kopf 1979: 161–2). Debendranath also set up a Tattvabodhini School to challenge Duff's own school, as well as a Tattvabodhini Press. In 1843 he started a monthly newspaper, the *Tattvabodhini Patrika*, with the rationalist thinker Akshoy Kumar Dutt (1820–1886) as editor, with the twin goals "of combating missionary propaganda and … of educating fellow Bengalis" (ibid.: 163).[4] The example of Dutt's aggressive evangelizing on behalf of Hindus served as a key motivating factor in Debendranath's composing of the *Brahmo Dharma*. Also in 1843 the Tattvabodhini Sabha became even more closely conjoined with the Brahmo Sabha, which Debendranath reorganized as the Brahmo Samaj. Like the Tattvabodhini Sabha, the rejuvenated Brahmo Samaj grew quickly, reaching several hundred members by 1847 (Borthwick 1977: 18).

The Protestant clergymen responded to the anti-missionary tracts of the Brahmo Samaj by deprecating the Samaj's basic scriptures, the Upaniṣads. Far from teaching a true monotheism, the missionaries proclaimed, the Upaniṣads inculcated only an impersonal monism. Debendranath's own study of the Upaniṣads revealed many verses supporting the idea of a personal God. Perhaps prodded by the missionary critique, he soon came to reject Śaṅkara's *Brahmasūtra Bhāṣya*, along with any Upaniṣadic statement tinged with monistic ideas. His own intuitive feelings balked at the Advaita idea of eventual mergence of the individual soul in the Absolute, and equally recoiled from the idea that God becomes the material universe.

In 1845 Duff published his *India and India Missions*, in which he severely criticized the Brahmo Samaj and charged them with believing in the infallibility of the Vedas (by which he meant primarily the Upaniṣads), despite their inconsistencies. These disparagements induced the Brahmos formally to review their position on this issue. A.K. Dutt was utterly opposed to the idea of Vedic infallibility, finding in the Vedic-Upaniṣadic texts errors and contradictions, and he eventually won Debendranath over to this view. The final outcome was a formal decision announced by Dutt in the Society's annual meeting of 1850, that henceforth the Vedic texts were not to be regarded as infallible guides, and that final authority rested not in scripture but in reason and conscience (S. Tagore 1994: ii–iii).[5] As the later Brahmo leader Keshab Chandra Sen commented on this momentous decision, the highest insights of the Vedic texts would be retained, but "their errors and absurdities were abjured" (Keshab 1940: 70).

Duff's critique also aroused Debendranath to collaborate with a young, English-educated student, Rajnarain Bose (1826–1899), to formulate a response to the missionary attacks. Their collaboration resulted in a book, written mostly by Bose, entitled *Vedantic Doctrines Vindicated*, published in 1845. It broached the notion of Hinduism's superiority to Christianity, going beyond Rammohan's general stance that Hinduism was merely its equal. Bose argued that the Vedānta did not espouse idolatry, but rather instilled the worship of God "as the supreme Regulator of this boundless universe and as the glorious and beneficent originator of all earthly good" (quoted in Kopf 1979: 170). He became a member of the Brahmo Samaj in 1846.

The ideals of Dutt and Bose, including their denunciation of missionary arrogance, are reflected in the *Brahmo Dharma*. As the opening passage of the *Brahmo Dharma* asserts, the knowledge of God inspired by study of the universe is open to all: "To be a theist or a professor of theism it is not necessary to belong to a particular country, age or nationality. The theists of all countries have the right to teach about God" (D. Tagore 1928: 1). Debendranath reflects here Rammohan's universalism and faith in the natural reason of humankind.

The *Brahmo Dharma Grantha* and the two sciences

One day in 1848, inspired by what he believed was God's grace, Debendranath felt the spiritual truths of the Upaniṣads dawn in his heart. He asked A.K. Dutt to write down the outpourings of his heart and in a three-hour sitting proceeded to compose the theological portion of the *Brahmo Dharma*.[6] This portion, representing the culmination of Debendranath's decade of study and reflection on the Upaniṣads, takes the form of commentaries on selected Upaniṣadic verses that he considered most illuminating, rearranging and modifying them in accord with his theistic ideals (Sarkar 1928: iv). Four major themes of the *Brahmo Dharma*'s theology are God's creative power, his providential benevolence, the subordination of reason to intuition, and the related idea of two sciences.

Regarding the first theme, Debendranath quotes a number of Upaniṣadic passages demonstrating that God creates all the material elements from which he fashions the universe. Near the beginning of the *Brahmo Dharma* (D. Tagore 1928: 2), he cites the famous *Taittirīya Upaniṣad* verse (3.1.1) that Brahman (God) is that from which all beings are born, by whom they are sustained, and into whom they shall enter in the end.[7] He quotes two more cosmogonic texts, *Bṛhadāraṇyaka Upaniṣad* (1.2.1) and *Chāndogya Upaniṣad* (6.2.1), stressing that God alone existed in the beginning, with nothing else beside him. Debendranath clarifies: "Before creation no other object besides the Supreme god existed. So He did not create [the universe] with the help of other materials like an artisan" (D. Tagore 1928: 11). He continues: "He alone has made this wonderful machinery of the universe by creating all these conscious and unconscious beings by his own innate power and intelligence" (ibid.). And he concludes: "That Almighty Perfect Person has created at His own will, water, air, fire and other materials for fashioning the universe and life, mind, and all the senses" (ibid.: 12). Such a view of creation, similar to the Christian idea of creation out of nothing, differs radically from the Nyāya-Vaiśeṣika idea of permanent atoms and souls, as well as from Rammohan's views.

As for the Advaitin notion that God is the material cause of the universe, Debendranath argues that the Supreme Spirit is distinct from the world, citing the *Kaṭhopaniṣad* (1.2.18) in support. Referring to classical Sāṃkhyan examples of substantive evolution, he argues that, unlike milk transformed into curds, clay formed into a pot, or gold into a ring, God "does not become anything … He is not converted into anything else" (ibid.: 57). Debendranath also denies the Advaitin idea that the world is only an appearance: "As the rope is mistaken for a snake, as the mirage is mistaken for water, as a shell is mistaken for silver, it is not that by some such mistake in Him this universe seems to exist. He has created this entire universe" (ibid.: 57–8). Debendranath thus cleanly dispenses with both classical Hindu theories of transformation (*pariṇāma*) and of appearance (*vivarta*).

Concerning God's providential benevolence, Debendranath sees the universe as devised for our comfort and aesthetic delight. He proclaims that by God's commands—or laws—the sun stands in the center of the solar system holding the Earth and planets in their proper orbits, shedding its vitalizing energy on every plant and animal. Similarly the moon, delighting all with its beautiful nightly changes of dress, follows its assigned path through space, obedient to the laws of God (ibid.: 19). By God's command the days, months, seasons, and years occur regularly, and rivers arising in the snowy mountains flow hundreds of miles to benefit innumerable creatures (ibid.: 20–1). We can trust in the ongoing stability of the universe, Debendranath insists, for God has established "such firm laws that the world can never be destroyed by their change. God holds up all like a dam, so that the worlds cannot be dispersed" (ibid.: 56; cf. 116–17).

For Debendranath all organic beings, from unseen atomic-sized creatures, ants, and worms, to birds, cattle, and men, are recipients of divine care. To each God dispenses food, drink, other necessities, and various enjoyments appropriate to their nature (ibid.: 39). God also endows creatures with suitable physical bodies, with limbs, organs, and senses matched to their needs. Debendranath declares that God "unites arms to the human body and wings to the body of birds. He has given such body and limbs to each creature as is necessary for its life and development" (ibid.: 62).

God has given special consideration to humankind:

> He has given us two hands, so we can catch hold of all things; He has given us organs of movement, so we can walk everywhere; He has given us the organ of speech, so we become happy by giving expression to all the thoughts of our mind …
>
> (ibid.: 81)

Debendranath goes on to remark that God has given humans a priceless gift, religion, which frees us from our sensual ties to the world and supreme self-contentment (ibid.: 82). The law of righteousness is benevolently enforced by God's dispensing reward and punishment to free us from delusion, sin, and death, in order to lead us to divine bliss (ibid.: 39).

Like most proponents of a benevolent designer, Debendranath feels the need to explain the great suffering in the world. His theodicy combines a traditional theistic emphasis on God's regulating and preserving role with elements of the classical Advaita ideal that the Supreme Spirit is an unperturbed witness of all actions, along with the Western ideal of progress. Thus God benevolently but fearlessly turns the wheel of the universe, clinging steadfastly to his role "as the witness and regulator of all" (ibid.: 93). What he witnesses is often fearful, indeed:

> By the current of rivers the towns and the villages on the bank are being demolished; countries are being washed away by the flood; millions of creatures are falling into the jaws of death by the devastating earthquakes and storms, but the All-knowing Beneficent God remains undisturbed steadfastly turning these apparently disastrous events into sources of future progress.
>
> (ibid.)

Debendranath lists other natural terrors, thunderstorms, erupting volcanoes that burn thousands of men and animals, as well as human-caused disasters like war whereby the earth becomes drenched with blood. But God remains calm and undisturbed as he pursues his benevolent purposes, apparently accepting the resulting suffering as part of some necessary evil for the sake of a greater good.

Debendranath's emphasis on the laws of God that regulate the natural world brings him inevitably to reflect on the role of science in revealing God. At times he sounds quite confident, reflecting Rammohan's views:

> Whether it be the astronomer's watching the movements of the heavenly bodies, or the geologist's searching the strata of the earth or the physiologist's studying the Laws of the body or the physicist's enquiry into the Laws of Matter or the philosopher's studying the subtle laws of the mind,—all declare His wonderful and inscrutable power.
>
> (ibid.: 50)

Such confidence was undoubtedly reinforced by A.K. Dutt's own rationalist inclinations. Dutt himself had studied astronomy, mathematics, and physical science as a teenager, and later became a strong advocate of investigating the laws of nature to understand God. But unlike Dutt, Debendranath had increasing doubts about the sufficiency of science to demonstrate God's existence.

Throughout the *Brahmo Dharma* Debendranath insists that God is beyond the senses, visible only to the eye of knowledge (e.g., ibid.: 88, 97, 119). God is hidden and difficult to know. Seekers must "penetrate behind the appearances" and then they "can see Him by their subtle intellect" (ibid.: 89). The only means for realizing God, Debendranath concludes, is intuition based on one's own inner conviction (*ātma-pratyaya*) (ibid.: 73). The term *pratyaya* evokes a sense not just of inner belief in God, but also of trust or reliance in God's benevolence. Debendranath explains:

> Truth is revealed in intuitive knowledge and our Self trusts in that Truth. So this natural intuition is the only (ultimate) means of proving the existence of God. When the Infinite Person is revealed in our innate knowledge through intuition, our understanding finds proof of His wisdom in the creation of the universe and makes known unto us His beneficent purpose.
>
> (ibid.)

Rational understanding of God through evidence of his benign design in the universe thus is subordinated to and supportive of the more immediate intuitive realization of God known in our innermost self.

As Wilhelm Halbfass observes, Debendranath's search for truth, however much inspired by "medieval Bengali mysticism," was also influenced by "European epistemological concepts and motivations" (1988: 396). Specifically, "Debendranath seems to have been influenced by the concepts of self-evident truths, intuitive certainty and common sense which we find in the works of Th. Reid, Dugald Stewart, W. Hamilton and other proponents of the so-called Scottish School of Common Sense" (ibid.). Halbfass also notes that Debendranath's use of the term *ātma-pratyaya* as an equivalent for the European notions of "intuition" and "self-evidence" represents a radical reinterpretation of the Upaniṣadic concept as found in the *Māṇḍūkya Upaniṣad* 7, where it stands for non-dual knowledge of the one Self (ibid.: 570n).[8]

For Debendranath, our innate intuition fully comprehends that, since we are dependent beings, there must be an Independent Person. Such intuitive knowledge requires no proof, being axiomatic. But if doubts arise about this intuition, then for Debendranath all reasoning is undercut. Doubt leads to delusion, fostering disbelief in cause and effect, in our own existence, as well as in the existence of God. God is realized only by those who trust intuition (1928: 119–20). At the same time, Debendranath sees "studying the purpose and wisdom" manifest in nature as a means of awakening divine knowledge (ibid.: 33). Such study strengthens one's intuition and trust in it. Intuition is further developed by shunning the pleasures of the world, controlling one's senses, eschewing sinful actions, renouncing the fruits of action, and uniting one's will with God's to perform his work in the world. One thereby attains God's grace and inner vision of his presence (ibid.: 121–40).

The disjunction Debendranath makes between reason and intuition culminates in his notion of two sciences. Commenting on a verse from the *Muṇḍaka Upaniṣad* (1.1.5) that expounds on a higher and lower knowledge (*vidyā*), Debendranath asserts: "The science by the study of which, that supremely desirable pearl of knowledge [God] can be attained is the true, the highest science [*parā vidyā*], all other sciences are inferior [*aparā vidyās*]" (ibid.: 15). The lower sciences, according to the *Muṇḍaka*, are the four Vedas and their ancillary limbs such as phonetics, grammar, rituals, and astronomy. Debendranath expands the list, indicating that all the lower sciences find their fulfillment in knowledge of God: "Astronomy, Geology, Medical Science, Psychology, Philosophy, Theology and all other sciences declare His infinite wisdom and goodness. One should attain the crown of all sciences, namely the knowledge of God from all these sciences" (ibid.: 33). The *Muṇḍaka* verse, incidentally, soon appeared at the head of the *Tattvabodhini Patrika*, pointing to the shift away from the idea of Vedic infallibility (N.S. Bose 1976: 150). Debendranath's appropriation of the *Muṇḍaka* verse in his *Brahmo Dharma* highlighted the hierarchizing of a higher intuitive science transcending all doubt and surpassing the lower empirical or rational sciences—a dominant theme in later Vedāntic interpretations of modern science and frequently used to subordinate Darwinism to a higher spiritual science.

Himalayan musings and a protest against a chance universe

Around the time that Debendranath composed the *Brahmo Dharma*, his family suffered a series of financial crises, distracting him from his spiritual quest for God. Growing estrangement between himself and friends and relatives due to his rejection of idol-worship further provoked a sense of the meaninglessness of worldly life, strengthening his desire for renunciation. He was partially alienated even from his own Brahmo Samaj due to an emergent rationalistic faction within the Society, exemplified by A.K. Dutt. Debendranath thus resolved to abandon the urban world of Kolkata and to roam freely through nature like the ancient rishis in pursuit of the direct experience of God.

Debendranath departed for the forest in 1856 intending never to return, leaving behind his wife and several children. Wandering for several months in the magnificent splendor of the Himalayan foothills, he repeatedly experienced what he felt were manifestations of the creative power of the infinite yet immanent God of the Upaniṣads, perceived as a nurturing mother as well as protective father. He was moved on many occasions by the natural beauty that revealed the artistic skill of the Creator. One day, for instance, moved by the sight of a creeping vine bearing lovely white flowers, he enthused: "I saw the Universal Mother's hand resting on those small white blossoms … with what loving care had she endowed them with sweet scent and loveliness, moistened them with dew, and arrayed them on the creeper" (1994: 101). God took loving care of nature even when no humans were present to enjoy it. Yet God also provided the occasional human inhabitant with everything necessary for a happy and contented life, as evidenced by a peasant family he observed living in a cave in an isolated valley. Debendranath was impressed by the simplicity of life in the hills, contrasted with his former hectic life in Kolkata.

While Debendranath saw God manifest in many facets of nature, he continued to stress that God is not directly perceived, elaborating upon this theme from the *Brahmo Dharma*. The senses can observe God's power, but not God himself. We can perceive the power of the life-force in a tree, but not the life-force itself, instilled in it by the will of God (ibid.: 108). The senses see only the outward reality, not the inner spirit, nor the Spirit dwelling in all things. Trust in inner intuition is necessary to realize God directly, as a self-evident truth. But some humans lack this trust, succumbing to doubt, and thus attribute regulation of the world to natural causes:

> Deluded by ignorance, some scholars say it is by the laws of nature,—by the blind force of matter; or some say that, without any cause, it is by the force of Time alone that this wide world goes round. But I say—it is the glory of that Supreme Deity alone, by whom this universal wheel is being turned.
>
> (ibid.: 107)

For Debendranath nothing happens in the world by chance. One day, while gazing on a turbulent river descending to the plains far below, he heard within the "command of the All-ruling One" to return, like the river, to the plains of his home, and to make known the truths discovered in the mountains. He arrived back in Kolkata shortly after the end of the Great Indian Rebellion of 1857.

Post-Himalayan paean to design

Soon after his return home Debendranath delivered a series of extempore sermons, taken down by his son Satyendranath and others, later published as the *Brahma Dharma Vyakhyan* (*Exposition of Brahmo Principles*).[9] He once again sets forth the design argument, now thoroughly infused with the notion of God's compassionate providence, and appealing as much to poetic imagination as to rational inference.

In a sermon entitled "God, the Creator," Debendranath presents a mystical rhapsody on creation, starting by paraphrasing the well-known cosmogonic verses from the *Bṛhadāraṇyaka Upaniṣad* (1.2.1) and *Chāndogya Upaniṣad* (6.2.1). He had utilized these same verses in the second chapter of the *Brahmo Dharma*, but there had emphasized God's creation of the elements from which he fashioned the universe, thereby distinguishing God from human artisans. In this sermon, Debendranath briefly reiterates that God alone existed in the beginning, but then expounds on how God willed out of the surrounding darkness all that exists, beginning with the sun whose light dispelled the darkness. God next brought the Earth into being and set it rotating around the sun. Debendranath notes that the Earth's original condition was not particularly pleasant, its surface covered with molten metals and its atmosphere permeated by gaseous vapors and massive clouds. He then muses:

> Who could have known … that such a world should eventually be adorned most wonderfully with life and joy, light and beauty, and be filled with numberless living creatures, and innumerable spacies [sic] of plant life? Who was it that sowed in this world the seeds of all these things? Who created it as the store-house of an infinite variety of mineral wealth, grains, flowers and fruits? There shines the sun millions of miles away from us, here rolls the earth in its orbit, and on its surface are all these animals and plants! But yet from that far-off sun comes the light which illumines the world, and makes the stream of life flow, and dispels our blindness. Who is it that has established such relationship between the earth and the sun? Is it the work of an insensate power? This life, this vitality, this wealth of various possessions, this happiness which we enjoy—are all these showered on us by a blind power? No, it is by the will of that Being who is All-wise and All-good, that all this that is, has come into being. God meditated, and then implanted in it those diverse and wonderful forces that gradually rendered the once hot, lifeless and desolate world a home and a place of comfort such as it is now.
> (1994: 128)

As throughout his life, Debendranath cannot believe that material and blind forces could create the order and magnificence seen in the universe.

Debendranath next sketches in the phases of terrestrial development, beginning with the appearance of rains to create the oceans wherein aquatics dwell, followed by the lifting up of mountains creating a division of land and water. Soon plants, trees, and animals are thriving on the land. Following this Genesis-like narrative,[10] Debendranath again raises the question of first cause:

> Did all these come to exist of their own accord? Is it all the work of an unthinking power? No, all this is the glory of that All-knowing Supreme spirit. It is He who created and constructed this universe so wonderfully. He gave us teeth to masticate our food, and before he gave us teeth He had poured milk into our mother's breast for us to drink and subsist upon. What marvellous art is this! How wonderful is His power to preserve and nourish His creatures! Are all these evidences of art the result of the action of blind

forces? Do they not manifest knowledge and wisdom? Do they not bear on them the proof of the will and thought of an intelligent being?

(ibid.)

Having indicated God's compassionate concern for our well-being, Debendranath briefly attempts another explanation for the suffering in the world. He argues that, while prosperity allows us to see God's "benign countenance," adversity "teaches us noble lessons and leads us unto Him" (ibid.: 129). He leaves unaddressed the problem of the massive tragedies that wipe out millions of human beings that he had struggled with in the *Brahmo Dharma*.

Conclusion

In many ways, Debendranath's emphases on an intuitive approach to reality and on the limitations of reason and rational analysis of nature parallels the Romantic response to the Enlightenment in Europe. Like the Romantic poets, he insists on God's indwelling presence in both the world and the human heart. Like them, he proclaims that the spiritual is not revealed directly through scientific investigation but through personal and intuitive response to the beauty of nature (cf. Barbour 1997: 41), but Debendranath combines his Romanticism with a Hindu renunciatory flare.

It is his heavy reliance on devotional intuition rather than on humankind's rational abilities and intellectual knowledge that separates Debendranath from the deistic Vedānta of Rammohan Roy (cf. Heimsath 1964: 76–7). It is also what eventually drove a wedge between Debendranath and his friend A.K. Dutt. From Debendranath's perspective, Dutt was much too influenced by the anti-Christian Deists in discarding the Vedāntic element of Rammohan's thought. Dutt firmly believed that only an intelligent being could create something as complex as the watch-like universe, but saw God as little more than the watchmaker. He came to interpret God as the Cosmic Atom or Seed (Visvavija), in effect identifying God with Nature (Chakraborty 1979: 45–6). He completely rejected prayer and meditation, as well as scriptures, as a means of attaining knowledge, advocating instead the pursuit of the natural sciences and the laws of nature (Kopf 1979: 50; Rambachan 1994: 260–1). Debendranath was fully aware of the differences between himself and Dutt, observing in 1858: "We were poles asunder, as I was seeking to know my relations with God while Dutt was seeking to know the relations of man with material objects" (quoted in Kopf 1979: 50).

Debendranath's disillusionment with Deism may also have been provoked by Alexander Duff's use of the design argument to convert Hindus to Christianity. In his *India and India Missions*, the missionary had argued that one effective way to convert Hindus to Christianity was by first teaching them science, in order to destroy their faith in the superstitious practices of Hinduism. No direct attack on the Hindu scriptures would be needed, for the collision between scientific truths and the teachings of the *śāstras* would be self-evident (Duff 1988: 580–92). Then, to avoid Hindus falling into skepticism, the missionary could use the design argument (very much in the style of Paley) to instill belief in the Christian God (ibid.: 656–67). Debendranath's distrust of the missionaries could easily have encouraged his cautious attitude towards natural theology and the design argument. At the same time, he saw in the design argument the surest rational proof against the idea of a godless and purposeless universe—an idea about to be severely challenged by the non-teleological explanation of nature found in Darwin's theory of evolution.

The year 1858 saw the first public appearance of Darwin's evolutionary theory, in a paper read at the Linnaean Society in London (along with Alfred Russel Wallace's paper propounding the same theory). It was also the year of Debendranath's return from the Himalayas. Upon his return to Kolkata, he met an energetic and talented youth, Keshab Chandra Sen, whom he took under his wing. Keshab was soon to become a major force in the Brahmo Samaj. He developed Debendranath's stress on intuition, assimilating it to classical Hindu notions of yogic intuition, and elaborated upon his mentor's notion of the two sciences. Keshab was also the first Brahmo leader to take up the challenge of Darwinian evolution. We shall examine his responses to and assimilations of Western science and evolutionary views in the next chapter.

9 Design and the religious evolutionism of Keshab Chandra Sen

Along with Rammohan Roy and Debendranath Tagore, Keshab Chandra Sen (1838–1884) was one of the most prominent figures of the Brahmo Samaj. Keshab joined the Brahmo Samaj in 1857 while Debendranath was wandering in the Himalayan foothills. Like Debendranath, Keshab came from a wealthy and influential Bengali family serving the British administration, and both attended Hindu College in Kolkata for part of their education. Concerning issues of social reform, however, Keshab was more radical than Debendranath, perhaps due to the influence of the Unitarian minister Charles Dall who had arrived in Kolkata in 1855 and quickly befriended the adolescent youth. While both Keshab and Debendranath had been mediocre students at Hindu College, Keshab was "a restless intellectual" far more interested in new ideas and forming new organizations (Kopf 1979: 254). Nonetheless, the two shared a number of philosophical and theological perspectives with regard to the relationship of science and religion.[1]

For Keshab the last few years at Hindu College marked a time of spiritual crisis. His ascetic inclinations, nurtured by a growing sense of personal sin imbibed from Christian missionaries, led to a deep depression, precipitated in 1856 by his arranged marriage to a girl of nine. His resentment over the marriage and the resulting alienation from his family was exacerbated by public humiliation over a charge of academic cheating. He found solace in prayer and in missionary friends with whom he was studying the Bible. At the same time his Hindu friends ridiculed him for his biblical studies and were increasingly suspicious that he was about to embrace Christianity.

Keshab turned to the Samaj after reading Rajnarain Bose's tract "What is Brahmoism?" which set forth a number of themes that resonated with Keshab: Unitarian universalism, the common theistic intuitionism found worldwide, and the science of religion. Included in the latter was the idea that Brahmoism, with its tolerance and emphasis on the essential unity of religions, represented a later and superior stage of religious evolution compared to Christianity (Kopf 1979: 67–8).

When Debendranath returned to Kolkata in 1858, Keshab soon found in him a sympathetic father figure and spiritual teacher. Gifted with considerable oratorical skills and with Debendranath's support, Keshab quickly rose to a position of leadership in the Society. One of his most important early activities was to set up a Brahmo missionary system modeled on the Protestant missions. His own missionary tours to the British Presidencies of Chennai and Mumbai led to the founding of similar societies in those cities, although they quickly became more conservative on matters of social reform than Keshab would have wished (Damen 1983: 110).

Among their shared theological ideals, Debendranath and Keshab held strong teleological views of nature combined with unwavering faith in divine providence. Both men were also deeply concerned with the skeptical faction within the Brahmo Samaj and emphasized the necessity for intuition in approaching God to correct and deepen the partial insights of unassisted reason.[2] And using the idea of intuition, Keshab further developed Debendranath's notion of two sciences.

Despite their similar theological outlooks, tensions between Keshab and Debendranath arose in the early 1860s involving their divergent strategies for social and religious reform. Keshab espoused a more accelerated course, attracting younger Brahmos less patient in their pursuit of social change than Debendranath's disciples. Keshab's party was also more oriented to a universalist perspective and less concerned with preserving a distinctive Hindu identity that was becoming increasingly nationalistic.[3] These tensions were exploited by Keshab's detractors during his missionary absences from Kolkata. The strains led in 1866 to a formal split in the Samaj, with Keshab's faction becoming known as the Brahmo Samaj of India, in contrast to Debendranath's Adi (Original) Brahmo Samaj.

In 1870 Keshab traveled to England where he was introduced to such influential people as Queen Victoria, John Stuart Mill, F. Max Müller, and Prime Minister William Gladstone (N.S. Bose 1976: 158). He also met with several Unitarian leaders and with members of the Swedenborg Society, where he may first have encountered the phrase *new dispensation*, which was to become the name of his universalist church established near the end of his life.[4] During his six-month stay in England, Keshab was deeply impressed by Western technology but equally disillusioned by the squalor of the masses in London and by the way Christianity was actually practiced. He concluded that India had much to teach the West regarding spiritual matters, while the West had much to give to India regarding science and technology—an increasingly common trope in the Hinduism and science discourse. Yet regarding society, he felt that England had a divinely inspired, providential role to play in India in helping to abolish caste, idolatry, and other social evils.

Shortly after his return from England in 1872, Keshab met Dayananda Saraswati, founder of the Arya Samaj, and three years later Ramakrishna, whose disciple Vivekananda was to play a crucial role in molding Hindu perceptions of modern science. The meetings with Dayananda and Ramakrishna further stimulated Keshab's growing appreciation of his Hindu heritage. Although Keshab never learned Sanskrit, he began increasingly to include Vedic and Upaniṣadic quotations along with his customary biblical citations in his sermons and lectures.

Some time in the early or mid 1870s it seems that Keshab first encountered the evolutionary theory of Darwin. Perhaps he had learned of it during his stay in England. His first explicit reference to Darwinism that I have found is in an 1877 lecture (discussed below). Keshab soon developed an evolutionary history of religion, building on Bose, and intertwining notions of organic, cultural, and spiritual evolution. His Church of the New Dispensation, which he formally announced in 1880, would fulfill God's progressive dispensations to humankind. The Church did not survive Keshab's death from diabetes four years later. It clearly had depended in large part on the force of Keshab's own personality, and there was no one to replace him after his demise.

While Keshab's Church did not survive his death, his impact on future Hindu responses to modern science and Darwinism—especially as filtered through Swami Vivekananda's Neo-Advaita—was immense, although most of the patently Christian elements became greatly attenuated. Three major, intertwining themes in Keshab's thought are of particular relevance: (1) the harmony of science and religion; (2) the spiritualization of science and its converse, the scientizing of religion; and (3) the evolution of religion. The first two themes reflect Keshab's elaboration of Debendranath's two sciences infused with European philosophical and theological ideals. The last, building on the first two and incorporating aspects of Bose's science of religion, represents Keshab's special contribution.

The harmony of science and religion

Keshab viewed the present age of science with optimism, for he saw science as doing away with a dark age of dreams, visions, and superstition dominated by a benighted and carnal priesthood (1901: 394). At the same time, he deplored the thoroughly mechanistic perspective of modern science, warning us that the machine-like order revealed in nature can all too easily lead to Deism and the rejection of an indwelling, personal God. God is not only Creator, but also Preserver, and thus more than the watchmaker of traditional natural theology: "Does the universe bear the same relation to God as the watch does to the watchmaker? Certainly not" (1979: 76). God is the living power sustaining the universe at every moment.

From Keshab's viewpoint, the ancient attitude that regarded science as religion's handmaid has given way to a very different and unfortunate perspective: "Surely it is a scandal to modern science that it conceals the Creator whilst it upholds the majesty of his laws" (1940: 282). Following the lead of the Scottish philosopher William Hamilton, Keshab argues that this development is due to an obsessive focus on the study of nature. He claims that an exclusive commitment to investigating the natural world leads to materialism, for the student "sees nothing but a series of secondary causes and the workings of blind necessity and mechanical laws" (ibid.: 124).[5]

Atheistic and deistic delusions, from Keshab's perspective, are understandable inasmuch as the natural world constitutes a veil cloaking the presence of God.[6] In an 1880 lecture on "God-Vision in the Nineteenth Century," Keshab develops a number of metaphors to illustrate this point. At the same time, he emphasizes the potential role of modern science in bringing about true knowledge of God. One such metaphor is that of nature as a tabernacle whose walls are inscribed with a mysterious code. While ancient priests struggled to interpret the obscure handwriting of God on the walls of Belshazzar's palace (Book of Daniel 5), the task of the modern priest in the Temple of Science is to decipher God's handwriting on the walls of his vast tabernacle of the universe. And those whom Keshab calls true scientists can perceive directly the Invisible God of Nature, can "see the Living God with the naked eye" (1901: 397). To justify this claim, Keshab says: "I have only to invoke optics. May the science of vision enable us to see the True God in the light of day!" (ibid.: 405). Keshab at this point leaves undeveloped the idea of the science of spiritual vision.

Shifting his metaphorical focus from the cosmic to the organic, Keshab goes on to describe the body as the Temple of the Living God, whose vital force is hidden to the casual observer but readily apparent to the "scientific" observer. He briefly describes the physiology of the body, its pulsating blood, respiratory process, and muscular action. He asks if it is only one's own "blood-force" that propels the fluid through the arteries. No, he replies, for "beneath that force I feel … the hand of the Living God plying the machinery of the blood system in the human body" (ibid.: 409). Similarly, Keshab continues, it is God who works the lungs and activates the muscles. The body is dead matter without the enlivening "God-force" that emanates from the Living God enthroned in the center of the bodily organism. Keshab here echoes another of the Scottish "Common Sense" philosophers, Thomas Reid, who proclaimed that the various involuntary bodily functions such as growth, digestion, illness and recovery, and the circulation of the blood are performed "by the power of some [supernatural] agent; but they are not done by our power" (1915: 109). Keshab further notes that the hidden vitality of a giant tree is to be discovered by uncovering its roots, wherein lies the root-force, another manifestation of the God-force.

Perhaps Keshab's most striking metaphor illustrating the hidden nature of God is the much-loved watch analogy of Western natural theology. He argues that, just as the gears and springs of a watch are hidden behind the dial, so the divine motive-force driving the universe lies hidden behind the façade of nature. Only clear vision (God-vision) can remove the covering dial of the universe to reveal the divine machinery within: "In clear vision you apprehend the Force of Forces. Lo! what was veiled has been revealed … You have removed the dial, and the beautiful machinery within, which was so long concealed before the atheistic eye, reveals itself in Theistic God-vision" (1901: 412). With the dial removed, one sees the secret workings of the universe:

> Lo! how beautifully those wheels work! You rejoice as you see those wheels revolving. What is the motive power, the primary agent? There is a force behind, a hand inside moving the hands outside, a mainspring moving the whole machinery. It is only by [re]moving the dial that you see within, and all things reveal themselves unto you.
>
> (ibid.: 413)

The point of the metaphor is clear, as Keshab explains: "View the universe as a vast machinery, the Lord is the mainspring … It is thus that science enables you to realize your God" (ibid.: 415). Keshab interjects here that science also enables one to realize God when the universe is regarded as a tree with God as its root, or as a child with the Lord as its mother, harkening back to one of Debendranath's favorite images of God.

The skeptic, of course, may ask if the machinery of the clock-like universe is perhaps self-regulating and self-moving, or if trees grow by their own innate vitality. What if the atheist's intuition leads to the conclusion that natural causation suffices to explain the ongoing movement and processes of the universe? Keshab simply pleads:

> Tell me not that an unknown Divinity, after creating the universe, left it to itself and went to sleep behind the clouds. Tell me not that mind and matter work with independent force, and have no connection whatever with the Creator. An absent deity is a fiction. A self-moving universe is a dream of sceptical minds.
>
> (ibid.: 419)

While the notion of God-vision as scientific tantalized Keshab's syncretistic mind, and while he enthused about its great powers of discovery, his explanation of what God-vision actually is remains vague in his 1880 lecture. We shall return to this motif of scientific "God-vision" in considering Keshab's scientizing of religion.

True science for Keshab, over against its modern manifestation, is science in the traditional sense, as knowledge in general, encompassing both the material and spiritual realms.[7] Keshab finds it astonishing that modern scientists proudly profess their ignorance of divinity, and equally bewildering that they are honored in the world. "These [scientists] are surely ignoramuses and they deserve to be pitied as such" (Keshab n.d.: 6). By contrast, the "true first class scientists possess a deep knowledge of science, the science of matter as well as the science of the mind, the sciences of heaven as well as the sciences of earth" (ibid.: 7). For Keshab, then, when science and religion are properly understood, they are in complete harmony. He elaborates upon three intertwined aspects of this harmony: complementarity, hierarchy, and ultimate unity.

Keshab often expresses the idea of complementarity between science and religion in terms of cooperation between the "West" and the "East." Shortly after his departure from England in 1870, in a letter to his new English friends, he implores England to bring to India her charities, industry, earnestness, and "respect for science," while the East will convey to the West her "sublime devotion, fervent faith, and deep spirituality" (1940: 270). Keshab further develops this model of West–East complementarity by evoking and modifying the famous European analogy of the two books of God, nature and revelation: "Thus shall the Scriptures of science in the West and the Scriptures of inspiration in the East constitute together the Word of God" (ibid.: 270–1). European science will rectify the errors and deficiencies of the East's intuitive communion with nature that had become encrusted with superstition and idolatrous rites; eastern spirituality will rescue the Western world from the "horrors of agnosticism" (1979: 287). Interestingly, he equates agnosticism with the "killing doctrine of an Unknowable fiction" (an apparent allusion to Spencer's Unknown), a doctrine that Keshab asserts "has ruined Europe, by removing the knowable and Holy God from all objects in Creation" (ibid.).[8]

Keshab's ideal of cooperation presupposes a hierarchy between the natural sciences and knowledge of God, harkening back to Debendranath's ideal of the lower and higher sciences. Science, wittingly or otherwise, contributes to faith—becomes once again faith's handmaid. With reference to the giants of Western evolutionary thought, Keshab thus proclaims:

> I honour Huxley and Darwin, and all other men, who by their skill are qualified to evolve the latent meaning of the universe. Let them all show the wonders of the Creator. They are aiding me and my work—the work of the world's salvation.
>
> (1940: 355–6)

At the same time, Keshab is concerned that his fellow Indians, many of whom have renounced the worship of visible idols of stone and clay, will now turn to "the invisible idols of modern scepticism, the superfine quintessence of an Unknowable, … impersonal Abstractions of Force and Law, unconscious Evolution, and mysterious Protoplasm, all shadowy and unreal" (ibid.: 388). It is the Living Spirit, for Keshab, which endows dead matter with life, opposing the notion of an "unconscious Evolution." He pleads with his disciples not to discount science, but to embrace the scriptures of science as God's own, more venerable even than the scriptures of individual religious traditions: "Ye shall respect science above all things, the science of matter above the Vedas and the science of Mind above the Bible. Astronomy and geology, anatomy and physiology, botany and chemistry are the living scriptures of nature's God" (ibid.: 389). The subordination of the Vedas and the Bible is not a reversal of Keshab's science–religion hierarchy, for above such parochial scriptures and the still higher scriptures of the sciences are those of the soul, including those of philosophy and logic, yoga, prayer, and inspiration.

The hierarchical harmony of religion and science is possible, Keshab argues, because of their shared quest for unity. Once again referring to the giants of Western evolutionary thought, he notes: "The Darwins and Huxleys, the Tyndalls and Spencers of modern times are all engaged in the work of unification. They find many species, many forces, and they try to reduce them to one" (1901: 406). These scientists are thus "unconscious labourers in God's vineyard, and lay ministers in Nature's tabernacle in so far as they are mightily endeavouring to evolve unity and advance the cause of science" (ibid.). We may note here that Keshab implicitly endorses the Darwinian notion of the common ancestry of all species as a key part of his evidence for the claim that science and religion constitute a unity. For Keshab both the ancient Vedic seers and modern scientists seek unity in nature, and ultimately the one primary force hidden behind nature. That one force underlying all forces, including electricity and light, Keshab enthusiastically calls

"God-force," a "personal creative force" that unravels all the mysteries of nature (ibid.: 409).

It is this ideal of unity that leads Keshab to proclaim that he finds as much science in prayer as in the modern scientific discoveries of the locomotive engine, the microscope, and the telegraph. He thus arrives at his grand conclusion: "All science is religion, and all religion is science" (1940: 356). Let us now see how Keshab brought about this spiritualizing of science and its converse, the scientizing of religion.

The spiritualization of science and the scientizing of religion

Near the end of 1883 Keshab completed a series of articles collectively called "Yoga—Objective and Subjective," published the following year after his death on January 8. In these articles he presents a grand synthesis of his major ideas, including particular attention to "the optics of yoga" that elaborates upon the idea of "scientific God-vision" developed in his 1880 lecture. In this context he emphasizes not the similarities between the ancient Vedic seers or rishis and modern scientists, but their differences: "The Rishis were Nature's poets, overflowing with imagination and gushing sentiment. How dry, unimaginative, cold and undevout the votaries of protoplasm and evolution!" (ibid.: 540). The ancient seers, in their child-like simplicity, "went far beyond the Huxleys and Tyndalls of our day in the search of the Unknown. They stopped not at dead material force, like modern pilgrims of science, but went further and reached the Temple of Living Personal Force" (ibid.). The rishis' discovery of an efficient and personal First Cause within nature was not the result of the tiresome processes of induction or deduction, but of intuitive insight, "the germ of scientific yoga vision" (ibid.).

This yoga vision is possessed only by a few who are endowed with "the favoured eye of the yogi" (ibid.: 541). This yogic eye allows them to see the presence of the living God in the celestial bodies as well as in the ordinary terrestrial elements of fire and water. To be sure, the primitive yoga vision of the Vedic rishis "is more poetical than philosophical, more mystical than scientific, more a matter of faith than of reason or thought" (ibid.: 542). Nonetheless, Keshab assures us, it is "a sound mental process, quite capable of psychological verification," and confirmed by the great spiritual leaders of the world's religions (ibid.: 542). He thus regarded both Moses and Jesus as true yogis with an exceptional degree of divine perception. Moses perceived God in a burning bush. Jesus with his "baptized eye"—going beyond those who rely on their uninitiated eye and see the sky only as sky—saw through the material veil of the sky to behold "the effulgence of heaven" and "the person of the Holy Ghost" (ibid.: 543). Through baptism, inspiration, and the "eye of faith," we can all "see God in nature as he [Jesus] did" (ibid.: 544).

Vedic yogic vision, despite its primitive and poetic character, still contains a fundamentally scientific disposition. When stripped of all local and accidental accretions such as polytheism and pantheism, the logical and scientific basis of yogic vision emerges. The nature and moral fruits of yogic communion with nature's God, Keshab claims, will become "evident to every believer, when he actually tries to realize it in his own consciousness, according to strict scientific rules and under proper logical tests" (ibid.: 545). Keshab leaves unclarified what constitutes the underlying scientific methodology or, in his words, the "strict scientific rules."

Such scientific yoga, Keshab tells us, can be practiced anywhere, in secluded forest retreats or in the bustling life of manufacturing, commerce, and politics. But the example he provides conjures up the ambience of a traditional forest retreat. The practitioner of an updated Vedic yoga first observes the beauties of nature surrounding him, oblivious to worldly concerns:

> Let us fancy a scientific yogi of the nineteenth century, freed from all superstition and versed in modern philosophy, quietly meditating on the works of nature … absorbed and entranced, he asks,—whence this beauty, this order and regularity, this growth and movement in nature? What is it, who is it that sustains and quickens the universe? Suddenly he starts, and almost instinctively he seizes the efficient cause, the prime force which is in all and through all … he feels that here faith and science harmonize, the seer and the scientist are one.
>
> (ibid.: 545–6)

In this manner the yogi joyfully attains a scientifically accurate picture of the universe.

Thus the scientific yogi perceives his God directly in or through the laws of nature, in the force of gravitation, in steam, electricity, and all other forces and objects. In this way, the presence of God is revealed in the astronomical observatory, the laboratory, the museum, the dissecting room, and through the telescope and microscope. The yogi, adept in every field of physical science, never fails to find the personal God in all his investigations:

> When the yogi scientist observes the rolling planets above, as an astronomer from his high observatory, every discovery he makes leads him into the very bosom of that Mighty Architect, who holds those stupendous orbs in his right hand … The more he studies the mathematics and the mechanics of physical forces, the more vividly does he see the great Mathematician and Mechanic who has constructed this wondrous mechanism. And if as a botanist or as a physiologist he deals with the countless forms of vegetable and animal life, his mind runs through all creature-life into the primary vitality of the living God.
>
> (ibid.: 548)

Keshab claims this science-God is indeed the "God of Natural Theology," but is alive, awake, and ever active, unlike the "unknown or distant or sleeping Divinity" of Deism (ibid.: 548–9). The above passage is reminiscent of Debendranath's paean to the investigations of the astronomer, geologist, physiologist, and physicist, all revealing God's wonderful power.[9]

In looking over the many pronouncements Keshab made about yogic science, we find that he utilizes a syncretic mixture of mainly Christian motifs with a smattering of traditional Hindu epistemological notions. Like a nineteenth-century Unitarian, he speaks frequently about the human capacity to discern God in terms of spontaneous reverence, immediate intuition, the favored eye, or eye of faith. These ideas conform in part to Debendranath's emphasis on intuition. They also merge with the notion of the spiritual eye of yoga, recalling the classical Hindu *pramāṇa* (means of knowledge) of yogic perception. Keshab assumes that yogic perception, being based on direct experience, is self-validating. Its experiential basis makes it "scientific," but such a claim conflates and confuses scientific empiricism with personal experience. By such means, Keshab not only affirms the scientific nature of Hindu yoga, but also thoroughly spiritualizes science—an interlinked process that becomes the hallmark of later Hindu approaches to modern science.

However scientific Vedic yoga vision may or may not be, it represents according to Keshab merely the first stage in the history of religion. In this stage the design argument featured prominently, although supplemented by intuition and yogic perception. The design argument for Keshab thus belongs to the childhood of humankind's religious history. In tracing out the further phases of yoga into polytheistic and pantheistic stages, he follows the path of several nineteenth-century Christian theologians and philosophers of religion. Let us then turn to his evolutionary reconstruction of religious history and the progressive self-revelation of God.

God's progressive self-revelation and the evolution of religion

The idea of successive manifestations of God in the world is not an alien concept to the Hindu tradition, as seen in the long-established doctrine of avatars or descents of the divine to help humankind in age after age, in the never-ending cycle of yugas. Quite different is the idea of linear and progressive self-revelations of God in history culminating in a future spiritual fulfillment of the historical process. Keshab introduced into the Brahmo Samaj just such a notion, with Christian ideals of salvation history clearly behind it.[10]

As indicated above, for Keshab humankind's earliest apprehension of God is in nature, where we observe his wonderful, orderly, and beneficent works. This primary revelation is available to all and compels our reverence for God (1979: 75). But God is manifest not only in the natural order of the cosmos:

> There is another revelation: there is God in history. He who created and upholds this vast universe, also governs the destinies and affairs of nations. The same hand which we trace in the lily and the rose, in rivers and mountains, in the movements of the planets and the surges of the sea, regulates the economy of human society, and works, unseen, amid its mighty revolutions, its striking vicissitudes, and its progressive movements.
>
> (ibid: 77–8)

Keshab eventually developed a theory of several stages in humankind's perception of the Divine, culminating in true theism and the discovery of God in history (1940: 280--5, 535–55). The earliest, primitive intuitions of ancient humanity regarding the existence of some divine force in nature came to be rationalized in polytheistic and pantheistic ways. Science attempted to avoid these two evils, but ended up throwing God out in the process. True theism also came to reject these two, positing one personal God distinct both from matter and from the impersonal "It" of pantheism. The progression from polytheism to pantheism to theism, incidentally, was a common notion among nineteenth-century European philosophers of religion.[11]

Keshab's particular evolutionary reconstruction of religious history may have been especially inspired by the evolutionary approach to religion of his friend, the great German-British Orientalist Friedrich Max Müller (cf. Killingley 1995: 185–6). Müller saw in the Vedas the polytheistic childhood of Indian religion, characterized by communion with and worshiping of the forces of nature. He characterized ancient Aryan (Indian) religion as "a worship of *God in Nature*, of God as appearing behind the gorgeous veil of Nature, rather than as hidden behind the veil of the sanctuary of the human heart" (1873: 159). Semitic religions, according to Müller, may be characterized as "a worship of *God in History*" (ibid.: 157). But unlike Müller, who questioned Darwin's account of human evolution, Keshab seemed to have had few doubts about the truth of Darwinian theory, subsuming it into his notion of the spiritual evolution of humankind. Keshab clearly was no Darwinian in a strict sense—he simply commandeered Darwinian ideas in developing his own theory of God's dispensational history.

Keshab's earliest mention of Darwin and Darwinism seems to be in his 1877 lecture, "Philosophy and Madness in Religion." Following a discussion of the relevance of meditation in today's world, he turns suddenly to a new and timely topic:

> Let me proceed to take up the great idea of the day—I mean Evolution. I am not going to discuss the details of the philosophy of evolution. Your protoplasm, your natural selection, I leave to be discussed by men like Huxley and Darwin. The question perhaps is not so serious after all, whether men have descended from inferior animals. But whether there is a progressive evolution going on in the individual life of man is a question in which we are all interested.
>
> (1940: 340–1)

Here we see the beginnings of a rhetorical strategy that often dominates Hindu discourse on Darwinism: glossing over the biological details of organic evolution and focusing instead on the individual soul's spiritual progress. Keshab immediately goes on to affirm unequivocally the truth of organic evolution, but insists that the process does not stop with the physical, culminating rather in a spiritual perfection that transcends the material realm. Remarkable in his argument is the facile transition from the biological struggle for survival to the moral struggle within humanity:

> It cannot be denied that man is developed out of the animal, and that humanity was originally a form of animality. How does human life originate? In the embryo. The embryo grew into a reptile, and a little biped, and then into a ferocious beast, with instincts and passions fully developed; and it is after many struggles that out of this animal humanity is evolved in the fulness of time … The animal lives in us still, and wars with incipient humanity. Now triumphs the animal, and now the man. If the lower passions repeatedly win, and if man wholly succumbs and yields to them, the man sinks in the brute. But if the war goes on, the ultimate result of this protracted series of struggles will be the evolution of pure humanity.
>
> (ibid.: 341)

Interestingly, in a sermon two months earlier, Keshab had spoken of this moral struggle in terms of humankind's development from an insentient "thing" into an animal and then into a person or human, but without any explicit reference to Darwinian evolution (ibid.: 329). Rather, the context there was to refute the Christian notion of original sin, in the sense of humans being born in a depraved state. For Keshab, since we begin as things without volition, we cannot at that stage be morally responsible and thus have no original sin.

In "Philosophy and Madness in Religion," Keshab posits four stages of evolution through which humankind must pass: the inorganic, the carnal or bestial, the human, and finally the divine. Scientific men—Keshab is alluding to European scientists—stop at humanity, but sages inspired by spiritual ecstasy or madness go on to recognize a higher stage of evolution that results in the transmutation of humankind into a new type: "The highest evolution is regeneration—the destruction of the lower type of humanity and the evolution of a new species of godly humanity—life divine instead of life human" (ibid.: 342). This doctrine of spiritual evolution, he claims, is "thoroughly scientific" (ibid.: 341).

Keshab provides his most effusive and elaborate interpretations of evolution in his 1882 lecture entitled "That Marvelous Mystery—The Trinity," in which he sketches out a grand evolutionary cosmogony. Prior to creation, he proclaims, only darkness and silence reigned. Therein lay hidden the supreme Brahman of the Vedas and Upaniṣads, which Keshab identifies with the mighty Jehovah of the Bible. A voice is heard, call it the Word, or *Logos*, or sound, or breath—Keshab thus encompasses both biblical and Vedic metaphors for the creative force. The voice spoke, "Let there be light," and creation commenced. Creation was not accomplished all at once, Keshab insists, but is an ongoing process: "It is nothing but a continued evolution of creative force, a ceaseless emanation of power and wisdom from the Divine Mind" (1979: 225). He then expands on the nature of the Logos: "What a grand metaphor is the Logos! The Hindu, too, like the Christian, believes in the continued evolution of the Logos, and its graduated development through ever-advancing stages of life" (ibid.: 226).

To explicate these ever-advancing stages, Keshab introduces the notion of an evolutionary "Indian Avatarism," an idea probably rooted in the writings of the Theosophist Helena Petrovna Blavatsky.[12] He briefly defines this concept as the temporal sequence of various incarnations of God, recounted in the Purāṇas, arising "from the lowest scale of life through the fish, the tortoise, and the hog up to the perfection of humanity" (ibid.). Purāṇic avatarism, Keshab informs us, is "a crude representation" not only of "the ascending scale of Divine creation," but also of "the modern theory of evolution" (ibid.). Here, by the way, is a clear example of the scientizing of a traditional religious concept, and Avataric Evolutionism continues to find sympathy among Hindus in the twenty-first century.

Keshab reiterates the four stages of evolution he had previously noted: vegetable life from gross matter, animals from perfected vegetative forms, humanity from the highest animal forms, and, finally, the culmination of the whole evolutionary process when, during the course of "progressive humanity," man "becomes the Son of God" (ibid.). Keshab next correlates the progressive stages of human evolution, that is, spiritual evolution, with successive revelations of God. He interprets these stages and successive revelations in terms of the Christian Trinity of Father, Son, and Holy Spirit. The Father, as creator of the physical world, from minerals to the first level of humanity, reveals his wisdom and power in the phenomena of nature (ibid.: 231). This creative wisdom and power, revealed in the Old Testament and in the Vedas, constitutes what Keshab calls the First Dispensation of God to humankind. The Son commences God's spiritual work, revealing the ideal of love and loyalty. The Son "was intended and designed by Providence to be … a pattern man, a God-man," that is, an exemplar for humanity and himself fully human (ibid.: 230). This exemplar, Christ Jesus, is revealed in the New Testament, the Second Dispensation (ibid.: 226, 231–2, 244).

The evolutionary process, however, does not stop with Sonship, for the work of Christ is incomplete until all humankind is saved. Otherwise, the universe is "an inexplicable enigma, without meaning, without a purpose. Creation with all its beauty and harmony, its laws and systems, is only wild force run mad, if it has no ultimate objective to achieve. It is nature's delirium" (ibid.: 227). That God could fail to fulfill his purpose in creation, for Keshab, is a blatant impossibility. Yet such purpose was not completely realized with the creation of Jesus. That millions perish in disobedience and sin indicates that something beyond the Son is required: the work of the Holy Spirit that brings humanity to the Father's heaven. From the one little seed that is Jesus Christ will evolve "a whole harvest of endless and ever-multiplying Christs" (ibid.: 228). The scripture of the Holy Spirit, unsurprisingly, is revealed "in the Church of the New Dispensation, which is in India," that is, in Keshab's newly established Church that seeks to harmonize and fulfill the preceding dispensations (ibid.: 244).

Keshab summarizes the history of humankind in the following terms: God as Father creates the world, descends as the Son to touch base with humanity while permeating the world, and finally ascends as Holy Spirit, dragging a regenerated humanity up to heaven. Even more briefly, Keshab notes: "God coming down and going up—this is creation, this is salvation" (ibid.: 228). This succinct formula is taken over and elaborated upon by later Hindu thinkers in their efforts to come to terms with Darwinism. Vivekananda and Aurobindo, for instance, use the terms *involution* and *evolution* for this twofold process, although adapting Keshab's ideas to a monistic rather than theistic framework inspired in part by Theosophical ideals, as we shall see.

Conclusion

In many ways Keshab's musings on the relation of religion and science reflected broader trends seen in Europe. As historian of science Ronald Numbers has noted with reference to Christian apologists of the Anglo-American world in the first part of the nineteenth century, "writer after writer celebrated the delicious harmony between science and religion" (2009: 18). Keshab enthusiastically engaged in the same apologetics, and like his Western counterparts, frequently invoked natural theology in his appeal.

At the same time, the shift away from heavy reliance on reason and natural theology that we see occurring in the Brahmo Samaj with Debendranath and Keshab paralleled a similar disillusionment with the design argument in Europe during the eighteenth and nineteenth centuries. By the end of the eighteenth century, the impersonal implications of the design argument could no longer be overlooked by theists (Hutchison 1998: 101–2). By the beginning of the nineteenth century, the view that natural theology and reason on their own could demonstrate the existence of God was on the wane. With reason increasingly distrusted, growing reliance on religious experience and intuition on the part of the devout was a natural development.

In the mid part of the nineteenth century, there arose in the West the suspicion that new scientific discoveries were taking over religious territory. Darwin's *On the Origin of Species* served to increase such fears. Still, apologists continued to see no real conflict between religion and science. Instead, "[r]eligious writers fearful of scientific aggression typically dismissed unwanted science as 'science falsely so-called' …, allowing them to insist on their continuing affection for genuine science" (Numbers 2009: 30). Again, we see in Keshab's lectures and writings a similar strategy, contrasting false, godless science with true science.

Keshab's thoughts on religion and science, while absorbed in large part from Western authors, were also deeply impacted by his mentor Debendranath, whose views as we have seen were likewise significantly influenced by Europe, although unacknowledged. With Debendranath we saw the development of a hierarchy of a lower and higher, a material and a spiritual science. In Keshab the hierarchy became a distinction between incomplete science and devotionally inspired science. Whereas Western scientists had stopped at the Unknown, Hindu scientists had gone beyond to discover the hidden truth. This theme, reverberating with Herbert Spencer's thoughts about the Unknown, was elaborated upon at length by Swami Vivekananda, the subject of Chapter 11.

Keshab justified his radical spiritualization of science in general and of Darwinian evolution in particular by what has come to be an oft-repeated mantra in the Hinduism and science discourse: the notion that in the East everything is sacred. What Western scientists regard as merely material forces and matter, for Keshab are immanent manifestations of the divine: "The Goddess of Force, Sakti, meets us at every turn. The beautiful Goddess of nature, Prakriti, is resplendent everywhere. There is nothing godless in the East, there is nothing profane. All is sacred" (1979: 286–7).

Keshab's scientizing of Hinduism, seen in his notions of yogic optics and Avataric Evolutionism, was clearly a product of the colonial discourse. Other participants in this discourse eagerly engaged in this scientizing process as well, as seen for instance in Vivekananda and his discovery of Darwinian ideas in the ancient *Yoga-Sūtras* of Patañjali. A different scientizing of tradition occurred with Dayananda Saraswati, who discerned in Vedic literature much of modern science. But Dayananda was an ardent champion of an unqualified natural theology, and near the end of his life, having just learned of Darwinism, rejected it outright. He thus could hardly be expected to discover any Hindu antecedents of Darwin. We turn to Dayananda in the next chapter.

10 The Modern Vedic Creationism of Dayananda Saraswati

In 1872 a wandering renunciate preacher and social activist from Gujarat, Dayananda Saraswati (1824–1883), arrived in Kolkata eager to learn about the missionary programs of the Brahmo Samaj. Inspired by his Sanskrit teacher in Mathura several years before, Dayananda had dedicated his career to re-establishing the Vedic way of life in India. He was particularly concerned with countering the idolatrous and superstitious practices he had seen everywhere in his travels and that had been on full display in Mathura, a town steeped in Vaiṣṇava ritual devotionalism. Naturally he was intrigued by the Samaj's reformist ideals and campaign against idolatry.

In Kolkata Dayananda met Keshab Chandra Sen, the liberal and cosmopolitan leader of the Brahmo Samaj of India, as well as the more conservative leaders Debendranath Tagore and Rajnarain Bose of the Adi Brahmo Samaj. Given Dayananda's revivalist zeal for Vedic ideals, it is hardly surprising that he was suspicious of Keshab's Westernizing and universalist agenda. He sympathized far more with the Bose-inspired Hindu revivalism and nationalistic perspective of the Adi Brahmo Samaj. Bose read to an appreciative Dayananda his lecture "The Superiority of Hinduism," emphasizing Hinduism's great antiquity compared to all other religions with their recent historical founders. In addition, Bose insisted that ancient Hindu learning encompassed all of human knowledge, a message Dayananda took to heart years later in reconstructing Hinduism's Vedic Age, adding technology and modern science to the store of ancient Hindu achievements (Jordens 1998: 73). Based on his knowledge of the Vedic texts whose references to fantastic weaponry and flying machines he took quite literally, he credited the ancient Indians or Aryans with such "modern" inventions as the airplane, steam engines, and telegraphy.[1]

Debendranath Tagore's book *Brahmo Dharma*, with its strong theistic viewpoint, also undoubtedly swayed Dayananda. Of special note is Debendranath's affirmation that the material universe is not a manifestation of God himself, contrary to the teachings of Advaita Vedānta (D. Tagore 1928: 57–8). Such a view must have disturbed Dayananda's long-held Advaita convictions. When Dayananda had first fled his home at age twenty-one in order to escape a planned marriage, he had sought out various teachers of Vedānta, and two years later was initiated into the strict order of Dandīs, followers of Śaṅkara teachings. Over the years he had studied the various classical schools of Hindu philosophy, including the Sāṃkhya and Nyāya-Vaiśeṣika, but he still retained much of the Advaita perspective when he arrived in Kolkata—a perspective that was to change significantly in the next few years. This change was due in large part to his meetings with theistic Brahmos like Debendranath (Jordens 1997: 105–7).

One major point where Dayananda departed from the Brahmo Samaj was in his steadfast belief in the Vedas as a unique, universal, and infallible revelation. He considered the Brahmo rejection of Vedic infallibility as the cause of their internal ruptures. He soon came to the conclusion that the Brahmo Samaj as a whole, with its modernist ideas, had gone too far in attacking traditional beliefs and customs, discarding the true with the untrue. He was thus inspired in 1875 to set up his own society, the Arya Samaj (Society of the Aryans), to represent the real Vedic faith, the true Sanātana Dharma or "eternal religion," as his later followers perceived it (Llewellyn 1993: 87–91). The Arya Samaj soon flourished in the Panjab, where Bengali Brahmoism was looked upon with skepticism.

Dayananda's dismay with the Brahmo attitude towards the Vedas also compelled him to begin his own analysis of the Vedic texts to spell out their model for society. In 1875 he published the first edition of his *Satyārth Prakāś* (*Light of Truth*), calling for a revival of the ideal society of the ancient Aryan or Vedic Indians. Two years later he completed his *Introduction to the Commentary on the Four Vedas*.

Both the 1875 edition of the *Satyārth Prakāś* and his *Introduction to the Commentary* still partially reflect Dayananda's early Advaita orientation, as he characterizes God as both the efficient and material cause of the universe. He thus insists that even the ultimate material atoms (*paramāṇus*) did not exist prior to creation and will eventually dissolve back into God. To be sure, Dayananda in the first edition of the *Satyārth Prakāś* already rejects two major Advaita teachings (or rather what he considered to be Neo-Vedānta misinterpretations of the ancient Vedānta texts) concerning the illusory nature of the world and the identity of the soul and God (Jordens 1997: 252). Within a few months following the publication of the *Introduction to the Commentary*, however, we find a marked change in his thinking in a debate during a "religious fair" in the spring of 1877, at Chandapur, Uttar Pradesh.

In the debate Dayananda asserts against his Muslim and Christian opponents that God made the world out of pre-existing matter, the eternal Prakṛti (Primordial Nature), the material cause of the world, also called atoms (Singh 1971: pt. 1, p. 129).[2] He especially attacks the notion that the world was created out of nothing, relying on the Sāṃkhyan *satkārya* doctrine that something cannot come out of nothing, and certainly not by someone pronouncing a few words—an obvious reference to the Genesis account of God's creation by fiat. He also pointedly rejects the Advaita notion that God is both the efficient and material cause of the universe, arguing that if God is the material cause, actually becoming the world, then he would be just like the world and subject to all its pains and limitations. Or conversely, if God is matter, then since matter is not creative in itself, the world could never be created, just as clay by itself cannot create a jar. However, if God is only the efficient cause, he is not subject to such imperfections, and his role is like "that of the potter who cannot fashion the pot without the earth" (quoted in ibid.: pt. 1, p. 131).

Dayananda further insists that there are three eternal substances, Material Nature or atoms, souls, and God, a doctrine derived from the Nyāya-Vaiśeṣika. This doctrine of three substances serves in the debate as the basis for a brief presentation of the atomistic design argument—already quoted in the first chapter—in which Dayananda alludes to the wondrous construction of the eye and God's distribution of karmic fruits to souls (ibid.: pt. 1, pp. 134–5). The three-substance doctrine also provides the fundamental ontology of the second edition of the *Satyārth Prakāś*, completed in 1882, the year before his death.[3] This edition contains the major elaborations of his design argument.

Design in the *Satyārth Prakāś* (second edition)

In the second edition of the *Satyārth Prakāś* Dayananda introduces the design argument in Chapter 7, "The Veda and God." An "imaginary objector" (the *pūrvapakṣin* or opponent in traditional Hindu argumentation) challenges Dayananda to prove the existence of God, of whom he has been speaking so glibly. Dayananda responds that there are various sorts of evidence, including perception, citing Gautama's *Nyāya Sūtras* (1.1.4) on *pratyakṣa* (perception) as a valid means of knowledge. He then expounds: "…the perception of God results from the perception of wonderous [sic] design [*racanā*] in the visible world and of the phenomena of knowledge [*jñāna*] and virtue [*guṇa*]," for the cause can be inferred from the effect (Dayananda 1970: 174).[4] The "phenomena of knowledge" apparently refers to the idea that all knowledge is taught to humans by God. Virtue as evidence for God is an inference from our moral feelings, since our conscience derives, according to Dayananda, not from our own egos but from the Infinite Spirit. Other evidence includes direct cognition of God by purified souls. All these evidences, incidentally, are endorsed in the classical Nyāya-Vaiśeṣika of Udayana.

Dayananda elaborates on the design argument in Chapter 8, "Creation, Preservation, and Dissolution of the Universe." The chapter commences with several scriptural citations supporting the view that the Vedas and Upaniṣads are monotheistic, and that God, souls, and matter are three distinct types of entities. Of special relevance is Dayananda's translation of *Taittirīya Upaniṣad* 3.1.1: "He is the Brahma or Supreme Being whose design [*racanā*] has brought forth all these planets and other material bodies, in whom the sentient beings live and die" (Dayananda 1970: 202).[5] Dayananda uses the word *racanā* to translate and modify the Upaniṣadic idea of giving birth. The original Sanskrit text reads more literally that it is Brahman "from whom all these beings are born (*jāyante*)." The notion of giving birth too readily suggests that God is the material cause of creation, a notion Dayananda now rejects, and so he substitutes design for birth. He immediately clarifies this point, having the imaginary objector ask if the world is "born of the Great God or of some other being." Dayananda affirms: "It is produced from the Deity, the efficient or intelligent cause, but its material cause is nature (prakriti) or matter" (Dayananda 1970: 203).

Another example of Dayananda's innovative understanding of scriptural texts concerns the frequent Upaniṣadic statements that Brahman is the universe and that nothing existed prior to creation except Brahman. Dayananda interprets these texts to mean that prior to creation matter was not absolutely non-existent, but simply in an invisible form—conforming to the Nyāya-Vaiśeṣika view of invisible atoms.

One final example of Dayananda's reinterpretation of traditional understandings of scripture regarding Brahman's relation to the universe occurs in response to the objector's citation of the spider analogy from the *Muṇḍaka Upaniṣad* (1.1.7). This is a favorite illustration of the Vedāntins to elucidate that Brahman is both material and efficient cause. Dayananda initially counters that if Brahman were the material cause of the universe then Brahman would be changeable, conditioned, and insentient like the world. Or conversely, the world would have the nature of Brahman, that is, endowed with infinite being, consciousness, and bliss, but in fact the world possesses the opposite of these qualities. Dayananda cites the *Vaiśeṣika Sūtras* (2.1.24), that "the qualities of the effect are seen in the preceding cause," in support of his argument. He then tackles the spider analogy directly:

> The example of the spider does not support your thesis, it rather refutes it. The material cause of the threads of the cobweb is the material body of the spider, and the living principle or soul in the spider is the efficient cause of the web. It is an evident instance of the wonderful works [*adbhuta racanā*] of God.
>
> (Dayananda 1970: 207)[6]

Having established the basic ontological framework of the three substances, Dayananda then weaves into his discussion of cosmogony various elements of the design argument. He incorporates and revises the various cosmological and teleological proofs for God found in Udayana's *Kusumāñjali*, often putting them to use in disparaging the teachings of other religious traditions. Udayana's proofs, we may recall, rest on the assumption that we live in an ordered, purposive, and moral universe, and may be summarized as follows: God is necessary to account for (1) the distribution of karmic fruits, (2) the formation of elementary compounds from atoms, (3) the formation of the cosmos, (4) the maintenance and dissolution of the cosmos, and (5) the religious and cultural instruction of humankind. Let us see how Dayananda handles each of these in the *Satyārth Prakāś*.

Dayananda begins his discussion of God's role in the distribution of karmic fruits by having his imaginary objector ask about God's purpose in creating the world—a world in which souls experience much pain as well as pleasure. Dayananda responds that, in the state of dissolution prior to creation, souls would be left in an unconscious state, like deep sleep. Such a state, he continues, is suitable only for idle and inactive beings, not for energetic and enterprising people—a clear rejection of the Advaita ideal of withdrawal from the world. Nor, he claims, could souls enjoy any happiness in the world, arguing that the happiness and pleasure found in the world far exceed the pain and suffering. Further, souls could not work towards salvation without the world, and God could not award souls their appropriate rewards. Dayananda concludes that God's inherent nature (*svābhāvika*), including essential attributes like knowledge and the power to make the world, would be useless if he did not actually create the world, just as the essential nature of an eye would be unfulfilled if it could not see (ibid.: 208).[7]

Regarding the formation of elementary compounds from atoms, Dayananda first quotes the famous verse from the *Bhagavad-Gītā* (2.16) affirming that the existent (*bhāva*) cannot arise from nothing (*asat*).[8] Dayananda caustically notes that ignorant people think it possible for the universe to be created without a cause—that is, out of nothing. He then proceeds to give an account of cosmogony incorporating a theistic Sāṃkhyan scheme emphasizing that God created the world out of pre-existing matter. When the time for creation arises, God brings together the invisible, subtle particles (*paramāṇus*) of Primordial Nature (Prakṛti). From this Prakṛti are generated the standard evolutes of the Sāṃkhyan scheme, beginning with Mahat (Great Element) and ending with the gross elements (Dayananda 1970: 216–17).[9] The further phases of creation outlined by Dayananda, dealing with organic bodies, I shall deal with in a moment.

The question of how God manipulates the subtle atoms is explained by Dayananda in response to a Materialist objection that the world by its own nature (*svabhāva*) arises spontaneously. Dayananda gives a classic Hindu theistic answer to naturalism, found in both Śaṅkara and Udayana. If the material elements by their own nature are creative, then they must ever be creating, producing suns and moons alongside the present ones, and thus destruction of the universe would be impossible, and if the elements are destructive by nature, then creation would be impossible.[10] Even if the elements by nature are simultaneously creative and destructive, then neither creation nor destruction is possible. Therefore all objects in the world, consisting of parts, must be brought together in the right way by a willful agent. He concludes: "In like manner, the various kinds of things and species of animals can not come into being of their own accord without God's uniting the different kinds of atoms of matter from His omniscience and transcendent design [*racanā*]" (ibid.: 213).[11] It is God's omniscience and his pervading all bodies that gives him the knowledge of atoms and the consequent ability to manipulate thosc atoms. And he is able to pervade all bodies only because he himself has no body, unlike a human artisan. If he had a body, not only would he be subject to pain, he would also be unable to penetrate the atoms. But "[s]ince he is more delicate than nature or matter and souls, He pervades them, and so is able to mould them in the form of the material world" (ibid.: 209).

Turning from God's atomic-creation process to the formation of the visible world, Dayananda proceeds to the production of herbs, plants, and food from the previously created gross elements. Food in turn produces the vital seed from which arises the body. He then launches into an expansive illustration of the design argument, beginning with the complexity and order of the body but soon extending to the whole of creation, praising God's providential ordering of the universe for the benefit and comfort of sentient beings. It is worth citing at length:

> Look at the plan of the body, how wisely is it constructed! Its observation fills the learned with wonder. Inside the body are found the system of bones, the system of nerves, the system of muscles, the covering of the skin, the spleen, liver, lungs, the valves of the chambers of the heart, the mysterious union of the soul and body. The structure [*racanā*] of the head contains the brain, the root of vitality, and hair. See the nails and other growths, the delicate organ of the eye, the arrangement of nerves like telegraphic wires, glands, the sensation in the organs of the senses, the provisions of special regions for the inducement of waking, dreaming and sleeping state of the soul,[12] the secretion and production of all tissues, mechanism, organism and other schemes of tempers, all of which is the most wonderful creation, and which can not be made by any other being but God. None but God Almighty can make the earth, stored with many kinds of precious stones and materials; the microscopic structures [*racanā*] of the various kinds of vegetable life; the phenomena of innumerable varieties of colors,—green, white, yellow, black and intermediate shades, which are painted on leaves, flowers, fruits, roots and the like. Except God, nobody can make the leaves, flowers, fruits, grain, tubes and roots, endowed with various kinds of tastes—sweet, salt, pungent, acid, sour, bitter; and diverse kinds of order [*racanā*]; innumerable worlds, millions of suns, moons and stars; construction, revolution, preservation, their subjection to laws, and other wondrous plans of nature's economy.
>
> (ibid.: 217)

Directly following this passage is Dayananda's allusion to the gold ornament found by chance in a forest that gives evidence of design by its beauty, quoted in Chapter 1.

With regard to an objection that perfected souls with their supernatural powers acquired through yoga could design and create the world, Dayananda responds that God is still necessary to create the bodies of yogis as well as the world that supports them, thereby allowing them to attain their marvelous powers. Further, whatever great powers yogis may attain, they will always be limited: "Look here, there has been none till now, who could change the order of nature set up by God … God … has made the law, that the eyes should see, and the ears hear. Now no yogi or occult powers can alter it" (ibid.: 213–14). In such manner Dayananda dismisses one of Rāmānuja's arguments against design as proof of a single creator.

As for the maintenance and dissolution of the cosmos, Dayananda, like Udayana, raises the question of who or what supports the Earth. Dayananda dismisses various mythological explanations such as the Earth's resting on a serpent or tortoise. He also rejects philosophical speculations that the Earth is supported by nothing, or by air, or that, being heavy, it is actually sinking downward in space. He accepts that the Earth is held in place by the attractive force of the sun, but concludes that ultimately there "is none but God to support the sun and other worlds" (ibid.: 222). God does so by his pervading all things, thereby supporting them. Dayananda pays little attention here to dissolution as such, merely attributing this cosmic phase, like creation and maintenance, to God's activity. At the Chandapur fair, he had briefly alluded to God's dismantling the assemblage of atoms at the time of dissolution, thereby rendering them once again invisible (Singh 1971: pt. 1, p. 130).

On the subject of God's original tutoring of humankind in all religious and cultural knowledge, Dayananda first considers the authorship of the Vedas. He proclaims that the Vedas are the infallible word of God, offering among his proofs that they contain no statements contrary to "the laws of nature, [or] the evidence of the senses" (1970: 197). The objector, however, claims that there is no necessity for the Vedas to have been revealed by God, since men are capable of gradually acquiring knowledge and composing books. Dayananda's response echoes Udayana's socioteleological argument that God taught all knowledge to humankind at the beginning of creation. Dayananda argues that "savages have not become savants by observing nature" but only when instructed by a teacher. Thus, "if God had not instructed the sages in the beginning of the creation, and if they had not imparted their knowledge to others: all persons would have remained ignorant" (ibid.: 198). In accord with his cultural-nationalistic zeal, he adds that all peoples outside India (Aryavarta), including the Egyptians, Greeks, and Europeans, remained ignorant and illiterate until instructed by the sages of Aryavarta in the arts of civilization. As for the indigenous inhabitants of America, they had remained ignorant for thousands or millions of years until the arrival of Columbus, which allowed for the dispersion of knowledge originating with the Aryans.

As for how God revealed the Vedas, Dayananda argues that God had no need of a mouth and vocal organs to impart knowledge. Rather, God revealed the Vedas to the ancient sages by being present in their interior. This is only possible because he has no material body and is all-pervading (ibid.: 196).

Before we leave the *Satyārth Prakāś*, it is worth noting that Dayananda's assumption of Vedic inerrancy, combined with his notions of atomistic design, lead him to several conclusions regarding the origin and history of life on Earth and in the universe that conflict sharply with Darwinian evolution. He proclaims, for instance, that the first organic bodies, including those of humans, were created at the beginning of the present cosmic cycle some 1.96 billion years ago (ibid.: 221). These were not the products of sexual reproduction, but rather of God's direct formation of male and female bodies, which he then conjoined with pre-existing souls (ibid.: 217).[13] According to Dayananda, God created humans originally in Tibet, from whence they emigrated to Aryavarta. God formed not just a single human couple but hundreds of thousands, since there were multitudes of souls of great merit waiting rebirth in the new creation and deserving of bodies created directly by divine agency (ibid.: 218). This, for Dayananda, is proven by the observation that humans are descended from a great variety of parents, not just from a single pair. As one translator of the *Satyārth Prakāś* notes, "This negatives the Biblical theory that only Adam was created in the beginning and from his body sprang Eve" (Dayananda n.d.: 313n). Further, Dayananda insists, God created the original humans and other primordial animals not in old age or in infancy but in youth, so that they could support themselves and propagate sexually (Dayananda 1970: 218–21).

As for the origin of diverse species, Dayananda attributes this to God's impartial distribution of different animal bodies to souls according to "their works done in the previous creation" (ibid.: 219). Interestingly, Dayananda asserts that the different species, including humans, exist on all celestial bodies, including the sun, moon, planets, and stars. He cites as evidence scriptural reference to these heavenly worlds as *vasus*, habitations. Further, since God does not create anything without a purpose, such worlds would be useless if not inhabited: "Hence, the human and other races are in them all" (ibid.: 224). And in each world all the species have the same basic design in limbs, organs, and the like, as they do in our world, with only minor variations. This uniformity of species throughout the universe occurs in all cosmic cycles (ibid.: 225).

Dayananda and the beginning of modern Vedic creationism

While the world view expressed in the *Satyārth Prakāś* is distinctly at odds with a Darwinian perspective, Dayananda never refers explicitly to Darwinism in this work, despite having known of Darwin for some years prior to completing the second edition. Nor does he refer to Darwinism in any other published work (Llewellyn 1993: 113). However, he did address the idea of Darwinian evolution on at least one occasion, in a public lecture in the town of Roorkee, Uttar Pradesh, in 1878. Dayananda's biographer, Ganga Ram Garg, indicates that in the lecture Dayananda quoted Sanskrit texts to prove that such modern scientific discoveries as Newton's law of gravitation were well known to the ancient Aryas, and that religion and science are in complete harmony. But regarding evolution, Garg writes, Dayananda raised an objection: "If man descended from monkeys, how is it that process had come to an end and monkeys no longer evolve into men" (G.R. Garg 1984: 70).[14]

Further details on Dayananda's anti-evolutionary arguments in the talk are provided by a follower of Dayananda who was present at the lecture. According to this eye-witness, Dayananda's discussion of Darwinism was part of a much broader refutation of several other "non-Vedic" world views, including Christian and Muslim metaphysics as well as the absurd views of the Purāṇas. The eye-witness then notes a number of "unanswerable arguments" that Dayananda brought forth against the Darwinian notion that "man was descended from a monkey" (quoted in Singh 1971: pt. 2, p. 4). One of the arguments concerns the apparent disruption of the evolutionary process so that monkeys are no longer evolving into humans:

> … if man was really a descendant of a monkey, then it was, as it were, a fact—a law which, according to the followers of Darwin and other philosophers of his school, was unceasing in its operations under any conditions whatsoever. If the law was of a constant and permanent nature, ever working itself out, how was it that for thousands of years past, no monkey's young one had developed into a human being?
>
> (Quoted in ibid.: pt. 2, p. 5)

Dayananda's misunderstanding of natural law is noteworthy, reflecting the ancient Hindu assumption that any unsupervised inanimate or eternal creative principle—whether relating to a metaphysical reality or law of nature—must constantly manifest itself and produce its necessary effects, regardless of spatial and temporal conditions. We saw how Śaṅkara and Udayana rejected the Sāṃkhyan notion of a self-transforming Material Nature on the basis that if it were eternal, as the Sāṃkhyas claimed, then it must always be transforming itself (and thus constantly creating the universe), and this would not allow for periodic cosmic dissolutions. Dayananda himself, as pointed out above, applied the same logic in raising the question why there are not multiple suns and moons around us if the elements are able to transform themselves in a creative manner.

His misunderstanding extends to his view of the mechanism of evolution, which he assumes involves inter-species hybridization. As Singh summarizes Dayananda's reasoning:

> If in times gone-by a monkey copulating with a fish (for instance) brought into being a strange creature, and if that strange creature copulating with some other creature brought forth another strange creature, and if the process of copulation on the same principle continuing, the last strange creature helped to bring on the scene, as the result of his copulation with a nameless creature, a human being: if all this happened formerly, why was it that a similar result had ceased to be brought about in the later ages? Did the last strange creature leave a will that no animal in future, especially a monkey, was to act after the fashion of his ancestor[?]
>
> (Quoted in ibid.: pt. 2, p. 5)

Dayananda then seemingly clinched his argument by noting that inter-species copulation does not produce offspring.

It is unclear how Dayananda obtained his knowledge of Darwinism. In any event, with Dayananda's Roorkee lecture we see the seeds of an anti-Darwinian evolutionism based on a "Vedic" world view that may be conveniently referred to as Modern Vedic Creationism. As noted in Chapter 1, Dayananda's rejection of Darwinian evolution did not preclude his acceptance of cosmic evolution.

Modern Vedic Creationism among early Arya disciples

As Dayananda lay dying in the town of Ajmer, Rajasthan, in October of 1883, he was attended by a number of Aryas. One of them was an intense young man of nineteen, Gurudatta Vidyarthi (1864–1890). Exposed in high school to Western philosophy and science and familiar with missionary onslaughts on idolatry, Gurudatta had begun to question his own cultural and religious heritage. He came across Dayananda's *Satyārth Prakāś* in 1880 and soon joined the Lahore Arya Samaj, but his skepticism remained. In 1883 Gurudatta initiated a science class in association with the Arya Samaj, and in the middle of the year delivered a lecture on religion espousing atheistic ideas on rationalistic and scientific grounds.[15]

Gurudatta's meeting with Dayananda in the fall of 1883 was his first and last encounter with the Swami. Seeing the Swami lying calm and joyful as death approached, Gurudatta felt all his doubts suddenly vanish. He determined to dedicate the rest of his brief life to advancing Dayananda's mission.[16] He completed his education with an M.A. in 1886 in Physical Science, and became an assistant professor of Science at the Government College in Lahore (Anon., "Biographical Sketch," 1902: 23; cf. Jordens 1997: 242; Chamupati 1937: 128–9, 146, 156–7). In the remaining four years of his life, he used his knowledge of modern science to support his guru's teachings.

At the same time, like his master, Gurudatta frequently cites scripture, providing updated reinterpretations to affirm his points. Perhaps the most ingenious of these is his explanation of the *Māṇḍūkya Upaniṣad.* This brief Upaniṣad correlates the three letters of the sacred syllable Aum with the three states of consciousness, waking, dream, and deep sleep. Gurudatta, in accord with classical commentaries, provides a cosmogonic rendering of the meaning of the syllable, but with a modern teleological twist in his interpretation of the first two letters. The letter *a*, representing the waking state, signifies the final phase of creation in which God—fully awake as it were—organizes the external parts of the universe, thus allowing souls enjoyment in the material realm. The letter *u*, the dreaming state, signifies the preceding phase wherein God "dreams up" or inwardly designs the structure of the universe (Gurudatta 1902b: 127–32).

Gurudatta then cites the classic watch analogy, noting the watchmaker's initial mental designing of the watch with all its wheels, pinions, escapement, and the like, and keeping in mind such principles as friction and the various properties of different metals and jewels. Only after seeing in his mind the proper arrangement of all the parts could the watchmaker then apply his knowledge in producing an actual watch (ibid.: 137). Gurudatta warns his readers that this analogy, like all such analogies, is limited—apparently to rule out any objections that the analogy suggests that the omniscient God has to ponder assiduously the features of the universe before he creates it. Nonetheless, the analogy is useful, Gurudatta argues, evoking Keshab's portrayal of the scientific yogi. By contemplating these phases of creation, Gurudatta proclaims, the yogi goes beyond the merely scientific apprehension of the mind and universe, until "there dawns a philosophical perception of the interior design of nature, with which perceptions the mind soars higher, till the design itself is found to be the outcome of constitutional and spontaneous tendencies of the Deity" (ibid.: 139).

Invoking ancient Hindu notions of macrocosmic–microcosmic correspondences, Gurudatta next asserts that the design in the universe, "with its mighty orbs rolling in their magnificent splendour," exhibits the same structure as the most wondrous and perfect structure on Earth, the human body (ibid.: 140). He cites as proof of such correspondence verses from the *Atharva Veda* proclaiming that the sun is the brain, the space between sun and Earth the stomach, the Earth and planets the feet, the sun and moon the two eyes, and so on. He then claims that the adaptations of parts to the whole, both in the universe and in the human body, reveal the divine maker. Focusing on the human body, and switching to the analogy of an architect, he concludes:

> Is this beautiful adaptation of parts merely the result of chance, or of mere "fortuitous concourse of atoms"? Have the blind forces of matter met unconsulted, and, after unexpected, unknown and unpredictable clashes, embraced each other and linked themslves [sic] into the apparently beautiful organisation of man? No, this adaptation of functions is not in vain. The edifice, constructed of the brain, the lungs, the stomach, the feet, the eyes, the ears, and the mouth, is but the building of a theatre. The adaptation of its rooms is the design of an architect. Surely, the architect made it for some one to act in.
>
> (ibid.: 143)

Near the end of his essay on the *Māṇḍūkya Upaniṣad*, Gurudatta uses a Paleyesque analogy, likening the eye to a camera obscura, in order to clarify the nature of the divine designer. He notes the standard correspondences between eye and camera, but then adds a new twist. The camera itself lacks the power of seeing, which resides in the faculty of vision possessed by the photographer. Thus, when a man dies, leaving his physical body, he shall still be able to see, just as the photographer is able to see when not looking through the camera lens. Such disembodied vision is already vouched for by the existence of the disembodied creator of the universe, for as the Upaniṣads declare, God is without physical organs or limbs, yet can grasp without hands, see without eyes, and hear without ears (*Śvetāśvatara* 3.19; *Kaivalya* 22). Gurudatta sums up his *Māṇḍūkya* essay: "So is it with the Divine Spirit. He is the true embodiment of all eternal, unchangeable principles, residing at the back of all form or organisation and independent of it, and standing at the foundation of all design" (Gurudatta 1902b: 143).

In another essay entitled "Evidences of the Human Spirit" (1902a), Gurudatta takes on the challenge of Darwinian evolution directly. Significantly, scriptural proof-texts here are largely absent, replaced by appeals to supposedly *scientific* evidences that point to the existence of a vital principle manifest in the "human spirit." Such evidence, according to Gurudatta, harmonizes perfectly with the classical Nyāya-Vaiśeṣika philosophy. He begins by lashing out in Dayananda-like fashion against the Christian and Islamic view of soul or spirit that it is merely the breath of God, created out of nothing. He includes in his condemnation of ignorant views the evolutionist theory—apparently alluding to the animistic theory of British anthropologist Edward B. Tylor—that the concept of spirit arose from the false interpretation of dreams, that is, out of nothing substantive. Such superstitions regarding the non-substantial nature of spirit, Gurudatta laments, have even begun to affect scientific speculation.

The mechanical theory of the universe, according to Gurudatta, ends up in materialism, as it attempts to reduce even life itself to mere matter and motion. He cites as examples the physiologist Wilhelm Wundt and the evolutionary biologist Ernst Haeckel, who seek to reduce all phenomena, from the stars to the consciousness of man, to the general laws of physics, that is, atomic mechanics (ibid.: 83). But in addition to matter, Gurudatta argues, there must be some "vital principle" to account for life.

At this point, he calls upon scientists themselves to make his case, citing the zoologist H. Alleyne Nicholson and the evolutionary biologist Thomas Henry Huxley. Gurudatta calls attention to Nicholson's observation of the absence of any organization in such tiny living organisms as the amoeba, "which is structurally little more than a mobile lump of semi-fluid protoplasm" (ibid.: 85). Yet it is still able to digest food without possessing a single organ for the purpose, so that there must be something beyond the merely physical or chemical to account for it. Similarly, Gurudatta notes, Huxley comments that the microscopist is unable to find any organization in such organisms, seeing only "a mass of jelly," yet they are able to metabolize, exert movement, and produce shell structures of great complexity and beauty. He further quotes Huxley's conclusion that the ability of such structureless particles "to give rise to those exquisite and almost mathematically arranged structures" is "a fact of the profoundest significance" (ibid.: 86). We see here in Gurudatta what has become a common creationist strategy, selective quote mining, to set evolutionists against themselves. For if we consult Huxley himself, we discover that he goes on to observe that, while such simple organisms are not permanently organized, they are not devoid of organs. And Huxley emphasizes that much was still to be learned about the amoeba (Huxley 1869: 10–11). Nonetheless, Gurudatta saw in these organisms scientific evidence for a vital principle, a principle known to the ancient Aryan philosophers and called Ātman. And as Nyāya-Vaiśeṣika demonstrates, in Gurudatta's view, the Ātman, or rather "ātmans," are eternal substances, not just breath or temporary creations out of nothing or misinterpretations of dreams.

With unshakeable faith in his guru, the Vedic scriptures, and a divine architect, Gurudatta saw no possibility of "Western" science ever unraveling the mystery of life. In late 1888, a little more than a year before his death, in a lecture for the Lahore Arya Samaj, he thus proclaims:

> The modern scientist might dissect every nerve and bone, subject every drop to a most searching examination under the most powerful microscope he could possibly have, but he was as hopelessly lost over his question as ever. He could not undo the mystery of life. He might go on for ages dissecting and experimenting, but he would be none the wiser for it on the question of life. That question could not be solved but by the aid of the Vedas. They alone could unravel that grand mystery, and to them the scientist must ultimately turn.
>
> (Quoted in Anon., "Biographical Sketch," 1902: 31)

Another early admirer of the Aryan anti-Darwinian arguments, interestingly, was not a theist but an Advaitin, Har Narayana. In his book *Vedic Philosophy*, Narayana observes that "Our Vedic theory of Evolution and Involution differs from the untenable theory of Naturalists" (ibid.: x). The naturalistic theory of evolution requires that progress goes on forever, thereby making the dissolution of the universe an impossibility. The Vedic theory, by contrast, views the world as gradually winding down as its energies wane, until the whole cycle recovers its strength and begins anew. Narayana utilizes this regressive view of time, combined with the *sātkarya* idea that effects pre-exist in their cause, to refute the theory of natural selection:

> The theory of natural selection is indicative of nature's imperfections. According to it Nature is still improving. This theory points out the absence in Nature at present of the best forms that it will produce in the future. We, on the other hand, believe in the pre-existence of all things in Nature, and in their later manifestation, not as an originality, but as a manifestation of a pre-existing substance … Our doctrine is, that out of nothing, nothing can come. Therefore, whatever comes into existence, must have had a pre-existence in Nature. We have thus little respect for a theory such as that of Darwin which propounds that human species are a development of an inferior animal nature. For, if we were to accept this, we would be at a loss to trace the origin of human language and the possession of Divine knowledge, which are peculiar to mankind only, and which … are not self-acquired, but can only be learnt from others.
>
> (Narayana 1895: x–xi)[17]

"No savage nation," Narayana continues, "has ever become self-civilized," but must be taught by other already civilized nations, which can ultimately be traced back to primordial India (ibid.: xi). While Narayana makes no reference to God as the primordial tutor, in accord with his non-theistic outlook, his argument echoes Udayana's and Dayananda's notion that all human language, customs, arts, and skills must be taught by the Supreme Being at the beginning of creation.

I will cite just one more Aryan supporter of Modern Vedic Creationism, Gurudatta's biographer, Pandit Chamupati (1893–1937). Chamupati developed a number of anti-evolutionary themes, insisting like Gurudatta that evolutionists shall never discover the mystery of life, since soul or life and matter are two distinct entities, alongside a third eternal entity, God. Thus it is impossible for soul to evolve out of matter, and vice versa (Chamupati ca. 1930: 11).

Chamupati also contends that scientific knowledge, while varying to some degree in different ages, is eternal. This, hc observes, is opposed to "the evolutionary hypothesis, that humanity has, from the day of its origin, been progressing in the discovery and application of the laws of science, [and] is belied by the testimony of excavations going on in different parts of the earth" (ibid.: 22). Intertwined with this contention is the idea that human beings are ancient, going back some 1.96 billion years. In support of such assertions, Chamupati cites a number of reports supposedly from the scientific literature. One example will suffice: "In Nevada … John T. Reid has discovered a human footprint and well made shoe-sole, which he claims to be five million years old … This shoe-sole was accompanied by footprints of dinosaurs and their bones" (quoted apparently from the *Theosophical Path* for August 1923, ibid.: 25).[18] The Nevada "print," interestingly, has been cited by Christian young-Earth creationists. It has also been used more recently in the Krishna Consciousness movement (Cremo and

Thompson 1993), but like the Aryas, in defense of Ancient Hominid Creationism rather than a young Earth. For Chamupati, it is also evidence of the ancient origin of language and science, revealed to humans by God in the Vedas, at the time of creation (Chamupati ca. 1930: 41–8).

Finally, Chamupati contends that the social virtues exhibited by the higher, non-human species cannot be accounted for on the basis of evolution, since these virtues "in the lower genera from which alone they could have evolved … are entirely absent" (ibid.: 79). Only karmic experience in preceding lives can account for such moral abilities of these higher animals.

Conclusion

While the outlook of Dayananda and his followers is firmly rooted in the classical traditions of Sāṃkhya and the Nyāya-Vaiśeṣika, their illustrative examples become increasingly those of the modern world with its watches, telegraphs, cameras, and detailed anatomical and astronomical knowledge. The eager espousal of modern science and modern technology reflected in the use of such examples belies, however, a fundamental distrust of scientific methodology and its disciplined skepticism, a distrust manifest in the rejection of Darwinian evolution. In the next chapter, we turn to Vivekananda and the Neo-Vedāntins, who at first sight seem much more open to the Darwinian perspective. But in the end, they too reject the naturalistic implications of a robust Darwinism—a rejection based in large part on the same *satkārya* viewpoint promoted by the Arya Samajists, a viewpoint that resists the idea of something truly new, like consciousness, emerging from mere matter.

11 The Modern Advaitic Evolutionism of Swami Vivekananda

Over the course of the nineteenth century Hindu intellectuals were increasingly discontent merely to discover in Indian culture select ideals like monotheism that were valued in the West, coming to prefer a more aggressive assertion of India's superiority to European civilization. Increasingly, the superiority was seen to include not only a spiritual but a scientific pre-eminence as well. Dayananda's revivalist Vedic Creationism, based on a theistic world view rooted in Nyāya-Vaiśeṣika, represents one such endeavor to establish Hindu superiority. The Modern Advaitic Evolutionism of Swami Vivekananda (1863–1902) exemplifies another important version of this growing Hindu self-assertion, but derived from the monistic idealism of Śaṅkara. Advaita for Vivekananda "is the most scientific religion" in the world precisely because it rejects the sort of intervening God proposed by theistic religions (Vivekananda 2003 [*Complete Works of Swami Vivekananda*, henceforth abbreviated as *CW*] 3:423). As he elaborates: "This universe has not been created by any extra-cosmic God, nor is it the work of any outside genius. It is self-creating, self-dissolving, self-manifesting. One Infinite Existence, the Brahman" (*CW* 3:423–4).[1]

Vivekananda particularly chides Christianity for its hostility to science, arguing that scientific progress in Europe has been made in opposition to the Church, concluding that "If Christianity had its old paramount sway in Europe today, it would have lighted the fire of the Inquisition against such modern scientists as Pasteur and Koch, and burnt Darwin and others of his school at the stake" (*CW* 5:533). Vivekananda ridicules Christian attempts to prove scientifically such legends as Moses' crossing the Red Sea, noting that "there is a craze in every country to find scientific evidence for the absurd religious myths" (quoted in Raychaudhuri 1988: 290).[2] Christianity, for Vivekananda, has been irreparably undermined by science and can no longer fulfill humankind's spiritual longings. But Advaita, being fully affirmed by the latest scientific discoveries, can provide just such satisfaction (Raychaudhuri 1999: 117; see *CW* 3:111).[3] Vivekananda's condemnation of Christianity for its scientific backwardness, combined with his insistence on the harmony of Advaita Vedānta with modern science, constituted a comforting reassurance to educated Hindus chafing under colonial domination and the onslaught of Christian missionaries.

In embracing modern science within his Neo-Advaitin framework, Vivekananda aggressively comprises Darwinian evolution, claiming that the ancient Hindus had already discovered the Darwinian idea. Vivekananda insists, however, that the ancient Hindus had a more complete version of evolution by incorporating the spiritual dimension. These two claims, that Darwinian evolution was anticipated in the ancient Vedic literature, on one hand, and that Vedic revelation demonstrates Darwinian evolution to be incomplete, on the other, are major characteristics of Advaitic Evolutionism.

Vivekananda arrived at the conclusion outlined above only after a long spiritual journey marked by repeated personal and intellectual crises. His journey, reversing in some ways that of Dayananda, began with a theistic world view that rejected Śaṅkara's monism but eventually embraced it, including its illusionary view of the world.

Vivekananda's spiritual crisis

Vivekananda, originally named Narendranath Datta, was born into a prestigious family of lawyers in Kolkata.[4] Exposed to the pious Vaiṣṇava devotionalism of his mother as well as to his father's dedication to social advancement and material success, the young Narendranath was torn between spiritual and worldly goals. His college years (1881–1884) precipitated a long-brewing crisis, raising major doubts about his religious ideas in the light of Western philosophy and science. Further, concerned about the apparent decline and stagnation of his own country compared to Western nations, and wondering how India had come under foreign domination, he studied the history of European nations seeking an explanation.

He joined the Brahmo Samaj under Keshab Chandra Sen, undoubtedly inspired by Keshab's proclamation of the harmony of religion and science, but he failed to resolve his growing spiritual doubts. His study of Western philosophy, from Plato to John Locke, John Stuart Mill, David Hume, Auguste Comte, and Herbert Spencer, merely served to intensify the "painful tension between reason and faith" (Raychaudhuri 1988: 229). His reading of Mill's *Three Essays on Religion*, with its refutation of the causal and design arguments for God's existence, took away his reliance on reason to bolster his flagging faith (Williams 1974: 19). With the impoverishment of his family following his father's death in 1884, and the consequent difficulty of further advancement in a legal career, Narendranath reflected on the lack of divine compassion: "'Does God really exist,' I thought, 'and if so, does He really hear the fervent prayer of man? Then why is there no response to my passionate appeals? Why is there so much woe in His benign kingdom?'" (quoted in Anon. 1948: 351).

While the European philosophers had sown seeds of skepticism in Narendranath's mind regarding a beneficent, omnipotent Creator, they also provided perspectives on religion and science that eventually helped him to resolve his intellectual doubts.[5] Among the most influential was Spencer with his insistence on the harmony of religion and science in their shared quest for the Unknown—the Ultimate Reality that unites the spiritual and material but remains forever beyond human comprehension. Just as all religions hide the final truth but point towards it, Spencer declared, so the individual sciences are partial, but point to that which unites all. Especially important for Narendranath was Spencer's emphasis on several interrelated ideas: the unity of nature, the uniformity of natural law throughout the universe, the conservation of energy, the unity of all force, and the indestructibility of matter that merely undergoes continual redistribution.

All these ideas are manifest in Spencer's non-Darwinian evolutionary views, encompassing cosmic as well as organic realms, and largely Lamarckian in tendency. The law of evolution, for Spencer, was thoroughly progressive in nature, and which "can end only in the establishment of the greatest perfection and the most complete happiness" (1958: 511). At the same time, evolution, also called involution by Spencer, is balanced by the reverse process of dissolution, these two phases of the cosmic process ever alternating. While Narendranath found much to admire in Spencer's thought, he was not satisfied with the idea that the Unknowable, even if it transcends reason, must remain forever beyond human comprehension. Nor was he content with the apparent lack of individual spiritual fulfillment in Spencer's cosmic scheme of evolution-dissolution, as it ignored any notions of karma and rebirth.

Another Western writer, Helena Petrovna Blavatsky, co-founder of Theosophy, was similarly dissatisfied with Spencer. Responding to the scientific challenges threatening to undermine traditional Christian teachings in the nineteenth century, Blavatsky developed a mystical world view supposedly in harmony with modern science. Adopting an extreme, Romanticist-Orientalist view of ancient Indian civilization, she regarded the early Vedic seers as having discovered all modern sciences, including Darwinian evolution and atomic physics (Blavatsky 1877: pt. 1, p. 620). Her mystical cosmology assimilated various Western evolutionary views of her day into a modified Advaitic framework. Her Theosophical teachings offered a spiritual path promising direct experience of the Ultimate, thereby surmounting Spencer's intractable Unknowable.

Blavatsky specifically took over from Advaita the notion of an eternal and immutable principle, Brahman, lying beyond human reason. But this Brahman, being identical with individual souls, is thus knowable within ourselves by inner awareness and intuition. Souls themselves are subject to karma and rebirth, by which they evolve towards ever greater understanding and experience of the Absolute. The universe is also eternal, being identical with Brahman, but undergoes cycles of evolution and dissolution. Evolution, Blavatsky claims, consists of two phases: an involution of consciousness into matter and an evolution of organic forms that allow for a reemergence of the primordial consciousness working through the mineral, vegetable, animal, and human levels, culminating in the attainment of the suprahuman.[6] This idea of spiritual involution and evolution, defined in opposition to Spencer's physicalist interpretation of evolution/involution and dissolution, was the cornerstone for the Advaitic Evolutionism of the mature Vivekananda.

It seems highly probable that Blavatsky's writings exercised considerable influence on Narendranath, beginning in his college years when he was closely associated with the Brahmo Samaj. He never acknowledged such influence, however, and later heaped considerable scorn upon her and Theosophists in general, coming to resent the idea that Theosophy had assisted the spread of true Vedānta in the West, or had ever provided anything of religious value to Hindus (*CW* 4:318). Nonetheless, the parallels between the major ideas of both thinkers are striking.[7]

While intellectually intrigued and stimulated by the Western writers he encountered in college, Narendranath still hungered for direct spiritual experience of God. His encounter with a nearly illiterate, local temple priest in the early 1880s radically transformed his life. With his Brahmo background, Narendranath initially rejected the priest's approval of idol worship, and considered the priest's Advaitic notion that one's own self was identical with the Creator of the universe as atheism (Anon. 1948: 342–3). Nonetheless, Narendranath was impressed by the priest's claim to have seen God personally, soon becoming his disciple, and eventually spreading the teachings of his master, the famous Ramakrishna, to the rest of the world. Shortly after his master's death in 1886, Narendranath took the vow of renunciation and assumed the leadership of the other disciples, in accord with Ramakrishna's wish. But his inability to experience God directly continued to frustrate him, leading him to doubt the existence of God. It also in all likelihood prompted him to engage more deeply in the study of Advaita Vedānta, and to deemphasize more theistic approaches to God-realization (Williams 1974: 32).

Given his self-doubts, exacerbated by his anxiety about teaching his charges what he had not himself realized, it was natural that Narendranth set about to find a new guru who might help resolve his problems. Of special note is a letter he wrote in 1889 to a pandit in Benares outlining his personal struggles and voicing many doubts. He questioned Vedic authority, noting the ignorance of supposedly omniscient seers regarding the physical world, including even the simplest of astronomical truths (*CW* 6:211–12).[8] The pandit's unsatisfactory reply was not to question and to give up doubt (Williams 1974: 45–6).

A few months later, however, the pandit sent Narendranath a pamphlet outlining his own views on the relation of Advaita and science. Its contents can be partially surmised from Narendranath's response to the pandit:

> A kind of scientific Advaitism has been spreading in Europe ever since the theory of conservation of energy was discovered, but all that is Parinamavada, evolution by real modification. It is good you have shown the difference between this and Shankara's Vivartavada (progressive manifestation by unreal superimposition).
>
> (*CW* 6:215)

This is probably the earliest surviving reference to evolution in Vivekananda's published works. His apparent preference for Śaṅkara's *vivarta* over Sāṃkhyan *pariṇāma* suggests that Narendranath was finally coming to accept Ramakrishna's Advaitic perspective. Narendranath's implicit equation of Sāṃkhyan *pariṇāma* with the modern, scientific theory of evolution, and the reference to the law of the conservation of energy, are noteworthy. His response anticipates his later teaching that Darwinian evolution was already known to the ancient Sāṃkhyan sages dealing with a lower level of reality, while Advaita Vedāntic seers had realized higher evolutionary-spiritual truths in accord with but transcending modern science.[9] Advaitic Evolutionism became for Narendranath the basic solution to his spiritual crisis.

The World's Parliament of Religions and Vivekananda's mission to the west

Following his espousal of Advaita in the late 1880s, Narendranath became increasingly inspired by the desire to combine the search for knowledge of Brahman with service to humankind, especially the poor and downtrodden.[10] Over the next few years, he traveled around India as a wandering monk, observing the impoverished conditions of India's masses. He eventually conceived a plan to visit America to represent Hinduism at Chicago's 1893 World's Parliament of Religions, hoping to raise money there for social work back home. But he needed the patronage of Hindu princes to underwrite his trip. One forthcoming patron bestowed on Narendranath the name Vivekananda by which he became known in the West.

In May 1893, Swami Vivekananda embarked for America, and after his arrival managed to secure a seat as a delegate at the last moment. His addresses at the Parliament emphasize the themes of the unity of religions and of religious tolerance. But he criticized the Christian missionaries in India for providing stones (to build churches) rather than bread to allay the hunger of millions, and requested aid for his impoverished countrymen. He defended Hinduism against missionary propaganda, asserting the sublimity of the Advaitic doctrine of unity over against crude notions of a creator god and a created soul. He confidently affirmed the scientific nature of the Advaita and its teaching of manifestation, contrasting it with the idea of creation—undoubtedly referring to the Christian notion of "creation out of nothing." He states: "Manifestation, and not creation, is the word of science today, and the Hindu is only glad that what he has been cherishing in his bosom for ages is going to be taught in more forcible language, and with further light from the latest conclusions of science" (*CW* 1:15).

The Swami's appeal to the authority of science to affirm his tradition and to criticize Christianity as superstitious was undoubtedly disturbing to the Christian organizers of the Parliament.[11] Such an "Occidentalist" strategy, turning the tables on the West "to defeat the Occident at its own game in its own terms" (Ketelaar 1991: 38), however, was not unique to him. It was adopted not only by the other Advaitin representative at the Parliament, Manilal N. D'vivedi from Bombay, but also by the Ceylonese and Japanese Buddhist delegates. They all rejected the idea of an extra-cosmic, personal creator, and the Buddhists insisted that there was no conflict between the non-theistic teachings of the Buddha and evolution. Such views paralleled those of the Theosophical delegate at the Parliament. Clearly the Buddhists, like the Hindus, had been impacted by Theosophical ideology, which had spread throughout much of India and beyond by the late nineteenth century.

After the Parliament Vivekananda embarked on lecture tours throughout the eastern part of America. His initial plans of seeking financial aid for projects back home quickly enlarged to encompass a vision of spiritual conquest of the West through Advaita Vedānta. Settling in New York for several months, he began teaching yoga classes, contributing articles to various journals, and writing books. He traveled to England and there as in America laid the groundwork for his Vedāntic mission in the West. After three and a half years in the West, he returned to India in 1897 with a few Western disciples to assist him in establishing the Ramakrishna Mission there. He visited the West again in 1899, but soon returned to India in ill health. Nineteen months later he was dead, expiring on July 4, 1902, having succumbed to diabetes. He was just thirty-nine. His lectures in the West, dutifully recorded by disciples, and his various writings on Vedānta, cosmology, yoga, and related topics, composed primarily during this time, constitute the major resources for interpreting his views on design and evolution.

Design: an argument for children and lunatics

From his college days on, Vivekananda was deeply suspicious of the design argument. Logical appeals to conscious design in the universe, he realized, have no more persuasive force than assertions of the creative self-sufficiency of the material universe. On one hand, he notes, "seeing the adjustment of nature of different parts, it is clear that it is the effect of intelligent work." On the other, with equal force, "it may be argued that intelligence itself being created by matter and force in the course of evolution could not have been before this world" (*CW* 8:145–6). Reasoning by itself, he concludes, cannot establish whether consciousness is the cause of matter or vice versa. Taking his cue from Śaṅkara, Vivekananda relegates reasoning to the realm of practical knowledge, adequate for the animal needs of humankind but incapable of dealing with the Absolute (*CW* 8:146).

Like Keshab, then, Vivekananda rejects the design argument as a means for ascertaining the highest truth and stresses the futility of relying on reason alone. Such reliance, he thought, is too easily undermined by the discoveries of modern science and by philosophy: "At the beginning of the nineteenth century man tried to find God through reason, and Deism was the result. What little was left of God by this process was destroyed by Darwinism and Millism" (*CW* 6:41). Vivekananda saw that the design argument readily reduces God to the disengaged deity of Deism. A popular Hindu metaphor for God's inactivity between successive creations is sleep. Perhaps with this metaphor in mind, Vivekananda raises the issue of an overly somnolent deity. He observes that the notion of an eternally recycling universe "is a better idea than that God created the universe in five minutes and then went to sleep, and since then has been sleeping" (*CW* 2:427). A deistic god, it seems, sleeps through most of creation, as well as through maintenance and dissolution of the universe.

But Vivekananda went beyond Keshab in relegating both the design argument and theism to a lower level of spiritual understanding. The higher science of Advaita deals with the One Brahman that transcends any notion of a personal

God, and of design. At the same time, Vivekananda found the design inference, appropriately tweaked, to be of some tactical use against atheism and scientific materialism.

Vivekananda (with total disregard for the views of schools like the Sāṃkhya and Nyāya-Vaiśeṣika) claims that the religions of the world all favor consciousness as the source of the material body, while various modern schools of thought hold the reverse. These latter maintain that the chemical and physical composition of the body and brain produce "the souls or mass of thought," but this is no answer at all, according to Vivekananda (*CW* 2:75). He asks, "What makes the body? What force combines the molecules into the body form? What force is there which takes up material from the mass of matter around and forms my body one way, another body another way, and so on?" (*CW* 2:76).

This force, for the Swami, should not be identified with an intelligent designer conceived in terms of a personal god. Nor should it be identified with any sort of individual mind, as such a mind is delimited in time and space and is thus constituted of parts or subtle particles that must be appropriately combined. Before its parts are conjoined, the individual mind cannot act to form itself, and thus must be assembled by something else. So the question remains, what is that force capable of combining particles in the appropriate way? That force, Vivekananda declares, "was called the soul, the Atman in Sanskrit" (*CW* 2:77).

Vivekananda regarded the intuitive realization of this priority of the intelligent Ātman in the universe as underlying the ancient notion of design: "It is intelligence, the adjustment of part to part, the display of intelligence, of which the ancient design theory was an attempt at expression" (*CW* 2:209). This attempt, from Vivekananda's perspective, while a promising beginning, fell short of true insight, being based on the mechanistic metaphor of contrivance and design natural to theism. But it failed to recognize the more organic notion of intelligence unfolding itself, an idea in harmony with, and necessitated by, the evolutionary nature of the universe. It was suitable for the childhood of humanity, but not for a more philosophically and scientifically enlightened age. Accordingly, today it is only suitable for small children:

> The idea that nature in all her orderly arrangements shows design on the part of the Creator of the universe is good as a kindergarten teaching to show the beauty, power, and glory of God, in order to lead children in religion up to a philosophical conception of God; but apart from that, it is not good, and perfectly illogical. As a philosophical idea, it is entirely without foundation, if God is taken to be omnipotent.
>
> (*CW* 6:97)

The illogicality of the design argument, Vivekananda points out, is apparent in its problematic analogy between human designers and a divine but disembodied artificer, an issue repeatedly discussed in classical Hindu works.[12] Citing the Buddhists, he remonstrates: "The idea of a God that is omnipresent, and yet is a person who creates without hands, and moves without feet, and so on, and who has created the universe as a Kumb[h]akara (potter) creates a Ghata (pot), the Buddhist declares, is childish" (*CW* 3:411). Vivekananda concludes that the idea of a disembodied personal God designing and creating the world is too incredible even for a child to believe today, and only a mad man could accept it: "Because a Kumbhakara creates a Ghata, therefore a God created the world! If this is so, then your Kumbhakara is God also; and if any one tells you that He acts without head and hands, you may take him to a lunatic asylum" (*CW* 3:413).

Vivekananda notes several other objections to the design argument, including the Sāṃkhyan idea that nature itself is sufficient to create the universe. He proclaims that the founder of Sāṃkhya, Kapila, had knocked the childish design theory "on the head" (*CW* 2:439), and that Kapila had also pointed out that liberated souls, being omniscient and omnipotent, could govern the universe—reminiscent of Rāmānuja's objection that exceptional souls may become world-designers.[13] Vivekananda further notes the Buddhist argument that karma, although inert, asserts its influence without oversight from a supreme Spirit, just as seeds, when watered, sprout by themselves without need of a superintending God (*CW* 3:414–5).

Other major objections in Vivekananda's mind concern the implications for God. The design analogy seems to restrict God's independence, for if he is like a human creator, then he is limited by the materials available (*CW* 1:353).[14] A designing God, for Vivekananda, also cannot escape the endless chain of cause and effect: "The argument from design claims that intelligence precedes form; but if intelligence is the cause of anything, it itself is in its turn an effect ... God creates us, and we create God, and this is Maya" (*CW* 7:58). The Swami here highlights the old dilemma confronting any theistic design theory: who designed the designer? Perhaps the most unacceptable consequence of the design argument for Vivekananda is the limitation it places on God's omnipotence: "If nature shows the power of God in creating the universe, (then) to have a design in so doing also shows His weakness. If God is omnipotent, He needs no design, no scheme, to do anything" (*CW* 6:97).

Probably the greatest challenge to the design argument prior to Darwin was the problem of evil. We have seen it was deeply disturbing to Vivekananda during his college days. His initial solution to the problem employs the traditional ideas of karma and rebirth, drawing upon Śaṅkara's theodicy that appears immediately preceding the latter's argument for design in his *Brahmasūtra Bhāṣya*.[15] Śaṅkara addresses two difficulties, God's seeming partiality (*vaiṣamya*) and cruelty (*nairghṛṇya*), given the vast discrepancies in happiness and misery that sentient beings experience. But for Śaṅkara the problem lies in us, in our karma, not in God's cruelty or partiality, just as different kinds of plants sprouting in a field depend on different seeds, not on any partiality in the rain (*BSB* 2.1.34). Vivekananda addresses the same two difficulties of partiality and cruelty, and utilizes the rain analogy, slightly modified, to indicate that our karma is responsible for our different experiences in life, not God. He supplements Śaṅkara by illustrating the world's cruelty in terms of the Darwinian struggle for existence: "One animal tears another to pieces, and every man tries to get the better of his own brother. This competition, cruelty, horror, and sighs rending hearts day and night is the state of things in this world of ours" (*CW* 3:123–4).

From Vivekananda's (and Śaṅkara's) Advaitic perspective, however, even karma provides no real solution. Stressing that "this world is a hideous world," he cautions: "there is no explanation—there cannot be any explanation. But the Vedanta shows the way out" (*CW* 2:110–11). One cannot justify evil by the good often alleged to arise from it (*CW* 2:112). Vivekananda elaborates:

> And if we suppose there is a Personal God like a human being, who made everything, these so-called explanations and theories which try to prove that out of evil comes good are not sufficient … Why should good come through evil? The question remains to be answered, and it cannot be answered. The philosophy of India was compelled to admit this"
>
> (*CW* 2:113)

For Vivekananda the Upaniṣadic sages showed the way out, past the ancient designer god, by going beyond both polytheism and monotheism, to the highest ideal of the Impersonal Absolute, the One that pervades and is all (*CW* 2:117).[16] Thus Advaita Vedānta shows the way out not by resolving the issue, but by transcending it: "He who sees in this world of manifoldness that One running through all, in this world of death he who finds that One Infinite Life, … unto him belongs eternal peace" (*CW* 2:117).

Vivekananda is especially concerned with the mechanistic conception of human beings implied in the design argument. While the West, he argues, has tended to reduce everything, including human beings, "to the level of a machine," thereby gaining great prosperity, it has in the process "driven away all religion from its doors" (*CW* 8:302). It is not surprising, then, that he favors a more organismic view of creation—a view consonant with his evolutionary understanding of the universe.

Advaitic evolutionism: evolution and involution

In his essay "The East and the West," Vivekananda proclaims:

> The theory of evolution, which is the foundation of almost all the Indian schools of thought, has now made its way into the physical science of Europe. It has been held by the religions of all other countries except India that the universe in its entirety is composed of parts distinctly separate from each other. God, nature, man—each stands by itself, isolated from one another; likewise, beasts, birds, insects, trees, the earth, stones, metals, etc., are all distinct from one another; God created them separate from the beginning.
>
> (*CW* 5:519)

India, in the Swami's view, was the fountainhead for Western scientific notions of evolution, for Indian religions alone among the world's religions recognized the interrelatedness of the universe and the unity of life. Thus, "the modern law of evolution" that explains everything "from inside" by self-transformation, for Vivekananda, is simply the Hindu notion of the effect residing in the cause (*satkārya-vāda*) which governs all evolutionary processes (*CW* 1:371–2). We can analyze these evolutionary processes on three distinct but interrelated levels: cosmic, biological, and spiritual. I will deal with each of these in order, before turning to his reflections on evolution and ethics.

Cosmic evolution and Vedāntic uniformitarianism

Fundamental to modern science is the assumption of uniformitarianism—that natural laws and processes are universal in time and space. This uniformity in nature, according to Vivekananda, was already known to the Upaniṣadic sages:

> … if it be true that the whole of this universe is built on exactly the same plan as the atom, if it be true that the same law prevails throughout the universe, then, as it has been said in the Vedas [Upaniṣads], "Knowing one lump of clay we know the nature of all the clay that is in the universe."
>
> (*CW* 2:204–5)[17]

Vivekananda interprets the uniformity of nature as establishing the cyclic character of the universe, for the vegetative life-cycle and the geological processes of destruction and renewal must also apply to the entire cosmos. There is no final destruction, only a return to an original state, an insight he insists that was realized by the Sāṃkhyan sage, Kapila (*CW* 2:205). Vivekananda explains:

> This universe must be resolved into its causes; the sun, moon, stars, and earth, the body and mind, and everything in this universe must return to their finer causes, disappear, be destroyed as it were. But they will live in the causes as fine forms. Out of these fine forms they will emerge again as new earths, suns, moons, and stars.
>
> (*CW* 2:206)

This cycle, he concludes, clearly expresses the ancient notion of *satkārya* that "this whole universe cannot be produced out of nothing. Nothing comes without a cause, and the cause is the effect in another form" (*CW* 2:207).

Regarding the underlying cause, or causal substance, from which the universe evolves and to which it returns, Vivekananda first posits two basic entities, matter and force. Matter in its most subtle, primal form is the *ākāśa* (ether), from which grosser forms evolve. It is motionless and unmanifest at the beginning of a cosmic cycle. Primal force is *prāṇa*, the life force that for Vivekananda includes the various forces discovered by modern science such as gravitation and electricity. It is *prāṇa* that stimulates the inert *ākāśa* to evolve into the universe. When the evolutes dissolve back into their finer forms, then once again only the primal *prāṇa* and *ākāśa* remain (*CW* 1:359–60).

But something transcends even these two: "Now there is something beyond Akasha and Prana. Both can be resolved into a third thing called Mahat—the Cosmic Mind. This Cosmic Mind does not create Akasha and Prana, but changes itself into them" (*CW* 1:360). And behind Cosmic Mind is Prakṛti, Nature in her unmanifest state. While Sāṃkhyan reabsorption ends with Prakṛti existing alongside Spirit (Puruṣa), Vivekananda insists that Spirit and Nature are ultimately one, call it Ātman, Brahman, or God. While modern science has rediscovered the ancient Hindu truths regarding the unity of force and the unity of matter, modern science has failed to find a higher unity. Only by following the same lines as the Upaniṣads will modern science discover their ultimate oneness, and thus understand the true nature of cosmic evolution, as well as biological and spiritual evolution (*CW* 3:398, 400).

Biological evolution: cellular transmission versus karmic inheritance

Vivekananda proclaims Hindu priority in discovering the principles not only of cosmic evolution, but also of biological or Darwinian evolution: "The idea of evolution was to be found in the Vedas long before the Christian era; but until Darwin said it was true, it was regarded as a mere Hindu superstition" (*CW* 8:25). The Swami found support for this view among Western scholars, citing for instance the great Sanskritist Monier Monier-Williams: "Indeed, the Hindus were Spinozists 2,000 years before the birth of Spinoza, Darwinians centuries before the birth of Darwin, and evolutionists centuries before the doctrine of evolution had been accepted by the Huxleys of our time" (*CW* 9:25).[18]

Vivekananda acknowledges that "[c]onstant changes are going on, no doubt, in nature, by its own processes. New species of trees and plants, birds and beasts are being always created in nature through changes, brought about by time, environment and other causes" (*CW* 5:521–2). But he is unhappy with purely naturalistic explanations and adjoins supernatural to natural mechanisms to account for transmutations, especially among the highest animals.

In an essay entitled "Reincarnation," Vivekananda contrasts the materialist theory of evolution with the Hindu doctrine of reincarnation. Invoking a classical Nyāya argument that "[o]ur experiences cannot be annihilated" (*CW* 4:270), he asserts that the psychological tendencies we are born with, like fear of death, result from experiences gained in former lives, the consequences of prior actions even if they remain temporarily unperceived (*adṛṣta*). He concludes:

> So far as explaining the tendencies of the present life by past conscious efforts goes, the reincarnationists of India and the latest school of evolutionists are at once [sic]; the only difference is that the Hindus, as spiritualists, explain it by the conscious efforts of individual souls, and the materialistic school of evolutionists, by a hereditary physical transmission.
>
> (*CW* 4:271)

The Swami here employs one of his common rhetorical strategies, first suggesting a basic agreement between traditional Hindu ideas and modern science, and then radically polarizing the two. The “only difference” between them, in this case, turns out to be a huge metaphysical divide.

Elsewhere, in arguing for the truth of reincarnation, Vivekananda invokes the *satkārya* notion that “something can never be produced out of nothing” and applies this principle to explain the psychological tendencies of an infant (*CW* 9:209). Its inborn tendencies must come from prior experience, not out of thin air. Thus, repeating the arguments of the *Nyāya-Sūtras* (discussed in Chapter 4), Vivekananda asks: “Why should it [an infant] have fear of death if it never saw death? If this is the first time it was ever born, how did it know to suck the mother’s milk?” (*CW* 9:209). He insists that habit alone provides knowledge of how to act in the world, dismissing instinct as “equivalent to saying that something came out of nothing” (*CW* 9:210). He also dismisses the idea that we are born “with a ‘tabula rasa’—a blank mind upon which nothing is written” (*CW* 9:210), rejecting Locke’s famous dictum.

Modern science, the Swami asserts, attributes the child’s knowledge to its father through hereditary transmission back through successive earlier generations. While this theory is better than the notion of instinct, in his view it is still insufficient: “We can understand a physical transmission, but a mental transmission is impossible to understand” (*CW* 9:210). A father may transmit “a certain hereditary experience” like “a drunkard’s character,” but it still takes karmic experience for a soul to take up a body with an inherited inclination towards drunkenness. Accordingly, the Swami concludes, “[a] simply hereditary transmission theory will only touch the physical man and would be perfectly insufficient for the internal soul of man” (*CW* 9:211).[19]

Vivekananda further explains the insufficiency of “mere physical transmission” by arguing that if “man” is merely “a bundle of atoms,” then when a father procreates children, his “soul … would decrease in ratio to the number of his children; and the man who has five, six or eight children must, in the end, become an idiot” (*CW* 9:211). While the reasoning here is not entirely clear, Vivekananda seems to be arguing that if the soul is material then the loss of semen entails a loss to the soul, including its mental powers—as though brain cells are split off from the father and passed on to offspring through the vital seed. In any case, the Swami goes on to explain that it is impossible for a “single cell of protoplasm” to contain the memory of all past human experience (*CW* 9:212). His claim that “[t]here is no difference between the protoplasm of a bird and that of a human brain” suggests that he envisioned the protoplasmic cell as some sort of amorphous fluid or seed, common to all species (*CW* 9:212). He then asserts that modern science supposes that the experiences of individual species are compressed and packed into these cells, two or three of which, when cut from the body of the father, constitute the mechanism of physical transmission. But such a view, he says, “is too tremendous a pill they [scientists] ask you to swallow with this little word *transmission*” (*CW* 9:212). Such a fantastic theory, for Vivekananda, is an example of modern “so-called scientific popery” (*CW* 9:212).

We should note that during the first several decades of the nineteenth century the scientific understanding of hereditary transmission was quite muddled, as Vivekananda points out (*CW* 9:212). Darwin himself had little idea of the genetic basis underlying heredity. Protoplasm was described by Thomas Henry Huxley in an 1868 lecture as "the physical basis of life" and argued for "a general uniformity in the character of the protoplasm … in whatever group of living beings it may be studied." We have previously noted Huxley's statement that the tiniest organisms like amoebas, that seem to consist of little more than semi-fluid protoplasm, reveal no structure or organization to the microscopist.[20] Indeed, until the invention of the electron microscope in the 1930s, the complex structure of cells (and protoplasm—a term no longer used by cell biologists) remained largely hidden.

Regarding Vivekananda's claim that only the karmic inheritance of habit can explain such psychological tendencies as fear of death in the newborn, we may note a rough parallel with Lamarckian notions of the inheritance of acquired characteristics. Huxley also points out a certain affinity between reincarnation and Lamarckian evolution regarding the hereditary transmission of dormant tendencies that constitute character:

> The Indian philosophers called character, as thus defined, "karma." It is this karma which passed from life to life and linked them in the chain of transmigrations; and they held that it is modified in each life, not merely by confluence of parentage, but by its own acts. They were, in fact, strong believers in the [Lamarckian] theory, so much disputed just at present, of the hereditary transmission of acquired characters.
>
> (Huxley 1894: 62)

Huxley concludes that, while the Lamarckian notion of the inheritance of acquired characteristics was controversial in the West, "Indian philosophy … did not admit of any doubt on this subject" (ibid.).

Vivekananda partially embraces the Neo-Lamarckian notion that transmutation is driven by the internal desire or needs of an organism. He argues that the Darwinian struggle for survival can never overcome all objective challenges in the environment. Rather, subjective change is required and drives evolution, culminating in the subject's "escape from all its environments" (*CW* 2:137). He elaborates:

> We cannot possible [sic] conquer all the objective environments. We cannot. The little fish wants to fly from its enemies in the water. How does it do so? By evolving wings and becoming a bird. The fish did not change the water or the air; the change was in itself … All through evolution you find that the conquest of nature comes by change in the subject.[21]
>
> (*CW* 2:137)

But Lamarckian-style evolution is problematic for Vivekananda, since the inheritance of acquired characteristics in the Lamarckian scheme benefits primarily successive generations, not the original animal. The Swami thus proposes an alternative theory, what I call the karmic inheritance of acquired characteristics, that allows for spiritual evolution through rebirth.

To explain further the insufficiency of materialistic evolution, Vivekananda argues that the "hereditary impressions of protoplasmic cells alone" cannot explain the transmission of thoughts (*CW* 9:213). He asserts that thoughts may propagate not only via speech, but also through the process of rebirth: "Now a father does not transmit thought. It is thought alone that transmits thought. The child that is born existed previously as thought. We all existed eternally as thought and will go on existing as thought" (*CW* 9:213). While Vivekananda grants that the theory of hereditary transmission need not be wholly brushed aside, still, "the main stress must be laid upon the transmission of thought" (*CW* 9:213). Vivekananda adds a key notion critical to his interpretation of biological evolution: "What we think, that our body becomes. Everything is manufactured by thought, and thus we are the manufacturers of our own lives" (*CW* 9:213). This notion of the primacy of thought, or consciousness, underlies his theory of spiritual evolution with its radical critique of Darwinian evolution.

Spiritual evolution and the involution of intelligence

Vivekananda readily concedes that modern evolutionary theory is beyond reasonable doubt. He defines the modern theory as the evolution of gross material forms out of the fine, but regards this as simply an incomplete view. We have to go further, he argues, and realize "[t]hat every evolution is preceded by an involution" (*CW* 2:207). And we need to know what undergoes involution. Vivekananda's answer is "God" (*CW* 2:208). Armed with this theory of involution-evolution, he dismisses materialistic evolutionists who deny God and claim that intelligence has evolved over millions of years. "This objection of the evolutionists," Vivekananda proclaims, "does not hold water, as we shall see by applying our theory. The tree comes out of the seed, goes back to the seed; the beginning and the end are the same" (ibid.).[22]

The above claims were made in a public lecture in New York in 1896. Two years earlier, Huxley had published his *Evolution and Ethics*, in which he had stated that evolution, as a natural process, was similar to "the development of a tree from its seed, or of a fowl from its egg," requiring no creative or supernatural intervention (1894: 6). There is sufficient reason to believe that Vivekananda was familiar with Huxley's book, as we shall see in dealing with his evolutionary ethics. In any case, Vivekananda uses the vegetative life-cycle to counter the linearity of Darwinian evolution, insisting on the endless round of involution-evolution. Huxley, in his turn, had noted that certain ancient philosophers fully acknowledged the idea of incessant change presupposed by evolution. But he concluded on the basis of paleontological evidence pointing to a common origin of all life forms that the ancients had "erred in supposing the phases formed a cycle" (ibid.: 5). If Vivekananda had read this passage, he ignored it.

Emphasizing the stability of species as seen in the vegetative life-cycle, Vivekananda circumvents the challenge posed by the idea of countless cumulative variations culminating in new forms of life, asserting that the beginning and end of the involution-evolution process are the same. He then soars to the grand Advaitic conclusion that intelligence, not matter, is the original reality:

> In the end we find the perfect man, so in the beginning it must have been the same. Therefore, the protoplasm was the involution of the highest intelligence. You may not see it but that involved intelligence is what is uncoiling itself until it becomes manifested in the most perfect man. That can be mathematically demonstrated.
>
> (*CW* 2:208–9)

Vivekananda's mathematical demonstration of involution-evolution invokes the law of the conservation of energy:

> We find then, that every evolution presupposes an involution. Nothing can be evolved which is not already there. Here, again, modern science comes to our help. You know by mathematical reasoning that the sum total of the energy that is displayed in the universe is the same throughout. You cannot take away one atom of matter or one foot-pound of force. You cannot add to the universe one atom of matter or one foot-pound of force.
>
> (*CW* 2:227–8)[23]

The law of energy conservation, for Vivekananda, not only refutes the idea of creation out of nothing, but also establishes the reality of spiritual evolution.[24] His claim rests on the assumption that the total spiritual energy or intelligence in the universe is constant:

> If it [intelligence] was not present in the protoplasm, it must have come all of a sudden, something coming out of nothing, which is absurd. It, therefore, follows absolutely that the perfect man, the free man, the God-man, who has gone beyond the laws of nature, and transcended everything, who has no more to go through this process of evolution, through birth and death … that perfect man who is at one end of the chain of evolution was involved in the cell of the protoplasm, which is at the other end of the same chain.
>
> (*CW* 2:209)[25]

As Vivekananda succinctly summarizes, using the Buddha to represent fully manifest intelligence, "If the Buddha is the evolved amoeba, the amoeba was the involved Buddha also" (*CW* 3:407).[26] The intelligence present in amoeba and Buddha alike, for Vivekananda, is the ultimate reality, God, "the involved universal intelligence unfolding itself" and "infinite cosmic intelligence" (*CW* 5:256).

Vivekananda felt that support for his theory of involution-evolution could be located in the classical *Yoga-Sūtras* of Patañjali, whom the Swami calls "the great ancient evolutionist" (*CW* 1:292) and "father of the theory of evolution, spiritual and physical" (*CW* 6:113). In his book *Raja Yoga*, Vivekananda provides a translation with commentary of the *Yoga-Sūtras*. The key passages are *Yoga-Sūtras* 4.2–3, where Vivekananda finds Patañjali's basic definition of evolution: "The change into another species is by the filling in of nature" (*CW* 1:291). Vivekananda elsewhere clarifies the meaning of the first phrase (*jāty-antara-pariṇāma*): "One species—the Jati is species—changes into another species—evolution; Parinama means one thing changing into another, just as one species changes into another" (*CW* 3:407). The second phrase, the "filling in of nature" (*prakṛty-āpūrāt*), explains Patañjali's evolutionary mechanism. In Vivekananda's translation: "Good and bad deeds are not the direct causes in the transformations of nature, but they act as breakers of obstacles to the evolutions of nature: as a farmer breaks the obstacles to the course of water, which then runs down by its own nature" (*CW* 1:291). As Vivekananda elucidates in his commentary, the farmer irrigates his crops by opening sluice gates on a canal, allowing water to flow by itself into the fields. In like manner, when obstructions in a human are removed, then "in rushes nature" (*CW* 1:292). But what is this nature?

Vivekananda employs two meanings of nature: cosmic force and universal spiritual essence. As he concludes in his commentary on *Yoga-Sūtras* 4.3: "It is nature [as cosmic force] that is driving us towards perfection … that perfection which is our birthright, our nature [as universal spiritual essence or Ātman]" (*CW* 1:292). Patañjali's "filling in of nature"—which in its original context seems to have referred to the reconstitution of the subtle and gross elements of the body in the process of rebirth—becomes for Vivekananda the self-manifestation of Ātman within the individual (Killingley 1990: 157–60). Vivekananda further explains that like the reservoir of water held back by the gates, our inner reservoir of "infinite power, infinite purity, infinite bliss, and existence infinite," that is, Brahman, is held back by the gate of the body (*CW* 3:408).[27] This body is overwhelmed by ignorance. Good deeds lead indirectly to the opening of the gate, but the direct cause is knowledge. Divine knowledge unlocks the body-ignorance complex, allowing the Ātman to flow forth and become manifest (*CW* 1:292).[28]

This divine knowledge for Vivekananda becomes the major driving force of evolution, both spiritual and biological. This is apparent in his discussion of evolution in relation to rebirth. The soul, Vivekananda declares, in its struggle against constraining environments, takes on new bodies, from amoeba to human, by its own inherent power (*CW* 2:136-7). And the soul's capacity to manufacture new bodies for itself can be realized even within a single lifetime. Vivekananda makes this point in commenting on Patañjali's discussion of supernatural powers (*siddhis*) leading up to the latter's definition of species change.

The third chapter of the *Yoga-Sūtras* describes various supernatural powers that a person can attain by means of yogic concentration (*samādhi*). Among these powers are the abilities to levitate, enter another's body, remember all past lives, and become invisible. The fourth chapter opens by elaborating upon the means of attaining these powers. Sometimes such powers are carried over from a previous incarnation, or developed through various alchemical means, mantric chants, ascetic practices, or by yogic concentration. In his commentary Vivekananda refers to an ancient Indian sect that emphasized strengthening the body to dispense with birth and death, thereby allowing for uninterrupted spiritual development. Vivekananda, accepting this possibility, concludes: “As soon as this body dies, we shall have to manufacture another. If we can do that, why cannot we do it just here and now, without getting out of the present body? The theory is perfectly correct” (*CW* 1:290). Species transformation thus becomes a yogic *siddhi*.

Such powers are part of the universal endowment of all beings, lying latent even in the lowest life forms. The Swami informs us, “Patanjali ... would tell you that even in the lowest worm that crawls under your feet, all the eightfold Yogi’s powers are already existing” (*CW* 3:334–5). However, up to a certain level of organic development the environment seems to play the major role in determining the material form of the organism. Once a sufficiently advanced level of purity and self-knowledge is attained, the soul’s power to manifest new organic forms abruptly becomes dominant as the *siddhis* are realized. Vivekananda provides this striking illustration, involving transmutation into a celestial life form:

> Generally the organism is weaker than the environment. It is struggling to adjust itself. Sometimes it over-adjusts itself. Then the whole body changes into another species. Nandi was a man whose holiness was so great that the human body could not contain it. So those molecules changed into a god-body.
>
> (*CW* 6:113)

This notion of spontaneous change from one animal form into another, specifically of a man into a god or animal, was perhaps suggested to Vivekananda by older commentaries on the *Yoga-Sūtras* (cf. Killingley 1990: 160). In any case, and in accord with his involution-evolution theory, Vivekananda contends:

> If we are developed from animals, the animals also may be degraded men. How do you know it is not so? You have seen that the proof of evolution is simply this: you find a series of bodies from the lowest to the highest rising in a gradually ascending scale. But from that how can you insist that it is always from the lower upwards, and never from the higher downwards? The argument applies both ways, and if anything is true, I believe it is that the series is repeating itself in going up and down. How can you have evolution without involution? Our struggle for the higher life shows that we have been degraded from a high state.
>
> (*CW* 2:173–4)[29]

While our present human life is the result of the degradation or involution of the supreme Brahman into apparent material forms, Vivekananda focuses on the ideal of spiritual progression. Thus, “[a]ll creation is moving onwards and upwards, obedient to the great law of spirit evolution, which is different from the evolution of matter” (*CW* 5:312). The single process of Brahman’s involution, incidentally, underlies two processes of evolution, biological and spiritual.

Vivekananda finds affirmation of his involution-evolution theory not only in Patañjali, but also in the writings of Rāmānuja, citing the latter’s ideas regarding the expanding and contracting nature of the soul.[30] Conflating spiritual and modern evolution, Vivekananda argues: "Indeed, Ramanuja's theory of contraction [*saṃkoca*] and expansion [*vikāśa*] is exactly what the modern evolutionists call evolution and atavism" (*CW* 3:239).[31] The contraction of the soul, he explains, is countered by performing good deeds and having proper thoughts. Of course, for Vivekananda, Rāmānuja's Qualified Non-Dualism, while proclaiming much truth, falls short of Śaṅkara's higher teaching of Non-Dualism with its doctrine of the immutability of the soul: "With the Advaitist the one difference is that he admits evolution in nature and not in the soul" (*CW* 3:239).

Elaborating on this theme, Vivekananda proclaims: "Evolution is in nature, not in the soul—evolution of nature, manifestation of the soul" (*CW* 6:92). In this manner he harmonizes organic evolution, *pariṇāma*, with the Advaitin notion of apparent devolution, *vivarta*, in hierarchical fashion: "This world is admitted as an apparent world, bound by time, space, and causation, and it comes to what is called the Vivarta-vada in Sanskrit, evolution [*pariṇāma*] of nature and manifestation [*vivarta*] of the Absolute. The Absolute does not change, or re-evolve" (*CW* 1:420). Such views clearly reflect those already expressed in his 1889 letter to the Benares pandit regarding "scientific Advaitism."

The inner perfection within all of us, then, is not subject to real change or evolution. Vivekananda illustrates this point with his beloved example of the amoeba and its latent spiritual perfection. The transformation from amoeba to the perfect man pertains only to the outer shell or covering. And even this covering is not ultimately real, for organic evolution, in the final analysis, is encompassed within the higher truth of *vivarta*, manifestation. Thus, despite the scientific demonstration of the involution-evolution cycle provided by the law of energy conservation, Vivekananda comes to the good Advaitic conclusion that there is "neither involution nor evolution," for both are simply "Maya, or apparent only" (*CW* 8:362). The affirmation of involution-evolution on the lower level of knowledge, and a denial of any real involution or evolution on the higher, constitutes Modern Advaitic Evolutionism in a nutshell.

However illusory involution and evolution may be on the higher level of reality, on the lower level we can differentiate three main sorts of transformation in Vivekananda's thought: involution of spirit, evolution of spirit, and evolution of matter. The first two succeed each other in endless cycles. The last, evolution of matter, runs for the most part parallel to and interacts with the two phases of spirit transformation. Material evolution in its organic mode, however, contrasts with spiritual evolution in lacking any meaningful progress in a moral sense. Interpreting evolution in Social-Darwinian terms, Vivekananda criticizes the view of those who see any principled progress in biological evolution, regulated as it is by the law of survival of the fittest. He observes that only the fortunate few who avoid most hardships and are not "crushed under the wheel of this so-called evolution" can believe that material evolution works to bring about greater happiness or justice (*CW* 2:95). This leads us directly to Vivekananda's view of ethics in an evolutionary universe.

Evolution and ethics: the struggle against nature

In his commentary on *Yoga-Sūtras* 4.3, Vivekananda elaborates upon the contrast between Darwinian and Yogic evolution. The former, with its mechanisms of survival of the fittest and sexual selection, serves only to justify the pursuit of economic success and sexual gratification. These are at best momentary goals, arising from ignorance and incapable of satisfying our longing for perfection. One unfortunate consequence of Darwinian evolution, according to Vivekananda,

> is to furnish every oppressor with an argument to calm the qualms of conscience. Men are not lacking, who, posing as philosophers, want to kill out all wicked and incompetent persons (they are, of course, the only judges of competency) and thus preserve the human race!
>
> (*CW* 1:292)[32]

Vivekananda's consternation with the notion of eugenics is prescient, given the later reaction in the West to such ideas.[33]

The superior Yogic explanation of evolution, he contends, is free from such rationalizations of evil and more spiritually satisfying. As he concludes elsewhere, "Patanjali says that these struggles remain only through our ignorance, and are not necessary, and are not part of the evolution of man" (*CW* 5:206). True evolutionary progress can only be measured in terms of progress towards God-realization (*CW* 2:359). The modern evolutionary mechanisms of competition inherent in natural and sexual selection not only justify tyranny, in the Swami's view, but also fail to provide any real explanation for one body assuming the form of another body. Patañjali, in Vivekananda's estimation, provides a "deeper" explanation, with his mechanism of "the infilling of nature" (*CW* 3:407).

Spiritual evolution, Vivekananda maintains, is at work in all sentient beings: "From the lowest form of life to man, the soul is manifesting itself through nature" (*CW* 6:35). Accordingly, he declares: "Taking for granted that Darwin is right, I cannot yet admit that it is the final conclusion about the causes of evolution" (*CW* 7:152). Even on the lower level of non-human, animal evolution, he insists that the "laws of struggle for existence, survival of the fittest, natural selection and so forth" are not the "last word," reserving this word for Sāṃkhya-Yoga and Patañjali's "in-filling of nature," that is, the manifesting of Ātman (*CW* 7:152).

While Vivekananda repeatedly affirms the unity of all life, at times he differentiates between higher and lower life forms, especially between human and non-human. It is rationality that especially distinguishes human from animal evolution in his view:

> In the animal kingdom we really see such laws as struggle for existence, survival of the fittest, etc., evidently at work. Therefore Darwin's theory seems true to a certain extent. But in the human kingdom, where there is the manifestation of rationality, we find just the reverse of those laws. For instance, in those whom we consider really great men or ideal characters, we scarcely

> observe any external struggle. In the animal kingdom instinct prevails; but the more a man advances, the more he manifests rationality. For this reason, progress in the rational human kingdom cannot be achieved, like that in the animal kingdom, by the destruction of others!
>
> (*CW* 7:154)

Thus, like many other interpreters of Darwin, Vivekananda wants to "restore the fence between man and nature which Darwin had breached" (Killingley 1990: 164).

Spiritual progress, especially "in the rational human kingdom," is a constant moral struggle against nature in order for the soul to manifest itself. Vivekananda specifically argues against complying with nature: "It is a struggle against nature and not conformity to nature that makes man what he is. We hear a great deal about living in harmony with nature, of being in tune with nature. This is a mistake" (*CW* 6:35–6). Such a view of humankind's relation to nature may well have been based on Huxley's essay "Evolution and Ethics" (Killingley 1990: 164). Huxley argues that the principle of survival of the fittest, an integral part of what he calls the "cosmic process," is amoral in character (1894: 81). He then cautions: "Let us understand, once for all, that the ethical progress of society depends, not on imitating the cosmic process, still less in running away from it, but in combating it" (ibid.: 83). These ideas may well have confirmed for Vivekananda the superiority of the Yogic theory of evolution (Killingley 1998: 154–6).

Vivekananda goes beyond Huxley in viewing the moral combat against nature as culminating not simply in social progress, but also in spiritual emancipation. Everything struggles to escape the slavery of natural law and the dictates of nature, but lower entities, such as minerals, plants, trees, and artificial objects like tables, are helpless and never progress. "Man" is different, for "eventually he conquers nature and realises his freedom. When he is free, nature becomes his slave" (*CW* 8:249). The Swami concludes:

> The awakening of the soul to its bondage and its effort to stand up and assert itself—this is called life. Success in this struggle is called evolution. The eventual triumph, when all the slavery is blown away, is called salvation, Nirvana, freedom.
>
> (ibid.).

From Vivekananda's Advaitin perspective, the moral struggle with nature is a contest between the true Self (Ātman) and the small self or ego. The former strives for the good of all, the latter for preservation of its own narrow individuality. Ego and Self correspond with two forces in the world, selfishness and unselfishness, or acquisition and renunciation. Vivekananda believes that Darwinism only gives play to selfishness, yet both forces are at work throughout the animal kingdom: a tigress may devour a human to survive, but sacrifice her own life to protect her young (*CW* 2:354). Humans may act selfishly, but the best of men, the ethical elite, recognize that love—loosely identified by the

Swami with renunciation—is the only real force in the universe. Selfishness is simply misdirected and limited love. The Advaita insistence on oneness leads Vivekananda to conclude that "the motive power of the whole universe ... is that one wonderful thing, unselfishness, renunciation, love, the real, the only living force in existence ... We insist upon this explanation because we cannot admit two causes of the universe" (*CW* 2:354–5).

In a grand Neo-Vedāntic Theory of Everything, Vivekananda reduces all causal forces, from atomic attraction to sexual attraction, to love. In a sweeping cosmic vision, he rhapsodizes:

> What is it that makes the atoms unite with atoms, molecules with molecules, and causes planets to fly towards each other? What is it that attracts man to man, man to woman, woman to man, and animals to animals, drawing the whole universe, as it were, towards one centre. It is what is called love. Its manifestation is from the lowest atom to the highest being: omnipotent, all-pervading is this love. What manifests itself as attraction in the sentient and the insentient, in the particular and in the universal, is the love of God. It is the one motive power that is in the universe.
>
> (*CW* 2:50)

The lover, he goes on to explain, transcends dogmatic religious beliefs as well as skepticism, realizing the Advaitic truth that the universe and the lover are one:

> The lover has passed beyond all these things, beyond rewards and punishments, beyond fears and doubts, beyond scientific or any other demonstration. Sufficient unto him is the ideal of love, and is it not self-evident that this universe is but a manifestation of this lover?
>
> (*CW* 2:50)

Vivekananda here radically subordinates scientific demonstration to the self-evident certainty of personal experience, raising crucial epistemological issues. His manner of resolving these issues is key to his reconciling Advaita and modern science.

Method and epistemology: yogic perception and science

Vivekananda's approach to scientific method and epistemology is based on his understanding of the term *science*. His frequent references to *bhakti yoga* and *karma yoga* as the sciences of love and work, respectively, point to his broad and inclusive definition of "science." Such a broad definition facilitated his harmonization of science and religion as "two sides of the same quest for truth" (Rambachan 1987: 284). Vivekananda's use of the term reflects the traditional meaning of the Sanskrit word *vidyā* that encompasses both spiritual sciences (*ātma-vidyā*, science of the Self, and *brahma-vidyā*, science of the Ultimate) as well as all other sciences (*sarva-vidyā*).[34] But this traditional view of *vidyā* is in tension with modern notions of science, for "the traditionalistic tendency to regard all 'sciences' (*vidyā*) as timeless, all-inclusive configurations of knowledge is incompatible with the ideas of progress and an open-ended empirical accumulation of knowledge" (Halbfass 1988: 186).[35]

Also underlying Vivekananda's understanding of science are two decisive notions anticipated by Keshab Chandra Sen. First, science, like religion, is a quest for unity, both seeking the unified reality within or behind the forces of nature. And second, science is any rational system with a goal, and a practical method for achieving it. But as Anantanand Rambachan points out, not all rational systems are scientific even if they are seeking some sort of unity, and science involves far more than just rational reflections and speculations, however systematic (1987: 285). Method is critical.

For Vivekananda, the preferred "scientific" method is yogic meditation, leading to superconscious awareness (*CW* 2:446–7). It is by such means that the ancient Vedic seers anticipated the discoveries of modern science, without the help of modern scientific instruments. Vivekananda thus says of the Sāṃkhyan sage Kapila and other yogis:

> How wonderful his perceptions were, and if there is any proof required of the extraordinary power of the perception of Yogis, such men are the proof. They had no microscopes or telescopes. Yet how fine their perception was, how perfect and wonderful their analysis of things!
>
> (*CW* 2:445)

Like Keshab, Vivekananda assumes that sage-philosophers, through disciplined contemplation, can penetrate the mysteries of both spiritual and material realms.[36] Contemplation of Brahman/Ātman encompasses complete understanding of the natural world, rendering scientific study unnecessary. As Vivekananda summarizes, "the Vedantic and other philosophers of the Indian schools hold that knowledge is not to be acquired from without … The human soul is the repository of infinite wisdom; what external agency can illuminate it?" (*CW* 4:431). The Swami thus proclaims with pride, emphasizing the mystical/experiential over against the empirical: "The Hindu does not go through nature to nature's God—he goes to the God of man through Man" (*CW* 3:459).

Superconscious perception, for Vivekananda, is not only infinite in scope, it is also infallible. But he warns us that not all claims of infallibility are legitimate: we should not be misled by hysterical trances or mere instinct, mistaking these for genuine inspiration (*samādhi*). Unfortunately, Vivekananda tells us, genuine inspiration cannot be externally tested—it can only be self-validated. Such self-validation may turn to reason to guide us along the way, but the truths of religion go beyond what reason can ascertain (*CW* 7:60).

Self-validation of personal inspiration is a principle eschewed by modern science. Validation in modern science is thoroughly empirical: "the ultimate judge of authority in science is nature itself … the natural authority of observation and experiment" (Miller 2008: 167). A clear distinction, then, must be drawn between personal experience and empirical confirmation. For Vivekananda, personal experiences of sages like Kapila are accepted as infallible without critical assessment of such experiences—without awareness that experiences are largely inseparable from personal and cultural interpretation.

In addition, the experiences of supposedly omniscient yogis are concerned with reconfirmation of truths already known and lack the "'empiricist' openness for future additions or corrections" (Halbfass 1988: 393). Accordingly, the conflating of yogic experience with scientific empiricism "appeals to the modern fascination with science, but rejects its commitment to objectification and quantification. It is a device of reinterpretation and cultural self-affirmation, which serves to defend the Indian tradition against charges of mysticism and irrationalism" (ibid.: 401).

Vivekananda provides one positive check regarding personal inspiration. Genuine inspiration presupposes extended moral training and devotion. He loosely equates the dedication and perseverance of scientists who make great scientific discoveries with the austerities practiced by yogis (*CW* 4:436). But to equate scientific dedication and insight with such yogic virtues as "purity of heart" and "devout contemplation" is problematic. Beyond insisting upon absolute integrity in carrying out observations and experiments, and in reporting results, modern science renders irrelevant, as far as possible, the personal moral qualities of the investigator.

The disjunction of morality from knowledge, the sacred from the profane, was the key axiom that created modern science according to Jacques Monod. With regards to Hinduism, he notes: "Westerners often have trouble understanding that for certain religions there neither is nor can be any distinguishing between sacred and profane: for Hinduism, everything comes within the bounds of the sacred; the very concept of 'profane' is incomprehensible" (1971: 175). The separation of morality from science often constitutes a significant point of contention and misunderstanding in the confrontation of traditional cultures, including the Hindu, with modern science.

Vivekananda's traditional epistemology leads him to reject conclusions of modern science that seem to him morally wrong, with little regard for supporting empirical data. Conversely, he invokes empirical data—often dubious in nature—to support moral and spiritual truths. For instance, he cites the increase in human population and decrease in animal populations as evidence, first, for confirming that there are a fixed number of souls in the universe—a corollary of the law of the conservation of (conscious) energy—and second, for establishing the transmigration of souls from animal to human forms. As he concludes, "It is a significant fact that as the human population is increasing, the animal population is decreasing. The animal souls are all becoming men. So many species of animals have become men already. Where else have they gone?" (*CW* 1:400). No evidence is cited for his empirical claims, and even if it were true that animal populations are decreasing, he gives no thought to alternative explanations.

Vivekananda's interest in science derived in large part from his "Occidentalist" strategy to affirm Hindu superiority by standards acceptable to the West. But his involvement with science was always secondary to his spiritual concerns, so that while embracing rationality and a scientific outlook, he strove at the same time to avoid the evils of "Western entanglement in *saṃsāra*" (Halbfass 1988: 242). Such entanglement had led the West into an unrestrained quest for domination resulting in foreign conquest and colonial oppression. The Swami aspired to a spiritual counter-conquest and liberation for all by spreading the message of Vedānta to the West.

Conclusion

As the father of Modern Advaitic Evolutionism, Vivekananda has played an extremely important role in shaping twentieth-century Hindu responses to Darwinism. Four themes become particularly prevalent among later Hindu thinkers, especially for those inclined towards a monistic rather than theistic perspective. First is the insistence that Advaita Vedānta completely harmonizes with modern science, and indeed is the most scientific of all world religions. A corollary of this idea is that all modern scientific discoveries are merely rediscoveries of what was first known by the ancient Vedic sages. Second, Vedic teachings, like modern science, teach the thoroughly evolutionary nature of the universe, both in its cosmic and organic aspects. Third, the ancient Hindu theories of evolution are superior to Darwinian-based theories, as the latter ignore both the involution of the spirit and the evolution of the soul through karma and rebirth. And finally, the physical universe can be understood and its laws discovered through superconscious awareness, or yogic intuition. The scientific laboratory with all its instruments can be replaced by the laboratory of the mind of the yogi, whose insights provide a more complete view of reality—encompassing both the physical and spiritual—than material science.[37]

In the next chapter we turn to the most renowned of twentieth-century Hindu philosophers of evolution, Sri Aurobindo Ghose. His Integrative Evolutionism is deeply indebted to Vivekananda's pioneering efforts while going beyond the Swami in assimilating and responding to the latest contemporary debates in the West about the means and meaning of organic evolution. Especially significant was the rise in the West of a teleological evolutionism that became the foundation for a new style of natural theology, based less on the ideas of a divine artificer manufacturing and assembling the parts of a complex machine and more on the ideas of emergent evolution. The new teleological view of nature embodied in these latter ideas resonated deeply with Aurobindo's own Vedānta meditations.

12 The Integrative Evolutionism of Sri Aurobindo Ghose

The development of Indian political activism in the waning decades of the nineteenth century was to play a significant role in Hindu responses to the modernizing influences of European culture, and to modern science and technology in particular. Two opposing political factions emerged in the 1870s, later viewed as the "soft" or liberal party, in contrast to the "hot" or extremist party. The former were instrumental in the founding of the Indian National Congress in 1885, motivated by the ambition to secure political and civil rights for Indians, who were being increasingly excluded from participation in colonial rule. The initial members were largely inspired by ideals of social reform and viewed working with the British administration through constitutional agitation as the best means of attaining their ends. The more revolutionary group of Indians, led by Gal Bangadhar Tilak, advocated resistance, by violent means if necessary, as the only viable strategy in dealing with the increasingly recalcitrant British. Given the meager results of the liberal group in securing concessions from the British, the extremists, with their goal of national independence, came to prominence within the Congress in the first few years of the twentieth century, leading to an eventual split in 1907. Among those in the extremist group at this time was a fast-rising star, the young, Cambridge-educated firebrand Aurobindo Akroyd Ghose (1872–1950), who may well have been responsible for the 1907 split.[1]

Sri Aurobindo and the two political strands

The liberal and extremist strands are variously reflected in the lives of Aurobindo's father and maternal grandfather, respectively. Aurobindo's anglophile father, the physician K.D. Ghose, took his seven-year-old son to England in 1879, to prepare him for the Indian Civil Service. The father had come to disdain all things Indian and saw the adoption of English language and ways as necessary both for personal advancement and for the improvement of India. In taking his son to England, the atheistically inclined father hoped to cut off his son from all things Indian.

Aurobindo's early academic achievement eventually led to a classics scholarship at Cambridge. But despite excelling there in classical studies, he failed to qualify for the Indian Civil Service, intentionally refusing to show up for the final qualifying examination—in horseback riding. He had become thoroughly disillusioned with the kind of life that awaited him in the Service, and had also met several Indian students interested in Indian liberation, instilling in him a strong desire to work for his native country's independence from British rule. He soon sensed that the moderate strategy of the Indian National Congress was unrealistic. Equally important for Aurobindo's intellectual development in England was his first exposure, via English translations, to the classical texts of Vedānta, and he became intrigued with the Upaniṣadic doctrine of the all-pervading Ātman.

Returning to India in 1893 just months after his father's death, Aurobindo began writing political tracts severely criticizing the moderate approach of the Congress. At the same time he continued his study of Hindu philosophy and began to practice yoga and meditation, seeing such practice as a means to strengthen himself physically and mentally for the upcoming struggle against the British. His deepening interest in his Hindu heritage and rejection of his father's disgust with Indian culture delighted his maternal grandfather, Rajnarain Bose, a former president of the Adi Brahmo Samaj and champion of the superiority of Hinduism over all other religions. In the 1870s Bose had "helped launch the Hindu revival movement which heralded the end of educated Bengal's century-long infatuation with Western ways" (Heehs 1989: 6). Sometimes considered the Grandfather of Indian Nationalism, Bose had also in the 1870s led a secret society in Bengal working for Indian independence by whatever means necessary. "He was thus a trailblazer in both the political and revolutionary movements that a quarter-century later would be led by his grandson Aurobindo" (ibid.).

Soon after the 1907 Congress split, Aurobindo became deeply involved in revolutionary plans to assassinate despised British administrators, and following one unsuccessful attempt that nonetheless resulted in the killing of two British women, he was arrested in 1908 for alleged involvement. After a year's imprisonment he was released for lack of sufficient evidence, but he emerged from jail a transformed man. He came to view the liberation of Indians, and of humanity at large, as something to be effected by a radical spiritual transformation through the practice of what he later called Integral Yoga.[2] Leaving the political work of Indian independence to others, he retired to the safe haven of the French enclave of Pondicherry in 1910, to devote himself to spiritual practice.

He was soon to become, along with Vivekananda, one of the major interpreters of Darwinian evolution for Hindus in the twentieth and twenty-first centuries. But whereas Vivekananda was inclined to assimilate Darwinism into a larger, metaphysical system of spiritual evolution, Aurobindo ultimately rejected Darwinism as an outmoded scientific theory.

Sri Aurobindo and Vivekananda

It was during his first few years in Pondicherry that Sri Aurobindo, as he is now known by his followers,[3] developed an evolutionary philosophy that in many ways represents an updated version of Vivekananda's Modern Advaitic Evolutionism. Both proclaim the priority of consciousness over matter and identify this consciousness with the absolute Brahman of the Upaniṣads. While they see this ultimate absolute as unknowable by the intellect or reason, they equally insist that it can be known by some form of supramental consciousness that Aurobindo refers to as "intuition," "self-vision," and "self-experience" (Aurobindo 1982 [*The Life Divine*, henceforth abbreviated as *LD*]: 43). Both are dismissive of design arguments that depend upon the notion of an extra-cosmic designer god, who in their view was increasingly called into question by scientific developments. Prominent in both is the idea that an involution or descent of Spirit into Matter precedes evolution.[4] In addition, for both thinkers, evolution itself consists of two strands. First is a physical strand that equates physical evolution with the hereditary transmission of traits through reproduction, thereby ensuring the stability of species. Evolution in this sense invokes the older, pre-Darwinian meaning of the unfolding of a pre-existing germ or essential form, like the blooming of a flower.[5] And second is a spiritual strand identified with the process of rebirth and evolution of the soul.

Sri Aurobindo, like Vivekananda, is perturbed by the materialistic and mechanistic presuppositions of Darwinian evolution and by what he perceives as its emphasis on competition rather than mutual cooperation and love. They both reject the Darwinian denial of cosmic meaning and believe that the true meaning of evolution is revealed by the eternal truth of the Vedāntic scriptures. As Aurobindo declares, "the ancient and eternal truth of Vedanta receives into itself and illumines, justifies and shows us all the meaning of the modern and phenomenal truth of evolution in the universe" (*LD* 114). Yet Aurobindo is far more hesitant than Vivekananda to grant, even on the material plane, any robust role for Darwinian evolution. His critique of Darwinism essentially dismisses it, gutting it of its crucial features.

Aurobindo also goes beyond Vivekananda in his own rendering of the involution-evolution process. We find in Aurobindo, for instance, a strong tendency to multiply the various phases or grades of involution-evolution. This is especially the case in his elaboration of the stages of spiritual evolution from the ordinary consciousness of Mind to the supreme consciousness of the Supermind, the intermediate grades consisting, in ascending order, of Higher Mind, Illumined Mind, Intuitive Mind, and Overmind.[6]

More significantly, while Vivekananda is inclined to see the involution-evolution process as cyclical and ultimately non-progressive, and further, as only an apparent manifestation of Brahman (*CW* 8:362), Aurobindo espouses a more progressive interpretation of evolution and rejects any notion that this world is merely an illusion of Māyā. For him, Māyā, in its lower role, hides from us in our limited mental state the ultimate truth of the Divine Unity. But there is a higher Māyā, the power of cosmic creativity belonging to the infinite Consciousness. It is the power to unfold the manifold phenomena of this world by which the divine "play of all in each and each in all" comes into existence (*LD* 115). It is the distinction between the lower and higher senses of Māyā that Aurobindo feels is missed or neglected by "pessimistic and illusionist philosophies" (*LD* 116), referring to the traditional illusionism commonly associated with Śaṅkara's Advaita.[7] Aurobindo gives credit to materialistic science for the corrective insight revealing "the truth and importance of the earth life and the human endeavour, its evolutionary meaning" (1944: 44).

Aurobindo insists that the goal of spiritual evolution, what he calls the divinized Man, can only occur in a divinized, spiritualized world. Such a development involves an ongoing reformation of the lower levels of existence by the intervention of higher consciousness, resulting in a radical transformation of the physical and organic worlds, thereby integrating Spirit and Matter.[8] For this reason I refer to Aurobindo's evolutionary philosophy as Integrative Evolutionism.[9] To appreciate more fully the innovative features of Aurobindo's Integrative Evolutionism, we need first to look at various philosophical developments in the West that helped to shape his response to Darwinism.

The emergence of a new style of natural theology in the West and the affirmation of Vedic evolutionary views

In the decades immediately following the publication of Darwin's *Origin of Species*, science in the West came to be strongly identified with materialism and was seen as a direct challenge to the Christian faith. Darwin's explanation of life as the result of the struggle for existence and of natural evolutionary processes unguided by any divine, purposeful intelligence was at odds with traditional notions of the benevolent, wise, and designing God proclaimed in natural theology. By the 1870s, however, various European evolutionary philosophers were emphasizing the need to incorporate the principles of love and self-sacrifice evident in nature as well as of struggle and competition. They affirmed evolution, but deemphasized or denied the Darwinian mechanism of natural selection as a creative force, regarding it as capable only of eliminating poorly adapted modifications and thus stabilizing species. Accordingly, there was renewed interest in vitalism and Neo-Lamarckism as a means to reinsert a sense of direction and meaning into the evolutionary process. For the devout, evolution was becoming God's method of creation, his means of fulfilling divine purpose (Lightman 2007: 41–94). Such teleological evolutionism led in the early twentieth century to a number of new and sophisticated natural theologies in which evolution was simply the unfolding of a divine plan, or of some seemingly conscious cosmic intention (Bowler 2001: 122).

At least seven important themes pervading early twentieth-century Western evolutionary philosophies and natural theologies were to resonate within Hindu views developing at the time, especially Aurobindo's Integrative Evolutionism. Perhaps foremost among these themes was the notion of inevitable progress, but now seen in less anthropocentric terms than formerly, with Mind or even Spirit rather than "Man" as the goal of evolution (ibid.: 131). Second, progress, while inevitable, was regarded as unpredictable in its specific manifestations. Third, mechanistic-deterministic explanations of life and mind were increasingly called into question as inadequate and outdated. Evolution, in sum, was viewed as open-ended, and pushed or directed by non-materialistic agencies—such as Hans Driesch's entelechy (vitalistic perfecting force within an organism), Henri

Bergson's *élan vital* (vital impulse), or Samuel Alexander's nisus (perfective drive). It should be noted, however, that such vitalistic theories were not well received by working biologists (Subrata Dasgupta 1999: 131–3).

Fourth, these non-materialistic agencies were often seen by the philosophers as immanent within or emerging from Nature, capable of supplanting an extra-cosmic, personal creator. Fifth, Nature was commonly regarded as layered, beginning with matter, then life, mind, and spirit, each successive layer with its increasing organization and complexity seen as emerging from the prior layer, perhaps abruptly. Higher levels, while in some sense dependent on the lower, could not be reduced to them, and could exert backward, or what today is called "top-down," causal effects on the lower levels. Sixth, while Nature may be layered, the whole idea of emergent evolution as seen in thinkers like C. Lloyd Morgan implied an underlying monism to the cosmos, rejecting any sort of spirit–matter or mind–body dualism. Such notions easily lent themselves to the idea that, if mind had emerged from matter, it must have subsisted therein from the beginning (Bowler 2001: 138–9, 181).

Seventh and last, the mystery of life was to be fully understood only by non-rational or intuitive means. Closely allied with this perspective was the idea of two sciences: metaphysics, on one hand, and all the special sciences like physics and chemistry, on the other. As Alexander argued, these two types of science simply had different aims: "Metaphysics … is dealing with the ultimates which the sciences leave over" (1950: vii).

Some of these themes like the two sciences already had close parallels in the writings of men like Debendranath Tagore and Vivekananda. Many Hindu cultural nationalists were gratified to see Western philosophers confirming the sort of non-rational epistemology that seemed akin to the intuitive, yogic insights of the ancient Aryan seers. But some Hindu thinkers, concerned with convincing the British that Indians were capable of self-rule, surmised that rationalistic Western critics of Hindu civilization would be little impressed by indigenous affirmations of ancient Vedic science based on claims of supramental perception. For instance, the Marathi Brahmin nationalist Narayana Bhavanrao Pavgee, while arguing that the Vedic sages had anticipated Darwinian evolution, insisted that their discoveries were the result of empirical observations, including the discovery of fossil trilobites and other ancient fossil forms (Brown 2007b: 719–27). Pavgee's indifference to the spiritual dimensions of evolution may account in part for the indifference of Hindus generally to his historical reconstructions, especially since intellectual developments in the West were making such an empiricist and materialist approach to evolution seemingly unnecessary.

The new European evolutionary philosophies also helped to shift the focus of Hindu appeals to the authority of modern science from physics to biology in justifying the evolutionary insights of the ancient Vedic seers. Rather than citing the latest discoveries in physics, especially in atomic and sub-atomic theory, as T.C. Rajan Iyengar had done in his 1908 book *The Hindu-Aryan Theory on Evolution and Involution or the Science of Raja-Yoga*, a new breed of Hindu evolutionary theorists increasingly invoked novel ideas in biology—as interpreted and proclaimed by philosophers like Bergson and not as expounded by biologists themselves. It was especially the monistic insights of the Vedāntic sages of the Upaniṣads that were seen as confirmed by the discoveries in modern biology.

One such Hindu theorist was K. Narayanaswamy Aiyer, a member of the Theosophical Society, who in his 1914 book *The Purāṇas in the Light of Modern Science* presents a contrast between ancient Hindu and modern—i.e., Darwinian—theories of evolution.[10] He prefaces his comparison with reference to Jagadish C. Bose's experiments demonstrating the presence of sensitivity in minerals and plants, thereby suggesting a general continuity in matter and life, as well as progress from minerals to man. He notes that the modern theories only deal with the material aspect of evolution, leaving out any sense of purpose. He then cites Bergson to corroborate the ancient Hindu theories:

> But if as the Hinḍūs do, we take a conscious unit or a life, as Professor Bergson says, working through matter and adapting itself to it, then there are two lines of evolution exhibited—matter evolving and spirit within unfolding its powers as the matter becomes more and more adaptable to its working. The Evolution of matter is therefore designed for the unfolding powers of the Spirit, till the latter reaches the *summum bonum* of evolution, that is to say, the Supreme Paramāṭmā.
>
> (Aiyer 1914: 36)

Aiyer goes on to invoke, rather loosely to be sure, Bergson's reference to descending and ascending movements in the universe, to confirm the Hindu idea of involution and evolution. And like Bergson, Aiyer notes the difficulty modern biology has in explaining the creation of complex organs like the eye, "a nut impossible for them [biologists] to crack" (ibid.: 37; Bergson 1911: 68–81).[11] For Aiyer, the Hindu Purāṇas have the answer: "Brahmā, the original creator, willed the form of eyes and other organs in the higher state, which archetypal form is taken up in the lower regions for formation" (1914: 37–8). This idea leads Aiyer to note another critical difference between modern biology and Vedic views: rather than the Western notion that man came from animals, Hindus view plants and animals as descending from man, at least during the present cosmic era. Once again, Aiyer invokes Bergson's authority—but quite misleadingly—in support of this reverse evolution (ibid.: 38–9).

The foremost Hindu evolutionary thinker of the twentieth century was Sri Aurobindo. Unlike Aiyer, Aurobindo rarely cites his authors and direct influence is often hard to trace. But the terminology of the Western evolutionary philosophers, like creative evolution and emergence, is common in his writings (e.g., *LD* 704–5), and his major work on evolution, *The Life Divine*, "bears striking resemblance to major works in the Hegelian and evolutional philosophical traditions" (McDermott 1974: 168). Nonetheless, some interpreters of Aurobindo's thought minimize or dismiss Western influence, in particular the ideas of Bergson, Hegel, and Nietzsche, arguing that Aurobindo's reading of the Upaniṣads and *Bhagavad-Gītā* and his early spiritual experiences based on those texts are the key formative influences on his later philosophy (Sethna 1981: 29–30; Heehs 1989: 109–10). While Aurobindo's system may well be derived from his spiritual experiences, this does not exclude the influence of Western philosophy on the manner in which he elaborates his system. As other exponents of Aurobindo note, his evolutionary philosophy represents a reconciliation of "Western" and "eastern" thought (Chaudhuri 1960: 178–9) and Aurobindo himself had a "thorough … grasp of the great Western philosophers of the present age, such as Kant, Hegel, Spencer and Bergson" (Maitra 1956: 69; cf. Singleton 2007: 134).

We are now ready to take up Aurobindo's Integrative Evolutionism in detail. I will first provide an overview of Aurobindo's notion of involution and the inverse movement of evolution that culminates in an integrative transfiguration of the cosmos, followed by a brief look at his explanation of the parallel and interlinked processes of physical and spiritual evolution. With this general summary of his cosmology in mind, we will be prepared to scrutinize his two major critiques of Darwinism: its alleged failure on the empirical level to deal with the missing links or gaps between different grades of organic beings, and its denial of any teleological sense in the cosmos—a denial he claims runs counter to the experiential dimension of life, especially as uncovered in higher contemplative states.

Sri Aurobindo's integrative evolutionism: involution, evolution, and transformation

Aurobindo describes the involution of Divine Being in our universe as the manifestation of seven (sometimes eight) principles or gradations of Being.[12] The first three principles constitute the tri-fold nature of Being or Brahman as "Sachchidananda": infinite Existence (*sat*), Consciousness (*cit*), and Delight (*ānanda*). These three, along with a fourth, the creative and governing supramental consciousness of Supermind, representing the power of self-determination, encompass the "upper hemisphere" of Divine manifestation (*LD* 663; cf. 262–4). This Supermind "rolls out the mighty rhythms of cosmic Law and views and governs the world as their immortal and infinite Seer and Ruler" and is identical with the "Sacchidananda Himself" (*LD* 266–7). It links the "upper hemisphere" with the remaining three principles of the "lower hemisphere." These lower three are Mind, Life, and Matter, loosely reflecting, in reverse order, the three first principles of the original Divine Unity, but now in an individuated or separated mode. Such separation results in a state of limited knowledge, of ignorance of the Divine Being underlying the cosmic and terrestrial manifestations evident to our ordinary consciousness. Supermind, Aurobindo tells us, along with mind and life, "are present in the atom, are at work there, but invisible, occult" (*LD* 711).

Since the idea of *seven* principles is only a convenient pedagogical model, Aurobindo at times posits an eighth, adding psyche or soul, "a projection and an action of the third divine principle of infinite Bliss" (*LD* 264). This allows for a complete correlation of the four higher principles with the lower: Existence with Matter, Consciousness with Life, Bliss with Psyche, and Supermind with Mind. Involution, in any case, is the descent of Spirit from Existence, through

Consciousness and Bliss to Supermind, continuing to the lower cosmic level through Mind, Psyche, and Life, ending in Matter (*LD* 264). In summary, involution is the result of the play of an infinite force producing and regulating all cosmic forms and movements under the influence of a directing cosmic Will or awareness, behind which lies infinite Consciousness (*LD* 265–6).

Evolution for Aurobindo presupposes the above involution, for evolution is the re-emergence or ascent of the involved Spirit, in the reverse order. As he states it, "we ascend from Matter through a developing life, soul and mind and the illuminating medium of Supermind towards the divine being" (*LD* 264). The human species as it exists is thus not the final step in evolution, being "too imperfect an expression of the Spirit" (*LD* 847). Humankind has an inner drive towards spirituality, whose fulfillment requires a further evolution of consciousness beyond the mental, through such levels as Higher Mind, Illumined Mind, Intuitive Mind, and Overmind, culminating in Supermind, stages describing various intermediate states of consciousness experienced by Aurobindo. Such evolution requires either that humankind be surpassed by "Supermind and superman," for if the human mind can reach beyond itself, "then there is no reason why man himself should not arrive at Supermind and supermanhood or at least lend his mentality, life and body to an evolution of that greater term of the Spirit manifesting in Nature" (*LD* 847).[13]

But the culmination of such evolution cannot be just an aimless and infinite circling of the involution-evolution process, nor an individual escape from such circling, but rather a transfiguration of all creation. The evolutionary ascent involves a "taking up of what has already evolved into each higher grade," thereby effecting a radical transformation of "the whole being and nature," an integration that "must be also part of the process, if the evolution is to be effective" (*LD* 703).

In clarifying this integrative process, Aurobindo describes a twofold "top-down" effect, depending on whether the active agent is a newly evolved principle impressing itself upon the lower substance from which it has just emerged, or is a higher, pre-established, supra-cosmic principle impressing itself onto an evolved principle on the lower level. The former involves an intra-cosmic action in which an evolved principle within the material universe, like Life, while constrained by the lower principle from which it has emerged, in this case Matter, will modify by its own intrinsic power that lower and prior substance. In the latter, involving a supra-cosmic top-down action, a superior principle residing on a higher plane "in its own full force above the field of evolution" may modify its own evolved and corresponding principle on the material plane by "pressing down into that field to possess it" (*LD* 704). Aurobindo justifies this supra-cosmic top-down action by appealing to ancient vitalistic and idealist notions:

> An original creative or evolutionary Power there must be: but, although Matter is the first substance, the original and ultimate Power is not an inconscient material Energy; for then life and consciousness would be absent, since Inconscience cannot evolve consciousness nor an inanimate Force evolve life. There must be, therefore, … a secret Consciousness greater than

> Life-Consciousness or Mind-Consciousness, an Energy more essential than material Energy.
>
> (*LD* 705)

The intra-cosmic and supra-cosmic top-down effects are critical for understanding Aurobindo's explanation of the two types of evolutionary process, spiritual and physical. He explains the two types as follows:

> This terrestrial evolutionary working of Nature from Matter to Mind and beyond has a double process: there is an outward visible process of physical evolution with birth as its machinery,—for each evolved form of body housing its own evolved power of consciousness is maintained and kept in continuity by heredity; and there is, at the same time, an invisible process of soul evolution with rebirth into ascending grades of form and consciousness as its machinery.
>
> (*LD* 825–6)

Physical evolution by itself, according to Aurobindo, ultimately goes nowhere, at least nowhere truly meaningful. It remains confined to the cosmic plane of existence, wherein knowledge of the universal Consciousness is concealed. And the individual—in the absence of spiritual evolution—would quickly perish, so that only the species would attain any benefit. He concludes that "rebirth is an indispensable condition for any long duration and evolution of the individual being in earth-existence" (*LD* 826). The various ascending grades of organic forms are thus a progressive series of "body housings" for the evolving soul, allowing it to manifest increasingly the higher Consciousness or Spirit. It would seem that the intra-cosmic top-down action is predominant in the process of physical evolution with heredity and birth as its basic mechanism. But this hereditary mechanism only maintains species in stability. In order for new organic forms to evolve, in Aurobindo's view, the extra-cosmic top-down type of effect is demanded. That is, the evolution of new physical, organic forms requires the assistance or guidance of the Spirit.

It is in this context of extra-cosmic top-down effect that Aurobindo utilizes the Avataric Evolutionism propounded by Keshab Chandra Sen and the Theosophists (Brown 2007b: 730–4). The divine descent of the Spirit in various animal forms makes possible the progressive stages of physical-spiritual evolution. Aurobindo argues: "The Avatar is necessary when a special work is to be done and in crises of the evolution" (1971: 401). He further clarifies: "If we admit that the object of Avatarhood is to lead the evolution, this is quite reasonable, the Divine appearing as Avatar in the great transitional stages" (ibid.: 402). The tenth and future Avatar, Kalki, will bring about the spiritualization of the physical realm "by bringing the Kingdom of the Divine upon earth" (ibid.). Aurobindo in the end regards Avataric Evolutionism only as "a parable of evolution," a parable intended not to endorse Darwinian evolution but to explicate involutionary and transformative phases of spiritual progress (ibid.: 403). For Aurobindo, these top-down processes are necessary to explain the major transitional stages seen in organic evolution. This brings us directly to Aurobindo's critique of Darwinism.

The problem of missing links

Today the problem of missing links is invariably brought up in relation to the incompleteness of the fossil record. Aurobindo likely had such evidence in mind when he commented on the great leaps "from one grade of consciousness … to another" seen in the emergence of Life from Matter and Mind from Life. Physical evidence from the past, he argues, provides no assurance that such an evolution between grades occurred, or how it might have occurred: "Even in the outward evolution, even in the development of physical forms where the data are clearly in evidence, there are missing links that remain always missing" (*LD* 708; cf. 829). He adds: "the scientist who does observe minutely the physical data, has been driven to believe in the continuity of evolution in spite of the gaps and missing links" (*LD* 708). Bergson had already pointed out that the fossil record reveals long periods of stability, suggesting that the evolution of complex forms required some sort of "internal push" or "impetus" (1911: 113).

But it seems that, for Aurobindo, the major evidence for gaps was more directly derived from his intuitive experience of and meditations upon the different grades of life, than from the geological record. These differences are especially apparent, he insists, with regards to the inner evolution of consciousness, even if there may be minute gradations that currently escape our powers of observation. He notes, for instance, with anonymous reference to J.C. Bose's experiments, that there may be "rudiments of life-reaction in the metal" that "may be identical with life-reaction in the plant." And yet, he affirms, there exists such a "vital-physical difference" between metal and plants and us: metal appears inanimate and plants unconscious (*LD* 708–9). Accordingly, his basic view is that

> there is still a real, a radical difference between grade and grade, so much so that the passage from one to another seems a new creation, a miracle of metamorphosis rather than a natural predictable development or quiet passing from one state of being to another with its well-marked steps arranged in an easy sequence.
>
> (*LD* 708)

The idea that newly evolved forms prior to their manifestation were unpredictable, even if somewhat inevitable, we may recall, is characteristic of Bergson's creative evolution model, as well as of Morgan's emergent evolution (Bowler 2001: 152, 380).

One possible solution to the problem of these gaps is to invoke an extra-cosmic creator who creates species all at once. But Aurobindo rejects this possibility. Alluding to the biblical and Platonic accounts of creation, he argues that "we can no longer suppose that God or some Demiurge has manufactured each genus and species ready-made … and left the matter there, having looked upon his work and seen that it was good" (*LD* 709; cf. 837–8). He claims, instead, that the transitions between life forms have been effected "by swift or slow degrees," either by "a secretly conscious or an inconscient Energy" (*LD* 709). Accordingly, though the demarcations between grades of being are sharp, they are not evidence for a creator god against evolution, but only against a purely mechanistic explanation of it. While natural mechanisms may bring a lower grade of being, via physical evolution, up to a point where it is ready to manifest a higher level of consciousness, a descent of Spirit—the extra-cosmic top-down action—is needed to effect the leap across the gap. When an organic kind is ready for a major transition, Aurobindo explains, then

> at that point a pressure from some superior plane where the new power is dominant may assist towards a more or less rapid and decisive transition by a bound or a series of bounds,—a slow, creeping imperceptible or even occult action is followed by a run and an evolutionary saltus across the border. It is in some such way that the transition from the lower to higher grades of consciousness seems to have been made in Nature.
>
> (*LD* 711)

Aurobindo regards heredity as the basis for the scientific concept of "life-evolution," but as noted earlier, he sees in heredity only a power for maintaining the stability of species. It is a conservative force allowing only for small variations:

> All the facts show that a type can vary within its own specification of nature, but there is nothing to show that it can go beyond it. It has not yet been really established that ape-kind developed into man; for it would rather seem that a type resembling the ape, but always characteristic of itself and not of ape-hood, developed within its own tendencies of nature and became what we know as man, the present human being.[14]
>
> (*LD* 828–9)

Aurobindo thus claims that there is no sufficient evidence for the evolution of matter into living organisms or of plants into animals by natural means. Even if it were shown that Life manifests itself under certain physical and chemical conditions, he says, such a coincidence does not prove that chemicals are constitutive of Life or its cause. The real cause of these different forms is the "Consciousness-Force" existent in Matter, that is, "the power of the Real Idea building its own significant forms and types for the indwelling Spirit's cosmic existence" (*LD* 829).

Aurobindo's abiding faith in distinct types that do not evolve one from another leads him to interpret similar embryological developmental patterns in different species as a sign that some Consciousness-Force is at play developing all sorts of possibilities from a basic template. By this means he also accounts for the existence of intermediate species that share characteristics of two fundamental types:

> Animal species in coming into birth may begin with a like rudimentary embryonic or fundamental pattern for all, it may follow out up to a stage certain similarities of development on some or all of its lines; there may too be species that are twy-natured [sic], amphibious, intermediate between one type and another: but all this need not mean that the types developed one from another in an evolutionary series.
>
> (*LD* 830)

All of this simply points beyond hereditary variation to "a consciousness and a force other than the physical at work making for variation in Nature" (*LD* 830). Such a non-physical force, he maintains, is needed to account for the working of natural selection, as well as for the diversification of species. Aurobindo's viewpoint reflects two common, interlinked assumptions underlying most creationist arguments today, that natural selection is a purely negative process eliminating ill-adapted species, and that random, non-planned variations are incapable of creating truly new and viable forms. In essence, Aurobindo, argues for the standard creationist conclusion: there is microevolution—to use contemporary creationist terminology—allowing for variation within a type, but no macroevolution that would account for speciation.

In relation to humankind, Aurobindo rejects the idea that the human body suddenly appeared as an "abrupt creation," arguing instead that "there was an evolution of humanity out of animal being, slow perhaps in its preparation and in its stages of development, but with strong leaps of change at the decisive points of the transition" (*LD* 838). He elaborates: "… it is certain that changes of characteristics in the type, though not of the fundamental type itself, can be brought about in species or genus" (*LD* 838). Aurobindo assures us that these minor changes have even been effected by human manipulation. But a major change requires the input of the superior Consciousness-Force working on the emergent principles struggling upwards from below.

Interwoven into and justifying Aurobindo's account of the diversity of life is the ancient notion of a great chain of being, a ladder or scale of nature, for "each grade of being exists in itself and by itself …, and the gradations above or below it are not origins and resultant sequences but only degrees in the continuous scale of earth-nature" (*LD* 829; cf. 837). Such a scale assumes that each kind of being manifests an inherent nature, a Platonic form if you will. Or in traditional Hindu terminology, as Aurobindo explains, humankind, like all other types of animals, "has his own native law, limits, special kind of existence, *svabhāva, svadharma*" (*LD* 831).[15] Such biological essentialism, whether Western or Indian, is "desperately unevolutionary," in Richard Dawkin's words, being radically opposed to a view of life that regards organic beings as capable of indefinite departures from ancestral forms (2009: 22). One need only recall Aurobindo's statement regarding "the power of the Real Idea building its own significant forms and types" (*LD* 829) to recognize his basic essentialism and antipathy to Darwinian theory. Indeed, for Aurobindo, anticipating current creationist rhetoric, Darwinian evolution was only a short-lived theory.

In Aurobindo's judgment, the facts discovered by science are trustworthy, but theories are another matter. He argues:

> It is true that Science now affirms an evolutionary terrestrial existence: but if the facts with which Science deals are reliable, the generalisations it hazards are short-lived; it holds them for some decades or some centuries, then passes to another generalisation, another theory of things.
>
> (*LD* 828)

Once again, he insists that heredity, with its power to maintain the stability of species, is a fact. But the generalization that heredity can produce "persistent and progressive variation" is for Aurobindo "very questionable" (*LD* 828). Scientific theories are like "shifting quicksands," and thus no basis for building a sound metaphysical view of life, something Aurobindo was committed to doing.

The keynote idea in nineteenth-century thought, according to Aurobindo, was the materialistic theory of evolution (1944: 1). From this theory flowed all sorts of disturbing social, political, and economic consequences, including "the breaking up of religious belief," "the theory of the recent emergence of the thinking human animal," "an exaggerated development of the competitive instinct," and "the increasing victory of the organised State … over the free individual" (ibid.: 1–2). Fortunately, however, Aurobindo reports, "The materialistic view of the world is now rapidly collapsing and with it the materialistic statement of the evolution theory must disappear" (ibid.: 2). The idea of the struggle for life was being questioned, even denied, in favor of moralistic interpretations of evolution adjoining the principle of love to that of egoism (ibid.: 4).

Perhaps most significantly for Aurobindo, the idea of a slow, incremental evolutionary process was being challenged (apparently by evolutionary philosophers like Bergson), giving way to "a new theory of evolution through sudden and rapid outbursts" (ibid.). As Bergson had already claimed, Darwinian gradualism "is tending … to give way to the opposite idea that a new species comes into being all at once by the simultaneous appearance of several new characteristics" (1911: 71). As a result, Aurobindo noted, "Theories of vitalism, idealistic tendencies of thought, which were supposed to have been slain by the march of physical Science, now arise, dispute the field" (1944: 4).

For Bergson, chance cannot account for the accumulation of slowly acquired random characteristics, since each, unless it were beneficial in itself, would be eliminated by natural selection. Nor could all the changes happen simultaneously, unless there is some "mysterious principle" (1911: 76) at work, a principle such as a "psychological cause" proposed by certain Neo-Lamarckians, but deeper than the mere conscious efforts of an individual (ibid.: 97). This deeper principle, for Bergson, is his famous "original impetus of life" (ibid.). Aurobindo similarly accepts that a "Life-Principle" is the cause for the development of life from matter (1944: 6). And beyond the Life-Principle there is an analogous Mind-Principle, reviving a modified form of Lamarckism: "The propagation of acquired characteristics by heredity was too hastily and completely asserted; it is now perhaps in danger of being too summarily denied. Not Matter alone, but Life and Mind working upon Matter help to determine evolution" (ibid.: 10; cf. Bergson 1911: 86). Such a view is more attuned to traditional Vedāntic ideals. As Aurobindo proclaims, the Vedic seers knew of the inconscient sea of matter out of which the worlds arose, but also posited "a governing and originating superconscient which accounts for the appearance of a hidden consciousness and knowledge pervading the operations of Evolution" (1944: 8). He concludes that new species result from the spasmodic intervention of this superconscient force: "… new steps in evolution are … effected by rapid and sudden outbursts, outbreaks, as it were, of manifestation from the unmanifest" (ibid.: 9).

It is clear that Aurobindo's stress on the problem of missing links is aimed at reintroducing the notion of Mind back into the evolutionary process. By appealing to the empirical facts of science, he hopes to reject a mechanistic determinism and to restore a sense of cosmic purpose. It is time to turn directly to the issue of purpose in Aurobindo's thought, and his attempt to reconcile a teleological view of the universe with traditional Advaitin notions of a complete and perfect Absolute that needs or requires nothing.

The problem of teleology and the purposelessness of the absolute

Certain of the evolutionary philosophers in the West at the beginning of the twentieth century were calling into question not just the idea of an extra-cosmic designer god, but even the notion of purpose or teleology. Foremost among these was Bergson, who rejected both the doctrines of mechanism and "finalism" (essentially, a variant of teleology), on the grounds that each restricts the freedom of the creative process (1911: 41–54).[16] Finalism or teleology for him is simply "inverted mechanism" (ibid.: 45). Thus Bergson could write: "… if the evolution of life is something other than a series of adaptations to accidental circumstances, so also it is not the realization of a plan. A plan is given in advance" (ibid.: 114). Evolution is constantly creative, and in essence unpredictable. If evolution were the realization of a plan, then it should reveal increasing harmony as it goes along, but this is not the case, since there are many deviations and turnings back (ibid.: 115). The harmony or unity of life we do see is thus not the result of a plan at the end attracting life forward, but of the vital impetus pushing from behind.

While we see some of these themes in Aurobindo, he refrains from an absolute denial of teleology. Behind the apparently inconscient matter, he feels, there lies an "evolutionary will of a secret Conscious Being," an "innate intention in the evolution," which constitutes "a teleological element" (*LD* 834). He elaborates:

> … the conscious or even the inconscient nisus arises from a truth of conscious being that has become dynamic…; the teleology, the element of purpose in the nisus is the translation of the self-operative Truth of Being into terms of self-effective Will-Power of that Being …
>
> (*LD* 834; cf. 1056, 1061)

Although details of evolution, such as speciation, natural selection, and the inheritance of acquired characteristics, may be debated, he argues, they do not affect the greater picture. The one self-evident conclusion of primary significance arising from all of this, he proclaims, is "the fact of a successive creation with a developing plan in it" (*LD* 836).

The term *nisus* that Aurobindo uses above refers in his system to the perfective urge that lies behind the successive evolution first of Matter, next of Life from Matter, then of Mind in living Matter, and finally of Supermind in or from Mind. The notion of nisus is reminiscent of, and may well be taken from, Samuel Alexander's use of the term in his *Space, Time, and Deity*. Alexander sees this perfective urge in the fundamental "stuff of the world" that he calls Space-Time (1950: vol. 1, pp. vi, xi): "There is a nisus in Space-Time which, as it has borne its creatures forward through matter and life to mind, will bear them forward to some higher level of existence [deity]" (ibid.: vol. 2, p. 346).[17] But whereas Alexander's notion of this higher level of existence, what he calls "deity," is not fully actual, but simply "the next higher empirical quality to mind, which the universe is engaged in bringing to birth" (ibid.: vol. 2, p. 347), Aurobindo's Supermind is a manifestation of the infinite Being, Consciousness, and Bliss of Brahman.[18]

Aurobindo recognizes that the facts of science, the observed nature of the universe, provide no evidence of any plan, or of soul evolution. But science for him deals only with insentient energy and mechanistic processes. It is incapable of affecting or addressing ideas of spiritual evolution, or of revealing a hidden or secret Consciousness directing the insentient energy (Brown 2010: 726).[19] More challenging than the scientific objection to teleology, for Aurobindo, is the one posed by traditional Advaita metaphysics, with its insistence that Brahman is perfect in itself without any need or desire to create the universe. The manifestation of the universe is due merely to the inherent nature of the Absolute to play, without ulterior motive. Aurobindo explains that the metaphysical objection "proceeds on the perception that the Infinite and Universal has everything in it already, that it cannot have something unaccomplished to accomplish … and there can therefore be in it no element of progress, no original or emergent purpose" (*LD* 833). He concludes: "The metaphysical objection is more serious; for it seems self-evident that the Absolute can have no purpose in manifestation except the delight of manifestation itself" (*LD* 834).

There are two ways that Aurobindo sees of reconciling teleology with the perfect Absolute of Advaita. First, an evolutionary teleology involves only the transformation of the material realm through the manifestation of the powers of Being itself filling the entire universe. Such a teleology "does not bring in any factor that does not belong to the totality; it proposes only the realisation of the totality in the part" (*LD* 835), a notion similar to Hegel's conception of history as the temporal development of a non-developing Absolute (K. Ward 1996: 98–9). And second, the nature of play or *līlā* itself, as a kind of game, admits of an intrinsic goal:

> All exists here, no doubt, for the delight of existence, all is a game or Lila; but a game too carries within itself an object to be accomplished and without the fulfilment of that object would have no completeness of significance.
>
> (*LD* 835)

From a somewhat different perspective, Aurobindo argues that only some aspects of Nature are teleological, that "purpose is only one minor part of existence," and that there is a higher principle that "will at once contain and exceed the teleological" (1972: 42). This higher principle is found in the play of divine imagination that expresses itself in cosmic manifestation. By this means Aurobindo rejects the idea of a grand designer, replacing it with the notion of the self-expression of the divine Bliss that still provides a sense of spiritual meaning, if not "purpose," to the evolved universe. Rather than rejecting all teleology, he thus embraces the higher non-purposeful principle of imagination that encompasses the lower "teleological element in Nature" (ibid.; cf. Srivastava 1968: 304). As he concludes: "Imagination is outside purpose … because she is an inherent energy not of some great teleological Master-Workman, but of Ananda, the Bliss of existence or Will to live; and beyond this delight in existence she has no reason for being" (1972: 42; cf. McDermott 1974: 173–4; Srivastava 1968: 433).

The interrelated notions of play and delight are frequently utilized in later Hindu elaborations of creative evolution as a means of reconciling an ultimately non-teleological cosmic process with a meaning-saturated universe. As Sarvepalli Radhakrishnan puts it in his 1929 Hibbert Lectures:

> The Indian figure of līlā makes the creation of the universe an act of playfulness. Play is generally the expression of ideal possibilities. It is its own end and its own continuous reward. The Absolute mind has a perfect realm of ideal being, and is free creativity as well.
>
> (1937: 344)

Similarly, Rabindranath Tagore, in his own Hibbert Lectures of 1930, appeals to the principle of joy and delight in his rejection of a meaningless, randomly produced universe. But unlike Aurobindo and Radhakrishnan, Tagore senses a more personal being or will as the guiding force behind evolution:

> Somewhere in the arrangement of this world there seems to be a great concern about giving us delight, which shows that, in the universe, over and above the meaning of matter and forces, there is a message conveyed through the magic touch of personality.
>
> (1931: 102)

Tagore near the end of his life, when he had become increasingly intrigued with the latest scientific discoveries, provides a striking metaphor for his guided theistic evolutionary views. In his 1937 book *Our Universe*, Tagore compares the balancing of positive and negative energies in the atom that enables a stable universe to a juggler-musician with his dancing bear whom he prevents from running amok through his steady beating of his drum. Tagore then proclaims: "There is also an invisible juggler who in his play and dance of creation is drumming a tune to make us follow its rhythm" (1969: 30–1).

Conclusion

Aurobindo's Integrative Evolutionism was to provide the basic model for various later Vedāntic (monistic) elaborations of evolution, especially those proposed in the post-colonial period. While reflecting in many ways Vivekananda's Advaitic Evolutionism, Aurobindo emphasizes a more progressive view both of physical and spiritual evolution, and his notions of integration and transformation of the lower chords of being manifest a more world-affirming perspective. In addition, Aurobindo ties in to his theory of involution-evolution the Western ideas of emergent and creative evolution, with their critiques of Darwinian gradualism and the notion of evolutionary unpredictability. Such Western ideas were not available to Vivekananda by the time of his death. Aurobindo was thus able to go well beyond Vivekananda in assimilating Western science and contemporary evolutionary philosophies into a traditional Vedāntic framework.

Before turning to post-independence developments, it is worth noting two major scientific developments that were unknown to Aurobindo but were later utilized in elaborating and updating his evolutionary model: quantum mechanics and molecular biology. Rabindranath Tagore provides a hint of things to come in his *Our Universe*. In the conclusion, Tagore notes, retrospectively, that the order and harmony of the planets orbiting the sun can hardly be "mere coincidence," recalling the old version of the design argument seen in both Paley and Newton. But prospectively, he points towards the phenomenon of cell differentiation that becomes a focus of attention for Hindu evolutionary and anti-evolutionary thinkers alike, by the end of the twentieth century. He refers to the appearance of a tiny life-cell particle on earth millions of years ago, that came to reveal a "rare, rich artisanship" (ibid.: 113). Within that cell, he proclaims, is an occult but conscious and intelligent creative force that explains cell differentiation, even if that force is beyond our comprehension:

> What this intelligence is that lies hidden in it, that is capable of amalgamating, readjusting, innovating and manipulating machinery of great complexity and how … it has made itself actively creative …, these are beyond all our comprehension. Extremely delicate and sensitive by nature these life-cells … are assembling at their destined places in the live body … How through sheer inner urge this miraculous division of labor has taken place in the functioning of the body, one never knows. The cell of the stomach has its specific duty, while the cell in the brain has its work of an entirely different nature. And yet, the atoms in the life-cells are fundamentally the same.
>
> (ibid.: 111)

Earlier in *Our Universe*, Tagore had noted the discovery that the rotating electrons around an atom jump from one orbit into another, and then back again—a reference to the non-continuous quanta of energy that are responsible for the "quantum jumps" of the electrons. The idea of quantum jumps, and of quantum mechanics more broadly, particularly in its "New Age" renditions that incorporate consciousness into the fundamental nature of reality, was eventually to be invoked by Hindu thinkers like Amit Goswami to explore and explain the actions of Tagore's mysterious intelligence underlying cell differentiation. These will be examined in Part III.

Part III

The post-colonial period

13 Design and Darwin in the post-colonial period

In the decades since Indian Independence in 1947, two major developments have helped to perpetuate and promote the various nineteenth- and early twentieth-century Hindu responses to Darwinism among educated Hindus. First is the growing influence within India of a Hindu majoritarian, cultural-political movement, militant Hindutva nationalism, which emerged in the 1920s. The Hindutva movement, while claiming early Hindu nationalist thinkers like Vivekananda and Aurobindo as their own, downplays the more tolerant, universalist teachings of the latter, adopting a supremacist agenda focusing on an ideal of Hindu identity in opposition to corrupting foreign elements. Among these alleged pernicious influences are secular Western culture in general and its modes of thought, including rationalist-reductionist science, as well as alien religions, especially Christianity and Islam.[1] Second is the accelerating growth of global Hinduism as increasing numbers of Hindus establish permanent homes outside India. In their adopted countries around the world, these generally prosperous Hindus espouse universal spiritual ideals, but are often heavily invested in preserving their cultural traditions and thus occasionally manifest the more exclusivist tendencies of Hindu nationalism (cf. Flood 1996: 265–7).

Thus, both groups share a quest for cultural self-affirmation that tends to view the two centuries of European colonization of India as having left indelible marks on the colonized minds of many Western-educated Hindus. Such colonized minds, in this nativist view, fail to appreciate the glory of India's ancient spiritual and scientific heritage. Especially important for our purposes are those among diaspora Hindus who have established academic careers in Western universities, chiefly in the fields of computer science, engineering, medicine, and physics. These commentators frequently contrast "Vedic science" with materialistic "Western science."

Ambivalence towards modern science

The prestige of modern science and Darwinian evolution are sufficiently great that they are seldom ignored by Hindu apologists, even by those who view modern science with deep suspicion. As the sociologist Eileen Barker points out with regard to the spiritually inclined members of any society today attempting to find confirmation of "old certainties" and faced with the challenges of religious pluralism and secularism, it is hardly surprising that they often "turn to science with both fearful suspicion and a desperate trust" (1986: 143).

An extreme example of this ambivalence may be seen in H.C. Mathur's *Text Book of Hindu Psyche (Scientific Rationale of Hindu Faith)*, published in India in 2002. On one hand, Mathur argues that "[s]cientific temper has no place in the Hindu way of thinking" (2002: 145). The myth of "scientific temper," he claims, has misled Hindu youth away from their spiritual values and into "materialism and consumerism," coupled with forgetfulness of the higher planes of consciousness that alone can provide true bliss. All this is due, not to the weakness of the Hindu psyche, but to a "deliberate design and conspiracy of the West" (ibid.: 146). The real Hindu psyche is so focused on the spiritual plane, Mathur asserts, that "the Hindu cannot really excel in any field in the material world" (ibid.). On the other hand, Mathur frequently invokes the discoveries of modern science to confirm traditional spiritual truths or to show parallels between them—after all, modern science for him is simply "a subset of the yogic thought" (ibid.: 125).

Such ambivalence manifests in predictable ways in Mathur's attitude towards Darwinism. In discussing evolution, he takes a jibe at Christian fundamentalism, dismissing the "anecdote of Adam and Eve" as a non-scientific explanation of the origin of life (ibid.: 16). And while Darwin's theory may be scientific, it is clueless regarding how life first appeared on earth, since it does not accept any outside, supernatural agency, in particular the soul, without which, Mathur claims, the phenomenon of life cannot be fully explained. He argues, for instance, that a single cell knows what to do metabolically by following its instinct for self-survival, but what gave it this instinct and its structure is unknown to science. These can be explained only by a guiding intelligence that vanishes at death, due to the departure of the superintending soul. Mathur concludes: "The Hindu belief of a soul, on the other hand, is logical and more scientific. It can be explained scientifically on the principle of interaction of vibrations" (ibid.). There ensues a lengthy "scientific" discussion of wave mechanics by which the entire array of physical, organic, and psychological or mental phenomena including divine beings is produced in the cosmic spectrum. The soul (ātman) is "a pure vibration … of very high frequency" that "explains all the mysteries of life" that materialistic science is unable to, for it is the soul that "organizes matter into specific shapes and sizes" (ibid.: 20, 22). But all vibrations ultimately have their source in the highest, quality-less (*nirguṇa*) Brahman, which cannot be perceived by the senses or scientific instruments, but only by "your intuitive feeling" and eventually through meditation (ibid.: 35). Ultimately, in accord with his Advaitic perspective, Mathur insists that "[t]he whole manifestation is an illusion" (ibid.: 36).

In the rest of this chapter I shall look at an array of post-colonial Dharmic responses to Darwinism that form part of the contemporary Hindu quest to affirm a cultural-religious identity. These responses fall generally into one of two broad modes: a theistic, creationist perspective that tends to reject Darwinism, often by utilizing design arguments, and a non-theistic, evolutionary perspective derived from Advaita Vedānta especially as reinterpreted by Vivekananda and

Aurobindo. Occasionally we discern a convergence of these two perspectives, as theistic thinkers increasingly accept limited aspects of modern evolutionary theory, while some recent Neo-Advaitins show a renewed interest in the design argument, which Vivekananda regarded as "childish," and Aurobindo radically subordinated to the idea of divine play. Common to theistic and non-theistic approaches is the tendency to cite the latest discoveries of modern science in support of their particular theological or philosophical perspectives.

Contemporary Vedic and Atomistic Creationism

We shall look at four contemporary Hindu creationist responses to Darwinism. Three rely heavily upon Vedic scriptural statements as interpreted by divinely inspired teachers for truth or truths about the physical world. But their specific "Vedic" sources and inspired sages vary, so that each provides a distinct version of Vedic Creationism. The fourth builds on the classical theistic atomism of the Nyāya-Vaiśeṣika.

Arya Samaj Creationism

Although Darwinism does not seem to be of much concern among contemporary Arya Samajists, when the topic is broached, evolution is summarily dismissed in accord with the creationist views of its founder, Dayananda Saraswati. Thus, on the now defunct online "Arya Samaj Forum," in response to a question regarding the theory of evolution, the Samajist administrator with the blog-name Shishya asserts: "It is vital to remember that it is and will forever remain a theory … any honest scientist will tell you that Darwin's theory is not the only one which could explain the available evidence. It is just the most widely publicized."[2] Shishya then contrasts the Vedic teaching that God created man with the evolutionary idea of "spontaneous and random mutations." He concludes:

> Do you think the world—as well all the creatures in it—is the product of blind chance or of intelligent design? Or in other words: Are you an atheist or theist? For only an atheist could believe that the world is the product of blind chance, and to believe in an intelligently ordered world is to presuppose the existence of a Divine Being who transcends creation.

Shishya's response parallels current Christian creationist claims that evolution is "only a theory," coupled with the radical dichotomizing of chance and design, correlating these with atheism and theism, respectively.[3] No consideration is given to evolution as an alternative between blind chance and intelligent design, or in Dawkin's terms, a process of "mutations" or "random changes in genes" evolving through "non-random selection" (Dawkins 2009: 35). At the same time, Shishya claims that "evolution *as a principle of creation* in no way conflicts with Vedic teachings. But God is always the guiding intelligence behind that evolution."

In another post to the Arya Samaj Forum, Shishya discusses the question of Avataric Evolutionism.[4] He dismisses the idea that the avatars were actual incarnations of God, merely "the stuff of pious imagination and wishful thinking," but does allow that the idea, as part of Purāṇic mythology, should be understood not in a literal fashion as Purāṇikas do, but as a parable or allegory only.

ISKCON creationism

One of the best known and earliest of the Hindu theistic movements to gain world-wide attention was the International Society for Krishna Consciousness (ISKCON), founded by A.C. Bhakktivedanta Swami Prabhupada in New York in 1966.[5] He based his anti-evolutionary views not primarily on the Vedas and Upaniṣads, but on the *Bhāgavata Purāṇa* celebrating Kṛṣṇa as the Supreme Personality of Godhead, and the *Bhagavad-Gītā*. Prabhupada considered such scriptural texts to be fully "Vedic" in their own right, containing infallible truths about the spiritual and physical realms. Prabhupada traced his spiritual lineage back to the medieval Vaiṣṇava ecstatic mystic Caitanya, whom he considered to be an incarnation of Kṛṣṇa.

Prabhupada generally insists upon the literal interpretation of the sacred Vedic texts, presenting them as accurate historical and scientific accounts of the universe. Such a view includes the idea that Kṛṣṇa created all species at the same time that he created the universe (in its present cycle), several billions of years ago (Prabhupada 1972: 456). The different species were intended as vehicles variously designed to benefit souls in different stages of spiritual progress via karma and rebirth—from aquatics through animal forms to the human. Prabhupada thus presents a Hindu version of the Great Chain of Being, radically in conflict with the Darwinian view. Not surprisingly, Prabhupada dismisses Darwin and his followers as "rascals" and Darwinian theory as "nonsense." One of his main arguments against Darwinian evolution is the old creationist canard that if evolution were true then lower species should not still exist, coupled with the query of why we do not see monkeys giving birth to humans (Prabhupada 1979: 48).[6]

A rather different approach to Darwinian evolution has been taken by two of Prabhupada's American disciples, Richard L. Thompson and Michael A. Cremo. Far better informed about modern evolutionary theory than their spiritual master, Thompson and Cremo avoid ridiculing Darwin and his followers, offering instead an alternative view of evolution that simply turns Darwinism on its head. This alternative view, called "inverse evolution" or "devolution," accepts the idea of "physical descent from a common ancestor and the appearance of new species by sexual reproduction," but identifies the common ancestor with "a superintelligent being, not a single-celled creature" (Thompson, Drutakarma, and Bhutatma 1984: 60). The devolution of Cremo and Thompson, although reminiscent of Aurobindo's involution, is distinct. Devolution characterizes a "fall" from Kṛṣṇa's heaven through a rebellious act of free will into the inferior realms of material energies and gross matter, rather than the creative and playful manifestation of the

Divine Absolute that ultimately spiritualizes the whole of the material, vital, and mental spheres, as enjoined by Aurobindo.

Thompson, in his 1981 book *Mechanistic and Nonmechanistic Science*, argues against Darwinism on the basis that it is not a scientific theory since it is "unfalsifiable," and that it was based on blind faith in the powers of organic forms to evolve by Darwinian mechanisms. He offers various creationist explanations for evidence often taken to support evolution, such as the "vestigial traces" of organs that seem to serve no purpose in the present form of the animal (Thompson 1981: 200–2). From an evolutionary point of view, such traces are remnants of once useful organs, whose genetic programs responsible for constructing them have not been eliminated by natural selection. For Thompson, such vestigial traces are evidence of a clever and efficient master engineer who creates one basic design or blueprint for creating a variety of animal forms, and then simply switches on or off certain regulative genes to create a multitude of related species from the original template. The odd designs that sometimes result from this strategy, Thompson explains, are due to the fact "that God does not necessarily intend the material world as an exhibition of His finest workmanship" (ibid.: 204).

In *Human Devolution: A Vedic Alternative to Darwin's Theory*, Cremo expands upon Thompson's ideas of divine genetic engineering, focusing on microscopic structures and functions that he claims elude the efforts of modern microbiologists to explain in evolutionary terms. According to Cremo's cosmology, we all begin as pure consciousness in a relationship of reciprocal enjoyment with the Supreme Lord, Kṛṣṇa. But then, if a conscious self desires independence from the Supreme Lord, it must descend into the realm of material energies, where it may take on a gross material body prepared by various demigod assistants of Kṛṣṇa. The demigods (mind-born creations of Kṛṣṇa's assistant creator, the god Brahmā) create gross physical bodies while engaged "in reproductive activities," making use of seeds, or *bijas*, that "contain the plans for the forms of the bodies" (Cremo 2003: 485).

Such *bijas* are not to be identified with genes or DNA, but rather with the divinely designed organizational plan that oversees the development of a plant or animal from a single cell. For Cremo, "[m]odern biologists have difficulty explaining exactly how the process of development takes place" (ibid.). Specifically, he argues, since each cell in an organism contains exactly the same DNA, "it is not easy to explain how and why, in the course of progressive division of a few cells into millions of cells …, that the cells sequentially differentiate and arrange themselves into complex forms of bodily tissues" (ibid.). The *bijas* or subtle developmental seeds containing the body plans for all species were generated by Kṛṣṇa and distributed to the demigods to use in creating appropriate bodies for conscious selves (ibid.: 486). In such manner Cremo answers the problem of cell differentiation noted by Rabindranath Tagore almost seven decades earlier.[7]

Cremo's criticism often seems directed at the shortcomings of older versions of evolutionary science. He would have been closer to the truth had he said that modern biologists have *yet* to explain embryonic development—although even at the time of his writing (2003), major discoveries had already contributed to an increasingly detailed understanding, on naturalistic principles, of cell differentiation and development. He makes no reference to the emerging field of evolutionary development—frequently abbreviated as Evo-Devo—which readily reduces the need for such divinely inserted design plans. Kṛṣṇa's role and that of the demigods become redundant.

Given Cremo's reliance on intelligently designed body plans to account for embryological development and the diversity of species, it is hardly surprising that he rejects the idea that natural selection can be a creative force (ibid.: 59–60). Like Thompson, he insists that some guiding intelligence is required. He admits that the results of guided natural selection may at times appear to be rather "weird" and "quirky," referring specifically to the infamous panda's "thumb," something one might not expect from a rational Creator. Nonetheless, however strange, it does not deny that the creative process producing it was "guided and intentional" (ibid.: 130). Agreeing with Alfred Wallace, the co-discoverer (with Darwin) of evolution by natural selection, that God and natural selection were not mutually exclusive, Cremo concludes: "The original Mind could have guided the path of natural selection in a certain direction to get human beings as an end result. And one of the byproducts may have been the panda, with its strange thumb" (ibid.).

While Prabhupada's comments on Darwinism seemed intentionally designed to promote conflict between religion and modern science, and especially evolution, Cremo clearly hopes to avoid such a scenario, promoting instead some sort of integration, or at least dialogue. In the conclusion to his book, Cremo notes that, while he has shown significant correspondences between the divinely revealed Vedic account of human origins and the scientific evidence available to our senses, he has not tried to use this evidence "to prove the Vedic account," which as revelation, he argues, "has its own epistemological certainty" (ibid.: 489). At the same time, he insists that all scientific evidence regarding human origins "can be evaluated in such a way that it does not contradict the Vedic account, and, indeed, is quite consistent with it" (ibid.). The whole purpose behind such an effort is to help humankind, with its growing respect for science, "become interested in what can be called re-evolution," returning to the original relationship of reciprocal enjoyment with Kṛṣṇa.

Before leaving ISKCON's anti-evolutionism, it is worth noting Cremo's rejection of Avataric Evolutionism, not on "scientific" grounds but on scriptural. As he notes, many of the traditional Purāṇic stories of the non-human avatars like the fish refer to the coexistence of human beings, thereby precluding any notion of linear ascent or evolution from animal to human. He correctly attributes the evolutionary interpretation of the avatars to nineteenth-century Westernized Hindu apologists.[8]

ISDL creationism

A parallel theistic movement to that of ISKCON is the International Society of Divine Love (ISDL), founded in India in 1975 by Swami Prakashanand Saraswati. His movement has spread around the globe, with a U.S. branch beginning in

Austin, Texas, in 1981. His anti-evolutionism, like ISKCON's, is steeped in Kṛṣṇa devotionalism, but with pointed critiques of the Hindu colonized mind. ISDL has yet to develop the more cautious and scientifically informed response to Darwinism that we find in ISKCON.

Prakashanand's anti-evolutionism is firmly grounded in his unquestioning acceptance of a literal interpretation of scripture, especially the *Bhāgavata Purāṇa* and its timescale of cosmic and terrestrial history. He claims that the total age of the Earth and sun is 155.52 trillion years, and that Hindu civilization has an unbroken continuity extending back 1.9 billion years, in contrast to other civilizations such as those in Europe and America that go back only 6,000 to 8,000 years (Prakashanand 1999-2001a; 1999-2001d). The eternal scriptures of the Vedas and Purāṇas were produced for this cosmic era by the god Brahmā at the beginning of the current cycle (155.52 trillion years ago) (Prakashanand 1999–2001e). It was Kṛṣṇa who at that time willed the universe into being, creating multiple chief celestial gods, Brahmās, who then set about creating the various galaxies and earth planets in each. Prakashanand's Ancient Cosmic Creationism is clearly at the other end of the cosmic time scale from young-Earth creationism, but it is equally problematic in terms of scientific findings regarding both the age of the universe and of the Earth.

With specific regards to Darwinism, Prakashanand acknowledges that scientists have discovered many biological facts, but claims that these scientists, on the basis of insufficient findings and drawing upon their own imaginations, formulated the incorrect theory of evolution. He then provides a caricature of modern evolutionary theory, substituting his own evolutionary explanations of speciation for standard evolutionary accounts. If his explanations are understood by his disciples as actual scientific explanations and not merely rhetorical ridicule, then his followers would be more than justified in rejecting evolutionary theory. As an example, he provides the following explanation for the evolution of dinosaurs from frogs:

> These frogs kept on jumping all the time and disturbing the atmosphere, so the nature pushed them down and they became reptiles like lizards, snakes and crocodiles. Tired of their slow motion they then evolved into running dinosaurs, from a miniature size to a real monstrous size. Feeling great they stamped the grounds of Asia, Europe and America when an unknown natural calamity happened and all the big dinosaurs died around 65 million years ago.
> (Prakashanand 1999–2001c)

The Neo-Lamarckian tendency of the above quotation becomes quite pronounced in Prakashanand's comments regarding the apparent lack of evolutionary progress in human beings over the last six thousand years:

> There are hundreds of questions to which the theory of evolution has no answer. When it says that the inner urge of a being to accommodate to the

> new situation causes a change in the body formation, then why has the evolution now stopped? The intellectual work load of a human being has increased to at least fifty times more than it was 6,000 years ago. But a human being still has the same size of brain as it had before.
>
> (ibid.)

Further, the Swami wonders why humans have not developed extra arms and hands, so that "the modern man could work on two computers at the same time, and could work better and faster in the kitchen or in the office" (ibid.). In such ways Prakashanand derides evolutionary theory.

Among Prakashanand's arguments against evolution, the following are worth noting:

1 Sense organs could never evolve from one-cell organisms, which can only grow larger. Sense organs must arise from the impulse of a being who already has senses and a developed mind—an objection based on the *a priori* assumption that mind is an independent, non-material substance.
2 Natural selection can only bring about slight changes in body color or slight changes in appearance, as seen among the various human races—the micro-evolution-versus-macroevolution objection.
3 Similarly, mutation can only effect a very small range of changes in DNA, preventing, for example, a tiger producing wolves, dogs, or bison.
4 Human uniqueness disproves evolution, for humans are capable of sins like premeditated robbery, murder, and same-sex mating. Animals are radically different, simply following instincts no matter how intelligent they may be, until their souls are reborn in the human family and begin acting on personal desires that create karma—an objection, incidentally, illustrating that notions of animal-human reincarnation are not necessarily more amenable to Darwinian evolution than the one-life perspective of the Abrahamic traditions.
5 The skulls of gorillas, chimpanzees, and various primitive tribes that scientists place in a sequence showing improvement do not illustrate human evolutionary developments, but only differences in brain capacities among different species, since they all exist at the same time.

Despite some familiarity with evolutionary terminology, Prakashanand demonstrates no genuine understanding of the details of modern evolutionary theory, nor even a familiarity with such basic facts as the widespread occurrence of homosexuality in the animal kingdom (Bagemihl 1999) and the demonstrated temporal sequence of hominid species, from *Australopithecus afarensis*, through *Homo erectus* and *Homo habilis*, to *Homo sapiens*. While the details of the exact nature of the human ancestral tree remain obscure, it is clear that the above species are not contemporaneous.

Like ISKCON's Cremo, Prakashanand rejects the notion of Avataric Evolutionism, but with a more acerbic condemnation of Darwinism. He protests that those who propose such a notion are injecting a materialistic and unscientific ideology (Darwinism) into the accounts of God's divine descents into our world. Such writers degrade the eternal Dharma by unduly elevating the status of the Buddha avatar while ignoring the supremely divine status of the two most important, Rāma and Kṛṣṇa. He concludes with a biting critique of the colonized Hindu mind:

> This is all the effect of the Western writers on the Hindu minds that collected the intellectual dirt of the followers of the English regime (knowingly or unknowingly) and tried to smear it on the face of Hinduism in their own intellectual style.
>
> (Prakashanand 1999–2001b)

A contemporary defense of classical Nyāya-Vaiśeṣika theistic creationism

On a more scholarly level than Prabhupada and Prakashanand, the comparative historian of philosophy, Kisor Kumar Chakrabarti, provides a critique of Darwinian evolution in his sympathetic interpretation of the classical Nyāya-Vaiśeṣika doctrine of psychophysical dualism. In his *Classical Indian Philosophy of Mind: The Nyāya Dualist Tradition*, Chakrabarti argues that the Nyāya-Vaiśeṣika dualism, while similar to Western, Cartesian dualism in asserting "that mind and matter are ontologically different and are not reducible to each other," differs in asserting that selves are not inherently conscious (1999: 1, 22). That is, although selves are eternal "spiritual" substances, according to Chakrabarti, they are merely the substrata of conscious states or qualities, like knowledge, volition, and desire (ibid.: 2). A self (*ātman*) thus differs from matter as it may become conscious; but this occurs only when it becomes conjoined with the inner sense (*manas*), itself also devoid of consciousness, and with the appropriate biological-neural mechanisms. Chakrabarti insists that such a dualism, in contrast to Cartesian dualism, is fully consistent with modern science since for Nyāya-Vaiśeṣika conscious states exist only in the presence of living bodies. Further, the Nyāya-Vaiśeṣika, unlike Cartesianism, allows for "the possibility that (ordinary) states of consciousness may have arisen through the evolution of animal life forms" (ibid.: 211).

While Chakrabarti acknowledges the possibility of organic evolution, he accepts Darwinism only as a partial explanation of conscious phenomena, arguing that there needs to be in addition some guiding divine intelligence. He notes God's role as the initiator of cosmic creation in his role "as the designer of the dyad [the first material effect or product resulting from the combination of two atoms], and indirectly of the whole world of change," as well as God's supervision of karmic fruits (ibid.: 169, 211). Chakrabarti explains:

> … the evolution of higher animals from lower animals and that of the latter from inanimate matter is evident and raises no problems for the Nyāya … Life forms are typically supposed to emerge later and evolve from the lower

> to the higher. But that does not imply that the Darwinian type of evolution can be the whole story. For that leaves the evolution of conscious states to a process of natural selection without the guidance of intelligence—and this is not acceptable to the Nyāya.
>
> (ibid.: 211)

Thus, Chakrabarti continues, selves may have existed for billions of years in unconscious states, until through their karmic merits and demerits, and according to the direction of divine intelligence, they assume physical bodies. In this sense "ordinary conscious states" may be said to evolve from matter. Chakrabarti distinguishes volitional and other internal states from ordinary states of consciousness, and these internal states have non-physical existence—and therefore are not the result of evolution.

Finally, Chakrabarti takes on the problem of theodicy, resorting to the classical Hindu solution of karma and rebirth. Then, to justify the idea of rebirth and invoking an argument already found in the Nyāya Sūtras—and reiterated by Vivekananda—Chakrabarti asserts that "the phenomenon of biologically inherited instinctive activities, such as that of suckling by a newborn mammal, provides evidence for preexistence" (ibid.: 217).[9] Chakrabarti contends that if "unconscious teleology" on the part of the newborn is ruled out, and if the principles of cause and effect are "uniform and objective," and if "the act of suckling is aimed at getting nourishment," then, since the newborn has had no opportunity in this life for learning that nourishment is beneficial, "the inference of a previous life becomes warranted" (ibid.). The possibility that such instinctive behavior can be the result of evolved genetic predispositions is not considered.

Chakrabarti's rejection of a robust Darwinism, unlike the Vedic Creationism of ISKCON and ISDL, is not based on scriptural authority but on reasoning and logical inference, and on the "methodological principles governing philosophical disputes" derived from Nyāya epistemology (ibid.: 5). These latter, I argue, are especially troubling when introduced into the scientific realm, for they include the criterion that examples and counterexamples brought forward in defense of a particular theory "should be generally acceptable or acceptable to both sides of an issue" (ibid.: 9), and the notion that common sense ideas should not be "summarily" dismissed (ibid.: 10–11). Such methodological principles are open to considerable abuse when applied to such common sense ideas as the design argument, and in the Hindu context, to notions of rebirth vis-à-vis instinctual behavior and epigenetic developments. Modern scientific methodology finds mutual acceptability and common sense as largely irrelevant, since confirmation ultimately rests on empirical, not logical, verification.

Before we leave Vedic and Atomistic Creationism, I think it is worth noting that not all Hindu creationists assume a radical mind–body dualism. One such creationist is the celebrated New Age promoter of Ayurvedic medicine, Deepak Chopra, noted for his insistence on mind-body unity. He accepts Darwinism in a very limited sense, as "an enormously fruitful theory" (Chopra 2005b), but finds it disturbing that "the current debate over evolution has become us-versus-them. To say that Nature displays intelligence doesn't make you a Christian fundamentalist" (Chopra 2005a). He thus decries the "hijacking of the whole notion of intelligent design" by Christian fundamentalists (ibid.). At the same time, he reiterates standard Christian creationist arguments against evolution, such as the one taken over from Fred Hoyle regarding the creative potential of a whirlwind in a junkyard: "To say the DNA happened randomly is like saying that a hurricane could blow through a junk yard and produce a jet plane" (ibid.). He goes on to list twelve common anti-evolutionary objections, from missing links in the fossil record to the supposed problems of simultaneous mutations, cooperation in nature, and the extraordinary beauty in the universe. As the example of Chopra suggests, we will find that non-theistic Hindus are increasingly drawn towards a creationist world view, but replacing a personal God with the transpersonal ideal of Pure Consciousness within an overarching evolutionistic framework.

Contemporary Vedāntic evolutionism

The evolutionist perspective of most post-Independence Vedāntic (monistic) thinkers reiterate and elaborate upon a number of characteristic claims of Vivekananda's and Aurobindo's Vedānta-inspired evolutionism. These interrelated claims include: (1) the anticipation of modern evolution by ancient Vedic evolutionary ideas, and the inverse notion of the confirmation of Vedic tradition by the latest findings of modern science; (2) the scientific nature of Hinduism, especially Advaita with its unitary view of reality, in contrast to the anti-scientific, anti-evolutionary nature of Christianity with its extra-cosmic creator; (3) the incompleteness of Darwinian and the superiority of Vedāntic evolution, due to the latter's inclusion of spiritual evolution, often expressed in terms of the involution and evolution of consciousness; and (4) an emphasis on the yogic-intuitive apprehension and experiential confirmation of scientific truths, combined with the alleged superiority of this methodology to the limited and fallible rational, sensory-empirical inquiry of modern materialistic science. This last claim authorizes the use of Vedāntic evolutionary cosmology to elucidate gaps in current scientific knowledge, especially in the fields of molecular biology and genetics that bear directly on questions of organic evolution. These elucidations frequently invoke recent discoveries in other fields of science such as physics, especially quantum mechanics.

One typical contemporary follower of Vivekananda already met in Chapter 1 is Dr. P. Joshi, a retired endocrinologist from Medunsa, the Medical University of South Africa, and one-time chairman of the Ramakrishna Center of South Africa in Pretoria. I noted earlier Joshi's affirmation of the harmony of Hinduism and modern science, as well as his differentiation of Darwinian physical evolution and Vedāntic spiritual evolution. He also asserts that "[t]he discoveries of Vedanta antedated science by thousands of years" through laboratory research, the laboratory being the mind of the Vedic seer (Joshi 1997: 118). He remarks upon the ancient Hindu discovery of the consciousness of the inanimate world (as in metals) and of plants, a discovery he feels was confirmed by the work of J.C. Bose, thereby providing scientific support for the Vedāntic idea of universal consciousness. While Joshi refers to the evolution and involution processes, he differs from Vivekananda in equating evolution with the rising of all things from Brahman, and involution with their disappearance back into Brahman. Vivekananda, by contrast, defined involution as the progressive involvement and manifestation of consciousness in the material realm, with evolution representing the ascent back to Brahman. Further, Joshi notes briefly, it is consciousness that creates external reality, a truth he feels is confirmed by quantum physics. Joshi, however, does not apply quantum mechanics to evolutionary theory in specific terms. This was a challenge taken up by the retired physicist Amit Goswami, whose ideas we shall consider in a moment.

One striking new trend among the monistic evolutionists is the notion that Vedānta can provide a bridge between the evolutionary model of life propounded by Western materialist scientists and the spiritually based creationist model encompassing the idea of Intelligent Design. Or in somewhat different terms, Vedānta can provide an alternative model to both Darwinian evolution and creationism that comprehends the insights of both, thereby resolving the contemporary conflict between religion and science in the West. This development is nicely exemplified in an article, "Vedanta and the Theory of Evolution," appearing in the *Vedanta Kesari* (January 2009), the monthly cultural and spiritual magazine of the Ramakrishna Order. The article's author, Chethan Belludi, according to an editorial footnote, "is an alumnus of Sri Ramakrishna Vidyashala (Mysore), pursuing higher studies in medicine." Belludi begins his essay with the almost obligatory reference to the antagonism between Christianity and science, citing Galileo and Copernicus, and noting that Darwin further exacerbated religious fanaticism against science in the West. Belludi indicates his sympathy with Darwinism, but laments its inculcation of materialism and support of atheism. The latest threat to Darwinism, he says, is Philip Johnson's Intelligent Design (ID) theory, and refers to the controversy in the United States regarding the teaching of ID in public high school biology classes. He sees ID as little different from the Genesis account, except that it has "scientifically altered the theory to sound more logical" (2009: 35).

The ID proponents and their evolutionist adversaries, Belludi argues, are like the proverbial blind men feeling the elephant, each realizing only a part of the truth. But an Indian scientist of the nineteenth century, he proclaims, saw the whole elephant, the full truth. This scientist, extrapolating from the organic cycle of seeds to trees and back, and realizing that the effect is the same as the cause, concluded that every evolution is preceded by an involution, so that an adult man has developed from a single cell, but the man (including his intelligence) was earlier involved in the cell. Needless to say, this "scientist," Belludi informs us, was Swami Vivekananda (ibid.: 36).

In the rest of this chapter, I shall look at five contemporary Vedāntic monists who either obtained advanced academic degrees in the United States, or established academic careers in the United States, in the fields of engineering, physics, and medicine. Except for the first, Subhash Kak, they all show a strong tendency towards reviving Vedāntic design notions within an evolutionary framework, with the designer specified as Brahman, or consciousness, that works from within the material realm, not from without. With Kak, I shall also consider briefly his collaborative work with colleagues.

Subhash Kak et al. and Vedic Evolutionism

Subhash Kak is professor emeritus in the Department of Electrical and Computer Engineering at Louisiana State University. In an article contributed to the online journal *Sulekha*, Kak refers to the "Cold War" between science and religion in the West ever since Darwin proposed his subversive theory that humans had evolved from apes. Noting the constant retreat of Western religion in the face of scientific progress, Kak claims that the Church, now having abandoned the story of Adam and Eve, currently insists that the origin of life can only be explained by reference to an intelligent designer. Evolutionary theory has thus caused a loss of traditional religious belief in the West with its anthropomorphic God that conflicts with the naturalistic explanations of science. Kak goes on: "Western and Indian thought is [sic] divided on the argument for design. In the West, thinkers from Aquinas to Newton maintain that nature manifests the design of a preexistent mind or the Creator" (Kak 2002). By this means the Westerner is able to define a "personal sense of purpose and meaning" (ibid.). In India the case is different, according to Kak, since, "In Indian thought, there is no separation between the Creator (the preexistent mind) and the universe. Consciousness is taken to be the fundamental characteristic of reality out of which material nature and individual minds emerge" (ibid.). Kak thus confidently asserts that "[e]volution is basic to this view. Life is seen to have evolved over millions of years in a manner that makes the cell mirror the cosmos" (ibid.). Kak clearly is selective here in his neglect of the rich design-argument traditions within Hindu thought, and the dualistic/pluralistic philosophies of the Sāṃkhya and Nyāya-Vaiśeṣika.

The compatibility of Hinduism and evolution stressed above is similarly highlighted in a book co-authored by Kak, Georg Feuerstein, and David Frawley: *In Search of the Cradle of Civilization: New Light on Ancient India*. In the book we find that "[t]he idea of evolution is integral to the Vedic theory of knowledge, which regards all of Nature as a manifestation of the Self" (Feuerstein et al. 1995: 225). In accord with Vivekananda's Advaitic Evolutionism, the authors insist that "[t]he Vedic understanding of evolution is not at odds with the Darwinian theory of evolution, though it has a different focus" (ibid.: 226). Reference is made to the Purāṇa-derived idea of Avataric Evolutionism, regarded as a striking expression of "ancient India's evolutionary ideas" (ibid.), as well as to Aurobindo's theory of evolution-involution, described as "a conscious attempt to connect the spiritually oriented Vedic evolutionism with the Darwinian theory of evolution" (ibid.). The ancient evolutionary theory was the product of "the Vedic scientific method" (ibid.: 227). This method involved a "system of analysis … supported not only by mystical vision but by skillful ratiocination," apparently creating for the first time "a science that went beyond mere observation" in constructing "a comprehensive understanding of the workings of the universe" (ibid.).

N.C. Panda's revival of Vedāntic design

One illustrative example of the Vedāntic revival of design is seen in Nrusingh Charan Panda's 2002 book *Cyclic Universe: Cycles of the Creation, Evolution, Involution and Dissolution of the Universe*. Panda received his Ph.D. in Veterinary Medicine at the University of Columbia, Missouri, and became Professor and Dean of the Veterinary Faculty at Orissa University in Bhubaneswar, India. The work summarizes the major cosmological theories of the ancient and classical worlds, Western and Indian, as well as modern scientific evolutionary theories and various philosophical interpretations of evolution, including those of Herbert Spencer, Lloyd Morgan, Henri Bergson, and Aurobindo. As the title of his work suggests, he dismisses the idea of creation out of nothing, preferring the idea of a cyclic reprocessing of the universe.[10] Also noteworthy in the subtitle of the book is Panda's substitution of evolution and involution for the traditional second phase of the cosmic cycle, preservation, implicitly equating the two.

Panda summarizes Darwinian evolution as "based on random mutation and natural selection" (2002: 717). He accepts the second idea, but insists that the idea of random mutation, or chance evolution, is controversial and problematic, setting up the common creationist dichotomy between chance and intention or design. The evolutionary "chance hypothesis," according to Panda, fails to convince the Vedic and Vedāntic sages. "The Vedic seers observe a Cosmic Blueprint," he declares, citing *Vedānta Sūtras* 2.2.1, which asserts that a non-intelligent cause, such as insentient matter, cannot be inferred as the cause of the world, because it cannot account for the design (*racanā*) seen in the world (ibid.: 545). It is this same aphorism in the *Vedānta Sūtras* that Śaṅkara used in elaborating his own design argument.[11] Panda thus argues that "Science should ponder over one suggestion made by Śaṅkara: If the observed phenomena (becomings) strongly suggest the existence of a metaphysical Being, without which the becomings cannot be satisfactorily explained, the Being is to be inferred from the becomings" (ibid.: 543).[12]

In Panda's mind, there is at least some accord between modern science and Vedānta:

> The Vedānta accepts the doctrine of evolution. According to Śaṅkara, the universe was created gradually, not simultaneously once for all.
>
> A second salient feature of the Vedāntic evolution is self-organisation and self-action. Nature evolves and God is a non-doer …
>
> Both of these above-mentioned concepts are compatible with modern theories in science.
>
> (ibid.: 731)

But as Panda immediately points out, there is a significant difference between Vedāntic evolution and modern evolution. The former recognizes a "Chairperson," the cosmic designer, variously identified as *Adhyakṣa* (Superintendent), *Puruṣa* (Cosmic Person), and *Īśvara* (God) (ibid.: 732, 545). In good Advaita fashion, Panda explains that the God of Vedānta is not the anthropomorphic, heaven-residing God of most religions (ibid.: 736). The Vedāntic God, or Brahman, Pure Consciousness, is within the universe, its substratum. Brahman manifests the universe through its power known as Māyā. When Pure Consciousness is reflected in Māyā, it then creates, preserves, and dissolves the universe. This Māyā, also known as Īśvara, the Chairperson, is thus the creative agent, while Brahman is only indirectly the agent, God the non-doer. Primordial Nature (*mūla prakṛti*), under the direction of the Chairperson, evolves into the Universe, the lower nature that science studies. Eventually, the evolution of Nature will stop, and involution, or dissolution, of the cosmos back into Māyā will occur (ibid.: 737–8).

On the individual rather than cosmic scale, Brahman is reflected in the individual self (*jīvātman*), which science has yet to explore (ibid.: 739). It is this *jīvātman*, the consciousness existing in all created entities from atoms to human beings, which in higher organisms directs the development of the fertilized egg into a complete adult in accord with the karmic impressions from previous lives. Without such impressions and the notion of an immaterial mind unconnected with the brain, according to Panda, "we face a serious problem in explaining embryogenesis and cell-behaviour" (ibid.: 537). He describes in considerable detail the processes of embryological development, beginning with cell division. He wonders, among other things, how the twenty-three human chromosomes are able to separate and come together in due order, without an intelligent guide: "Why the idiot chromosomes behave intelligently is a question which baffles scientists unless they are determined to avoid the issue" (ibid.: 539).

Likewise, regarding cell differentiation in the early stages, Panda asks, "How could the same genes (DNAs) direct the embryonal cells to differentiate meaningfully so as to specialize for some actions to be performed in the future?" (ibid.: 542–3). Since the "DNAs" are insentient macro-molecules, they could not be the cause. Vedānta has the solution: "The Vedānta states that the light of consciousness of the self (*ātmā*) seated in the blissful sheath … is reflected on the cells of the blastula and the gastrula, resulting in the induction of consciousness and translation of guided action" (ibid.: 542). The body as a whole, with its ordered brain, spinal cord, eyes, ear, nose, taste buds, and skin, is "very intelligently designed and engineered" (ibid.). The individual self (*jīvātmā*), then, is "the hidden guide" within the body regulating cell division. Panda acknowledges that science has found many biological mechanisms that regulate embryological development, but science will never be able to discover the hidden guide (ibid.: 543).

We may recall that, for Cremo, a special mantric seed was required to provide the necessary guidance for cell differentiation. Thus, the Vedāntic monist Panda ends up like the theistic dualist Cremo, rejecting organic evolution in any meaningful Darwinian sense. Panda, while rejecting the personal designing god of theism, brings in common creationist arguments against Darwinian chance evolution in defending Vedāntic design. He invokes, like his theistic counterparts, information theory and the second law of thermodynamics, insisting that the latter rules out any increase in genetic information due to the increase in random disturbances or noise (ibid.: 719–20).

As for some common Hindu objections to the design argument, Panda takes note of the Vedāntic claim that Brahman has no need or purpose, and thus no reason to design and create a universe. He explains in typical Advaita fashion that it is not Brahman, but only its reflected consciousness as Īśvara, that desires to be many, in order to play: "The One became many through the projecting and concealing power of *Māyā*, through the empirical evolution, *vivarta*. *Īśvara* wanted *līlā* (sports)" (ibid.: 546–7). Evolution as mere manifestation (*vivarta*) seems to dissolve the whole problem of suffering in the world cycle, for Panda. He claims, with little argumentation, that empirical diversity, involving "the coexistence of knowledge and ignorance, and of ecstasy and agony," is justified, for it allows the Cosmic Person to play (ibid.: 547). Further, with the manifestation or emergence of life and man, God's *līlā* attains a greater "charm." Panda concludes: "The evolutionary process ascends to higher and higher levels; the becoming of the Being is a continuing process for the accentuation of the *līlā* of the Cosmic Person" (ibid.). Here we see the modern notion of emergent evolution interpreted in Advaitic terms of *līlā*, thereby allowing for the traditional resolution of both the problem of divine completeness and of the apparent suffering in the world. But this traditional Advaita resolution runs the risk of trivializing the sorrow and pain experienced by sentient beings.

One other common objection to the design argument that Panda acknowledges is the atheist's query: If God created the world, then who created God? And if the theist asserts that God is self-existent, the atheist can equally assert that the universe is self-existent. Panda sees these two conflicting perspectives as irresolvable in their own terms. The only resolution, he claims, lies in the unique Vedāntic concept of Brahman, which transcends the dichotomy between a creator God (subject) and a created world (object), for Brahman is neither subject nor object (ibid.: 751–2).

Regarding Panda's epistemological approach, his cosmological speculations rely very little on scientific method. He frequently quotes authoritative texts and Śaṅkara, as well as modern scientists, as secondary support for his views. At the very end of the book, Panda confesses his dependence on the light of the great seers and realized souls. He also acknowledges his attempt, with the help of these guides, to go directly to the source of the light through practicing meditation to "try to link up his mind with the Supermind" (ibid.: 818). As to whether he attained such linkage, whether he attained the fourth and final state of meditation known as *turīya*, Panda is not entirely sure; perhaps he has only made "wild speculations" (ibid.). But then, he makes an intriguing parallel between himself and the Vedic "Chairperson" (*adhyakṣa*) of *Ṛg Veda* 10.129: "He who is the Chairperson of this universe in the highest heaven perhaps knows it (the creation and fate of the universe) or He may know not" (ibid.).

Gopala Rao's two-dimensional model of evolution

Gopala Rao is a retired professor of medical physics, formerly affiliated with Johns Hopkins University, Vanderbilt University, and the Medical College of the University of Virginia. In an online article entitled "A Two Dimensional Model of Evolution," Rao takes many cues from contemporary Christian theologians and combines them with aspects of traditional Vedānta metaphysics in an attempt to bridge the gap between science and human religiosity.[13] Rao thus reflects something of Vivekananda's universalistic perspective, but avoids the latter's claims of the non-scientific nature of Christianity.

Specifically, Rao utilizes the notions of bottom-up and top-down processes in nature, suggested by theologians and scientist-theologians like Arthur Peacocke, John Polkinghorne, George Ellis, and Nancey Murphy. Bottom-up processes are those long employed by modern science, characterized by Rao as reductionistic, versus the holistic nature of top-down processes. The latter include such examples as volcanoes disrupting ecosystems, human intention initiating hand movements or speech, but also, according to Rao, divine interventions from above. As he states it: "In religious parlance, one might say that God allows evolution to progress naturally on its own through 'Bottom-Up' processes, and at every stage, He, She, or It endows the evolved entities with properties not present in their predecessors" (2001a). This divine endowing of new properties is a top-down process involving an element of choice. It denies that bottom-up processes can "satisfactorily explain the evolution of such things as consciousness, self-awareness, mutual awareness, and the associated spectrum of emotions among biological species" (ibid.).

Rao asserts that biologists have failed to recognize that top-down interactions may be involved in Darwinian evolution—apparently overlooking the tremendous amount of work being done on complex systems, ecology, feedback mechanisms, and the intricate interactions of wholes and parts in living organisms. Not surprisingly, biologists have indeed failed to recognize the divine source behind certain kinds of top-down processes, that according to Rao pervade all evolutionary developments: "The evolution of each and every observable entity, phenomenon and concept in the Universe and each and every perceived law of Nature are the mechanisms for the dynamic realization of the omniscient and omnipresent potentiality of the Ultimate Source [i.e., God]" (ibid.). This source, he says, is what physicists sometimes regard as a Virtual Domain, a pure potentiality "devoid of space and time, matter or energy," and sometimes as a Quantum Domain, an actualized realm "where creation manifests, energy exists as a unified field, time begins and space expands" (ibid.). Elsewhere in the essay, Rao also suggests that the Quantum Domain "may very well be the domain of angels and departed saints, the purgatory in Christian terms" (Rao 2001b). The Virtual Domain is akin to Amit Goswami's potential state of genetic mutations, to be discussed below.

In support of this overall model, Rao quotes scriptures, scientists, various enlightened persons, Nobel prize winners, and New Age gurus such as Deepak Chopra. Among "scientists" he cites William Dembski, the well-known proponent of Intelligent Design. Among the scriptures, he quotes from the Bible, the Qur'an, and from that favorite of Modern Vedic Cosmologists, *Ṛg Veda* 10.129. He sees this verse as expressing the idea that "evolution is a continuous process involving a dynamic realization of the potential of the Ultimate Reality, symbolically referred to as Brahman. Creation begins with an impulse by Brahman to realize Himself and His own potential" (Rao 2001a). Rao interprets the first two verses as referring to the time before creation when there was neither existence nor non-existence, when "[a]ll this world was unmanifest energy," although the verses literally refer only to the idea that "[t]he one breathed, without breath" (Rao's literal translation, to which he has added the explanatory phrase regarding "unmanifest energy") (ibid.). Such a rendering seems to evoke the ideas of the Virtual Domain devoid of all matter and energy.

Varadaraja V. Raman's ambivalent evolutionism

Varadaraja V. Raman is an emeritus professor of physics at the Rochester Institute of Technology. In an essay entitled "Traditional Hinduism and Modern Science," he remarks upon his commitment to science and his sympathy for, even attachment to, his tradition, while noting that he has "experienced moments of dissonance" between them (Raman 2003: 185). He ends up rejecting certain aspects of science, as well as of tradition, while living in both worlds. He asserts that the traditional Hindu view of nature anticipated that of modern science, yet notes that the modern scientific view is based on a different empirical methodology and sees humankind as a merely incidental phenomenon. Nonetheless, new scientific discoveries can extend and expand ancient visions, he claims, suggesting for instance that *Ṛg Veda* 10.129 (cited by both Panda and Rao above) can be elucidated by the Big Bang. Similarly at the other end of a cosmic cycle, "Hindu mythopoesy" offers Shiva, god of destruction, responsible for the ultimate dissolution, "the mythic symbol of what in thermodynamics is called the heat death" (ibid.: 188).

With specific reference to biogenesis and evolution, Raman finds the Vedic imagery of sun, air, and water as relevant to the emergence of life—a rather general point left unelaborated. He also views the ten avatars of Viṣṇu as having an "uncanny parallel with our current views on biological evolution, and they transcend the current scientific picture" (ibid.: 189). However, he adds that to interpret organisms at various levels of complexity as "in different *stages* of an *evolutionary process* is a huge leap" that would require "considerable observational studies" in order to confirm (ibid.).

Raman is unsatisfied with the alleged Darwinian explanations of genetic traits like athleticism and illness as being due merely to chance, arguing that we humans "crave for clearer answers to the question of the uneven, and often unfair, distribution of fate and fortune" (ibid.: 190). Raman then provides the classic Hindu response to the problem of theodicy: "Here the Hindu view presents a picture that has provided answers as satisfying as any. This is the *law of karma*" (ibid.). Raman does not explore whether answers satisfying human desire have any relation to actual, empirical truth.

Turning to epistemological considerations, Raman justifies the typical mystical epistemology of Neo-Vedānta with reference to quantum physics and the idea of reality as a unified whole, transcending subject and object. The interconnectedness of mind and physical reality means, from his particular Hindu perspective, that the physical substratum can be discovered not only by science with its mathematics and instruments, but "can also be apprehended through the mystical mode via meditation, prayer, or yogic exercises" (ibid.: 192). Raman notes the need felt by many Hindus today, occasioned by the intrusion of European influences, for an apologetics to bridge the gap "between science and sainthood," and to confirm ancient spiritual truths by modern scientific discoveries such as quantum physics and the Big Bang (ibid.). But many other Hindus like himself, he declares, find such apologetic efforts "quite unnecessary, and sometimes even awkward" (ibid.).

We may recall that Rabindranath Tagore, in his rejection of a random and meaningless universe, effused that "in the arrangement of the world there seems to be a great concern about giving us delight," which suggested to him that some "magical touch of personality" lies behind the universe. Near the end of the essay, Raman engages in a similar poetic outburst:

> If there is splendor in the perceived world and pattern in its functioning, and if it can all result in the grand experiences of life and thought, then even prior to the advent of humans, there must have been a *purusha* of a vastly superior order, an Experiencer who spanned the cosmic range in space and time.
>
> (ibid.: 193)

That Experiencer for Raman is Brahman, the "primal effulgence" from which we have all emanated, "like photons from a glorious galactic core" (ibid.: 193–4). The urge to lyricize and scientize tradition is clearly irresistible for Raman.

Amit Goswami's quantum evolution

Amit Goswami, a retired professor of physics at the University of Oregon, has written a number of books utilizing ideas derived from quantum mechanics to integrate Neo-Advaita spirituality with modern science. Goswami's basic thesis is presented in succinct form in his *Science and Spirituality: A Quantum Integration* (2006).[14] Especially relevant for our purposes is his chapter on "How Consciousness Creates Biological Order," in which he explicates the idea of "quantum evolution." His more recent book, *Creative Evolution: A Physicist's Resolution between Darwinism and Intelligent Design* (2008), elaborates upon his earlier thesis invoking the idea of "quantum creativity" to critique Neo-Darwinian accounts of evolution and to offer an alternative model.

Both of Goswami's books attempt to find some sort of reconciliation between science and spirituality, and specifically to resolve the tension between evolution and creationism. A tentative truce between creationists and evolutionists in terms of an independence model for science and religion, in which each deals with a separate realm, has not worked well according to Goswami, as encroachments from both sides regularly occur (2006: 45). However, spiritual traditions around the world, by questioning the legacy of Western secularism and materialism and positing a transcendent consciousness as the ultimate reality, provide a way out of this impasse (ibid.: 2, 167). Foremost among these spiritual traditions is the idealist monism of Vedānta, encompassing the insights of purposive, conscious design and of creative evolution that allows for a new model of evolutionary theory. Hindus have long "recognized the cogency of evolution," he adds, noting with approval the idea of Avataric Evolutionism (ibid.: 60).

The title of Goswami's 2008 book evokes Henri Bergson's *Creative Evolution* written a century earlier. Goswami reiterates some of Bergson's anti-Darwinian arguments, but bolstered with an injection of quantum mechanical theory to explain the mechanism of creative evolution. Far more influential on Goswami than Bergson, however, are Aurobindo and other more recent Western evolutionary philosophers like Pierre Teilhard de Chardin and Ken Wilber. Goswami explicitly acknowledges, with reference to Aurobindo and Teilhard, that his own *Creative Evolution* "incorporates the revolutionary ideas of these two great thinkers" (2008: 6; cf. 72, 79–80). Like Aurobindo, Goswami attempts to undercut naturalistic evolution by bringing in the idea of reincarnation to account for the "propensities, habit patterns, and learned contexts" of one life that are carried into a new birth via our mental and vital bodies that survive the death of the physical body (2006: 86). He claims that such ideas are being confirmed by "our new science," arguing that the reincarnation studies of researchers like Ian Stevenson and Satwant Parsicha provide "clear scientific reliability" (ibid.).

Perhaps the most important Neo-Advaita idea Goswami utilizes is the complementary process of involution and evolution, a concept he indicates that was emphasized by Aurobindo (as well as Ken Wilber) (ibid.). In this dual process, he explains, Brahman or consciousness, "in order to know itself," enters into ever grosser levels of reality, progressively forgetting itself until it arrives at the fully unconscious, material level (ibid.: 86–7). Then, in an evolutionary reversal, consciousness gradually re-emerges, corresponding to the increasing levels of material complexity, thereby realizing the pre-existent potentials for life and consciousness.[15]All this, however, is only an appearance, for there is no genuine separation of matter and consciousness, in accord with Śaṅkara's monism: "Consciousness forgets itself for the sake of play—it pretends to forget, so to speak" (ibid.: 87). Esoteric traditions with their "idealist science" explain how "forgetfulness or *māyā*" arises, but lack details of exactly how this occurs (ibid.). Quantum mechanics, Goswami argues, supplies these details.

Before going into the details of Goswami's quantum evolution, let us first summarize his critique of Neo-Darwinian theory. Neo-Darwinian evolution, for Goswami, is fraught with major difficulties. Perhaps most basic is a philosophical objection—and the reason underlying almost all anti-evolutionary thinking. Neo-Darwinian evolution, by relying on the dual process of random mutation and natural selection, and thus on chance and causal necessity, denies all purpose and design in the universe (ibid.: 61–2). Quantum evolution, however, is the creative play of divine consciousness, Brahman or God, and the purpose of the universe is to manifest creatively the ideas within this consciousness that guides evolution to ever greater complexity and meaning.

Goswami notes on several occasions the conflict between creationist (especially of the young-Earth variety) and Neo-Darwinian explanations of the diversity of life. Scientists, he argues in good Neo-Advaitic fashion, rightly reject the idea of an extra-cosmic creator as a violation of energy conservation (ibid.: 4, 60). But creationists, he claims, have many points in their favor that call naturalistic accounts of evolution into serious question. These include: (1) the problem of gaps in the fossil record, (2) the puzzle of embryological development, specifically cell differentiation, (3) the inability of blind processes to construct complex organs like the eye, and (4) the implausibility of "hopeful monsters," as Goswami refers to the issue of speciation. I shall deal with each of these in turn.

Regarding gaps in the fossil record, Goswami claims in his *Science and Spirituality* that there are no transitional forms "between plants and animals, between the appearance of birds and the earlier reptiles, between primates and humans" (ibid.: 46). This last, I may note, is a bit imprecise, as humans are primates. Later on, Goswami also denies that there are any "fossils that show fish traits gradually being replaced by amphibian traits" (ibid.: 62). Such assertions ignore or misrepresent the fossil record as known today, in which there are admittedly many gaps between any number of species, but also many well-documented transitions, including between larger groups such as those mentioned above.[16] The discovery of *Tiktaalik roseae* in the High Arctic of Canada in 2004 is only one of the latest findings clearly disproving that there are no transitional species between fish and amphibians (Shubin 2008: 3–27).

To be sure, Goswami in his later book (2008: 10, 14–15) acknowledges that a few intermediate fossils have been found, but insists—without critical analysis or argument—that, according to Darwinian theory, "thousands upon thousands" of such intermediates should have been found by now but have not. He does not indicate what number would be sufficient, and ignores the rich trove of what Jerry A. Coyne calls "fossils galore" that fulfill all the predictions of Darwinian theory (2009: 26). And like many creationists, Goswami dismisses any possible "so-called intermediate" by classifying it in one or the other of the two classes between which it is supposed to form a link. Thus, *Archaeopteryx*, he insists, is simply a bird because it could fly (even if not very well, as Goswami admits), and thus not really an intermediate. But the reptilian features of *Archaeopteryx* that Goswami acknowledges, such as the "tooth-equipped jaw and trailing tail," are not explained (2008: 184).

Turning to the puzzle of cell differentiation, we have already seen Cremo's solution in terms of the Kṛṣṇa-created mystic *bīja* and Panda's Vedāntic resolution invoking the hidden guidance of the Ātman residing within the cell. Goswami follows Panda's general lead, but incorporates details from quantum mechanics (ibid.: 9–10, 57–8, 214). Goswami regards Consciousness (God, Brahman) as a programmer "using downward causation" to instill information for making a biological form into a one-celled zygote (ibid.: 56). These instilled programs then direct undifferentiated cells in the developing embryo to become liver, brain, and other specialized cells as needed. Goswami notes that contemporary biologists see these built-in programs as simply residing in the DNA. But he sees a problem in such explanations. He quotes from the contemporary biologist Richard Lewontin, who emphasizes that cell differentiation can only be understood by knowing about the position of molecules within a cell and the position of a cell in relation to other cells within the developing embryo. But how do molecules and cells know their positions? Goswami dismisses developmental biological explanations that molecules and cells know their positions through "fields" and "gradients," seeing these only as "metaphors" (ibid.: 58). His conclusion is that Lewontin really seems to be saying that the transfer of positional information from one molecule or cell to another must be nonlocal—an idea adapted from quantum physics—and thus non-physical, implying the role of consciousness in the process.

As for the evolution of complex organs like the eye, a not-yet-exhausted anti-creationist staple, Goswami observes:

> Consider for example, the evolution of the eye, an important component of a new species. It must have taken hundreds, maybe thousands, of gene mutations. But what good is one mutation, one-thousandth of an eye, even one-hundredth, or half? See the problem?
>
> (2006: 59)

The "one-thousandth-of-an-eye problem" has been repeatedly resolved on Darwinian grounds, for it is clear that 5 percent, or 1 percent, or even .001 percent vision is better than no vision at all (Dawkins 2007: 82–5). Creationists of all stripes, Neo-Advaitin or Christian, also overlook the evidence documenting the repeated and independent evolution of the eye, "the forty-fold path to enlightenment" in Richard Dawkins' term, from light-sensitive tissues to the various sorts of compound and camera-type eyes found in different animals (Dawkins 1997: 138–97).

And finally we come to the crux of the entire controversy: the problem of speciation. Like other creationists of today, Goswami accepts microevolution but balks at macroevolution. The basic argument against macroevolution, that is, speciation by natural means, is that a new species allegedly requires a large number of genetic mutations, each of which must be of some immediate use (or at least not deleterious) to the evolving organism. But most individual mutations are not beneficial, and thus will be selected against *if* they are immediately expressed in the observable traits of an organism (phenotype). And since Neo-Darwinian evolution, in Goswami's view, presupposes such immediate expression, most single mutations, being selected against in their unfavorable phenotypic expression, will not be passed down to future generations. (Such immediate expression, I note, assumes that the genes involved are dominant and not recessive.) Several such mutations, while individually harmful to an organism, may, as a group, produce a new trait that is advantageous to the organism. But Neo-Darwinian evolution,

Goswami claims, does not allow for the accumulation of individual mutations to form these beneficial groups over several generations. It allows only for the rare beneficial genotypic mutation that may produce a small phenotypic change in the organism, resulting in the variations in a species that we observe. Speciation is thus essentially the same problem as the evolution of the eye (Goswami 2006: 58–62). The probability of several advantageous mutations occurring simultaneously is vanishingly small. To illustrate the problem of macroevolution, Goswami cites biologist Richard Goldschmidt's idea of "hopeful monsters," creatures that, as characterized by Goswami, are the "surviving monstrosities of simultaneous genetic variation" (2008: 164).[17]

Goldschmidt originally floated the idea of the hopeful monster in a 1933 lecture, during the final working out of the Modern Synthesis of Darwinian evolution that took into account key developments in biology since Darwin's day, including Mendel's laws of inheritance and population genetics. The hopeful monster soon came to be seen as a direct challenge to a strict Darwinian interpretation of gradual speciation, arguing for a different sort of mechanism that would account for sudden leaps in evolution. These leaps, according to Goldschmidt, were due not to individual gene mutations, but to pattern changes in chromosomal arrangements. The hopeful monster soon became ridiculed by mainline biologists as increasing evidence accumulated for the role of gene mutations versus drastic systemic modifications of chromosomes. But as Stephen J. Gould points out, Goldschmidt's hopeful monster has been misinterpreted and much maligned, for it was originally based not on the idea of chromosomal re-patterning, but on "a separate, older and more important theme" (2002: 457). This older theme, inconsistent to be sure with the systemic mutation explanation, accounted for macroevolution by single, small genetic mutations that affect the early stages of embryological development, with cascading consequences that result in major changes in the finished anatomical structure. Most such single genetic mutations produce only monsters, incapable of surviving. But if a genetic change does not disrupt the ongoing regulation of embryological development and the resulting organism can take advantage of a new environmental niche, then the monster is hopeful. Properly understood, the hopeful monster accords readily with the Neo-Darwinian system. In brief, Goldschmidt's hopeful monster anticipates the important discoveries in evolutionary development—coupling evolutionary theory with embryological development—that form part of the "Extended Synthesis" that is currently under way in modern biology.[18]

Goswami, not surprisingly, ignores Goldschmidt's developmental ideas, as well as his ideas regarding chromosomal re-patterning—which after all still provides a naturalistic explanation for sudden evolutionary leaps. Instead, Goswami turns to another prominent biologist from the heyday of the Modern Synthesis, Gaylord Simpson, who provides Goswami with a suggestive hint.

Simpson, in his 1944 book *Tempo and Mode*, proposed a hypothesis to explain the discontinuities in the fossil record by presenting an alternative or complementary view to the idea of gradual phyletic change—that speciation occurs only through slow, measured adaptations. Simpson called his alternative model "quantum evolution," involving the mechanism of genetic drift in a population, a process that could carry the population through an "inadaptive" phase, resulting in an "all-or-none reaction" in which the intermediate and unstable (non-adapted) forms would either quickly perish or through selection rapidly find a new stable form.[19] Such a process would leave little evidence of the intermediary forms in the fossil record.

Simpson's hypothesis, it should be noted at once, is only loosely related to quantum mechanics. Simpson used the discontinuous quantum jumps in energy states of electrons within an atom to describe, metaphorically, the relatively abrupt, major changes of organic forms seen in the fossil record—an attractive point for Goswami. Goswami takes over Simpson's terminology of quantum evolution, but gives it a mystical reinterpretation infused with his own Neo-Advaita metaphysics. These quantum evolutionary leaps for Simpson, of course, had nothing mystical about them, genetic drift being a thoroughly naturalistic mechanism fully in accord with modern genetics. It is perhaps ironic in light of Goswami's mysticizing of Simpson's ideas that Simpson later (1953) radically redefined quantum evolution: it no longer represented a different mode of evolutionary change, but only an extreme type of phyletic, adaptive evolution operating "at a maximal rate" (Gould 2002: 531).

Let us now turn to the specifics of Goswami's Advaitic quantum evolution. According to Goswami, it commences like its Neo-Darwinian counterpart with random single mutations in the genes of an organism (genotype), or possibly in larger chromosomal rearrangements. But quantum evolution recognizes that single mutations or chromosomal rearrangements need not be expressed at once in the phenotype. They may remain in a state of potential expression only, which cannot be selected against, and thus they can gradually accumulate over several generations. The collection of mutations present in the genotype represents various possibilities in the phenotypic world. These non-manifesting mutations persist, then, in what quantum mechanical theory calls a state of superposition, in which an object can exist in various possible virtual states until it is observed, at which point the various possibilities collapse into one manifest state. Thus, single and potentially harmful genotypic mutations endure from generation to generation as mere possibilities, without manifesting on the phenotypic level—this non-manifestation or non-collapse being due to their not being observed (2006: 66–7).

The collapse of these states occurs only when enough mutations have come about so that, by acting in a coherent, mutually cooperative manner, they can produce a major new trait capable of forming a significantly altered and viable phenotype, and possibly a new species. At that point, the group of mutations may become an object of consciousness—apparently the divine consciousness that is the all-pervading Brahman—which then chooses among various possibilities, collapsing them into a new form. This creative choosing and resultant collapse is "in accordance with the purposiveness or grand design of consciousness" (ibid.: 67). Drawing upon ideas from Rupert Sheldrake, Goswami proposes that this collapse occurs "when a 'resonance' takes place between the archetypal form and its manifestation" (ibid.: 68). That is, when a long-awaited and coherent set of mutations defining a new form or species and perhaps present in a "morphogenetic field" finally becomes ready for phenotypic manifestation, the divine consciousness then observes it, bringing it into actual existence.

In one final co-option of quantum principles in the service of Neo-Advaita, Goswami suggests that the coordinated action of genes "far removed structurally on the spatial body of the DNA" to produce a new trait "seems … to be evidence of quantum non-locality" (ibid.: 72). He claims that such action at a distance is not permitted in deterministic (e.g., Darwinian) biology. Such a claim is distinctly at odds with current biological explanations involving chemical molecules, "messengers," that travel within the cell nucleus conveying the necessary information via distinct molecular structures that activate or interact with other genes. Further, such supposedly nonlocal influence according to quantum theory is relevant only at much smaller scales (the sub-atomic) than are involved with the large molecular structures of DNA.

Like Cremo's criticism of Darwinian evolution, Goswami's often seems to deal with older evolutionary ideas, although both attempt to incorporate the latest scientific findings. In *Creative Evolution*, Goswami notes the new work in "extended Darwinism," the Extended Synthesis that goes beyond the Modern Synthesis. He notes such new developments as evolutionary developmental biology (Evo-Devo) and the notion of the development of evolvability. But he dismisses these as rather desperate and unconvincing attempts to fix Darwinian theory, like the futile attempts to salvage geocentric models of astronomy by adding (fictional) Ptolemaic epicycles (2008: 147–54). Goswami, for instance, refers to biologist Sean B. Carroll's explanation of evolutionary development and its attribution of speciation to the natural and material characters of the genes. But Goswami lumps together all the attempted fixes, including Carroll's, concluding that "unless one is a true believer, these many epicycles of Darwinism do not satisfy" (ibid.: 154).

While Carroll's explanation of Evo-Devo and his extensive evidence for a naturalistic accounting of speciation may be unconvincing to creationists like Goswami, the Advaitin physicist makes no real effort to show why. But as Carroll points out, one of the great discoveries of Evo-Devo is that there are an enormous number of homologous genes, including regulatory genes, from flies to humans (2005: 9). Such discoveries make clear the pathways for evolving complexity from simplicity, from single genetic changes affecting the early stages of embryonic development. As Carroll summarizes:

> Indeed, the new facts and insights from embryology and Evo Devo devastate the lingering remnants of stale anti-evolution rhetoric about the utility of intermediate forms or the probability of evolving complex structures. We now understand how complexity is constructed from a single cell into a whole animal.
>
> (2005: 10)

Regarding the old microevolution–macroevolution controversy, Carroll remarks: "For half a century since the Modern Synthesis, this specter of a 'hopeful monster' has lingered. The facts of Evo Devo squash it" (ibid.: 291). He later adds: "Evo Devo reveals that macroevolution is the product of microevolution writ large" (ibid.).

Goswami's basic motivation for dismissing Evo-Devo seems to be more philosophical or metaphysical than empirical. Evo-Devo (and other approaches of the Extended Synthesis) does not recognize the need for quantum leaps to get over the gaps supposedly manifesting in all the alleged anomalous data, because there are no gaps, rendering Goswami's own quantum version of the hopeful monster irrelevant.

I have spent as much time as I have with Goswami because he is representative of many contemporary Hindu creationists who employ a number of Christian creationist anti-Darwinian tactics that have been largely discredited: the frequent quotations of scientists either out of context or in irrelevant ways, casting aspersions on the integrity and honesty of evolutionary biologists, and a basic refusal to engage with the details of rapidly emerging Darwinian explanations. Recourse is frequently had to irrelevant estimates of probabilities, complemented by speculative elaborations of a theological or metaphysical evolutionary model, spiced with a smattering of recent scientific discoveries and facts. When the creationist is also a scientist, the speculative models are commonly constructed upon the foundation of the scientist's field of expertise, regardless of its possible irrelevance.

In this context it is instructive to compare Goswami's *Science and Spirituality* with T.R. Anantharaman's *Ancient Yoga and Modern Science*.[20] Both are part of the monograph series on the "History of Philosophy, Science and Culture in India." Anantharaman, a former professor of metallurgy, like Goswami, regards speciation as consisting of "sudden and distinct jump[s]" or "qualitative leaps" that cannot be explained by modern science (2007: 69, 73). For Anantharaman, these leaps are actually sudden soarings "in the level of *consciousness*" (emphasis in the original), this consciousness being that of adept yogis capable of directing the "evolutionary or radial energy" undetectable by science but responsible for both material and spiritual evolution. Such terminology as "radial energy," incidentally, comes from Teilhard de Chardin (1975: 64–6). As a metallurgist, Anantharaman finds the phase transformations, or evolutionary reconstructions, in material solids and alloys to resemble those of spiritual transformations in the yogi, although he insists that the metallurgical and mental transformations, while both internal, are of two distinct sorts. Nowhere, however, does Anantharaman appeal to quantum mechanics to explain the phenomena under investigation. Anantharaman's and Goswami's quite divergent explanations seem underdetermined by the data they are attempting to explain, and overdetermined by their particular fields of expertise.

Conclusion

To summarize post-colonial developments in Hindu responses to Darwinian evolution, we find that both theistic and monistic approaches are strongly inclined to cite modern science as confirmation of their particular perspectives. Both approaches increasingly accept many features of Darwinian or Neo-Darwinian evolution, especially involving microevolution, but they balk at the notion that speciation or the production of complex organs and organisms can be due to random variation and natural selection. They both absolutely reject the idea that consciousness is something that is emergent from matter as an epiphenomenon, regarding consciousness as pre-existent to the creation or manifestation of this world. We also find in both approaches a marked tendency to revitalize notions of design, and to harmonize design with evolution, at least with microevolution. And both subscribe to the idea of two levels of science: a higher, supernatural science, and a lower, material one.

The theistic and monistic approaches still differ in standard ways: in their epistemological assumptions, in their accounting of the specific mechanisms to explain (non-Darwinian) evolution, and in their interpretations of the ultimate purpose of cosmic manifestation. Theistic approaches rely primarily on scriptural authority, with secondary reliance on inferential reasoning when in harmony with scriptural authority. Monistic approaches appeal to yogic-intuitive apprehension and "experiential" confirmation of scientific truths. Modern Hindu theism tends to develop design arguments similar to those of Intelligent Design, looking at molecular processes that, in this perspective, require an external designer or engineer. Modern Vedānta emphasizes the internal intention and will of organisms, indwelled by the transcendent-immanent consciousness that is Brahman. And finally, theism views the purpose of the created world as merely a training ground for individual souls, enabling them to return to a superior divine world. Vedānta, however, looks upon the manifest world as the product of a creative and playful evolutionary process allowing consciousness to delight in ever new configurations of spirit and matter, to attain supramental states that do not involve an escape from this world but a transformation of it.

In the next chapter, I will challenge the common perception that Hindus generally are relatively accepting of modern evolutionary theory. My discussion will focus on an online survey I conducted in 2009–2011, in which I asked Hindus around the world about their attitudes towards evolution and design, as well as related issues in the Hinduism and science discourse.

14 Design and Darwin in the contemporary Hindu world

A common perception in the West is that Hinduism and evolution are largely compatible. For instance, in their book *Evolution and Religious Creation Myths: How Scientists Respond*, Paul F. Lurquin and Linda Stone compare and contrast the general attitude of five major world religions towards evolution. Regarding Hinduism they conclude:

> One can see that in Hinduism there is little, if any, conflict between creation and evolution. For one thing, Hinduism can easily reconcile the idea of creation (through Brahma the Creator) with that of change (evolution) as represented by the 10 incarnations (Dash avatars) of Vishnu, the Protector. In addition, as a colleague from Hindu India pointed out, the idea of humans descending from a long line of animal ancestors would not be a problem in Hinduism, where there is such a high value on animal life and some animals (cows, for example) are considered sacred. One great god, Hanuman, is a monkey. Hindu mythology is also replete with animals taking human forms and vice versa. In addition, the Hindu idea of reincarnation includes humans reincarnating into animals and animals reincarnating as humans in endless cycles.
>
> (Lurquin and Stone 2007: 26–7)

Lurquin and Stone also see Buddhism as consonant with "evolutionary science" (ibid.: 27). They claim that, in contrast to the more fundamentalist factions of Christianity, Judaism, and Islam, "neither Hinduism nor Buddhism is monotheistic or revealed, and neither sees evolution as a threat. No supernatural force within polytheistic Hinduism reveals one divine Truth" (ibid.: 28). They finally note that "Hinduism and Buddhism contain strong ideas of change, transformation (for example, reincarnation), and impermanence that make evolution less of a new and unfamiliar view of life" (ibid.).

As I hope to have demonstrated in the preceding chapters of this book, such conclusions regarding the alleged harmony between Hinduism, in particular, and evolutionary science are highly misleading. Lurquin and Stone, on one hand, essentialize the Hindu Dharmic traditions in very problematic ways, specifically their characterization of "Hinduism" as polytheistic and as rejecting any notion that a supernatural force reveals one divine truth.[1] Rammohan Roy, Dayananda

Saraswati, and Swami Prabhupada, among many other monotheistic Hindus, would be quite perplexed by these simplistic generalizations. And there is no acknowledgment of the monistic strands of Hinduism exemplified by Neo-Vedāntins like Vivekananda and Aurobindo that have played by far the greatest role in Hindu accommodations of Darwinian evolution. On the other hand, Lurquin and Stone accept as unproblematic a supposedly self-evident concordance between Darwinian evolution and notions of change or transformation. Nowhere do they scrutinize what Hindus or Buddhists really mean by evolution. Such scrutiny is critical for understanding Hindu and Buddhist attitudes towards Darwinism, for as in the West, Hindus and Buddhists have used the term *evolution* in a variety of ways, some uses being distinctly anti-Darwinian.[2] Most importantly, Lurquin and Stone do not question whether the idea of reincarnation, or spiritual evolution as Hindus frequently refer to it, is truly compatible with Darwinian evolution. While a detailed examination of such issues is clearly beyond the scope of their very brief consideration of Hinduism and Buddhism, their failure to note any of these problems displays a lack of critical inquiry that encourages misunderstanding.

At the same time, their conclusion that Hinduism and evolution, including Darwinian evolution, are compatible is quite understandable, for it is frequently affirmed by Hindus themselves—as we have seen in the case of Avataric Evolutionism alluded to by Lurquin and Stone. For instance, a 2008 survey of various religious groups in the United States by the Pew Forum on Religion & Public Life (2008) found that Hindus (at least for the 257 respondents in the Pew survey) and Buddhists were far more accepting of evolution than most other Americans. Specifically, most members (over 70 percent) of Christian conservative or evangelical groups "reject the evolutionary account as the best explanation for the development of human life, while large majorities of Catholics, Jews, Buddhists, Hindus and the unaffiliated agree that evolution best explains the development of life on earth." Buddhists and Hindus top the list of religious groups, with 81 percent of Buddhists and 80 percent of Hindus agreeing with the evolutionary account (followed by Jews at 77 percent). Only atheists at 87 percent were more accepting of evolution. An obvious problem with these results is that what Hindus and Buddhists mean by evolution is not clarified, and is surely different from what Mormons, Jehovah Witnesses, and Catholics mean by it.

A few other surveys have been conducted that polled Indians in general, rather than Hindus specifically, without providing the religious affiliation of respondents. For our purposes, then, their usefulness is considerably lessened, even if their results are suggestive. One of the more interesting is an international survey covering ten countries including India conducted by the British Council on the occasion of the 2009 bicentennial of Darwin's birth.[3] In India some 909 adults were interviewed. One question of direct relevance concerned "the origins of species and development of life on earth." Twenty percent of Indians chose as closest to their own view a purely naturalistic explanation: "Life on earth, including human life, evolved over time as a result of natural selection, in which no God played a part." Thirty-two percent chose a guided-evolutionary account: "Life on earth, including human life, evolved over time in a process guided by a God." And

43 percent chose a creationist account: "Life on earth, including human life, was created by a God and has always existed in its current form." Since 80 percent of Indians are Hindu, these results may indicate that in India, at least, a fairly significant number of Hindus ascribe to creationist ideas.[4]

A tantalizing glimpse into Indian Hindu views of evolution is provided by David Gosling's investigation conducted in the early 1970s regarding the beliefs of "Indian scientists."[5] His survey focused on the general issue of the impact of scientific study on religious belief and was confined to "Indian scientists" at five major Indian colleges, universities, and institutes, including faculty, graduate, and undergraduate students.[6] Gosling's analysis includes occasional references to the religious background of the respondents. One question specifically asked whether biological evolution conflicts with religious belief. Gosling notes that, for those Hindus who saw religion as very important in their lives, a majority (unspecified) "consistently saw no conflict between science and religion with regard to biological evolution" (2007: 118). Incidentally, for the Christian Indian scientists in the survey with the same degree of religiosity, 62 percent perceived conflict between religious belief and biological evolution (ibid.: 118–19).

With a partial exception in Gosling's case, none of the surveys attempts to differentiate between various Hindu perspectives ranging from those of theistic fundamentalists to monistic idealists. Imagine the usefulness of surveys of Christian attitudes towards evolution that did not differentiate between liberal and evangelical Protestants, Catholics, Eastern Orthodox, and Seventh-Day Adventists. Or that did not differentiate between the various interpretations of evolution, from theistic evolution and progressive evolution on one extreme, to young-Earth creationism on the other—after all, even young-Earth creationists, as Ronald Numbers points out, "have no trouble with 'microevolution' by natural selection within biological families."[7]

After studying the wide variety of responses to Darwinism on the part of Hindu writers and intellectuals over the last 150 years, I became interested in knowing what Hindus today think about evolution. Given the lack of empirical data that addressed this question in any depth, I decided to conduct my own survey focusing not only on questions directly related to evolution but also on relevant historical and epistemological issues. The survey was conducted between October 2009 and January 2011.

Author's 2009–2011 international survey of Hindu attitudes towards evolution

I wished to address two general questions in the survey:

1 What exactly are contemporary Hindu attitudes towards evolution and related issues?
2 What is the extent to which contemporary Hindus around the world reflect the various perspectives on evolution articulated by the major Hindu thinkers discussed earlier in the book?

I developed the survey in consultation with Hindu friends, as well as with a statistician at Trinity University.

As is inevitable with such surveys, the wording of questions must be simplified and cannot deal with all the nuances involved in complex issues. Accordingly, several recipients of the survey were not happy with its construction, e-mailing me about their concerns or adding comments in the survey itself. Some, right at the start, did not like my use of the term *Hinduism*, arguing that there is no such thing and that I should have used *Vedic religion* or *Sanatana Dharma*. I was aware of the problematic nature of *Hinduism* but chose to go with it as each suggested alternative has its own issues. And many Hindus today have come to accept this term for their own religious tradition; it is widely recognized, if not always fully endorsed.

Others felt that the survey was "just a tool for Westerners to grasp Hinduism in terms of their own beliefs" and should have been designed by someone more knowledgeable about the tradition. One respondent argued that the survey "equated Hindu religion with the Abrahamic religions" and that in future I should "work with some Hindu experts" in order to "elicit a true 'Hindu' response." As an example of my Abrahamic bias, he cited my question about whether the Vedas contain all knowledge concerning both spiritual and scientific truths. He concluded: "As you say you are an expert on Hinduism you will know that the Vedas say no such thing." Indeed I do know that, but as the survey reveals, 64 percent of the Hindus responding disagree with me and this commentator. In many ways I agree with this commentator that a number of the questions reflect an Abrahamic perspective. But a major reason for that is that the questions are derived from the leading Hindu evolutionary thinkers of the nineteenth and twentieth centuries who were themselves responding to Abrahamic (specifically Christian) ideas and values, and thus they inevitably reflected and assimilated part of the Abrahamic notions into their own Hindu (or Vedic) perspectives.

On the other side, several respondents thanked me for the survey, saying it was "thought-provoking" and "priceless." One commentator remarked: "I'm glad someone is trying to determine Hindu attitudes by actually reaching out to Hindus, instead of self-interpreting old texts." One of the more articulate respondents elaborated:

> Thank you for seeking a Dharmic opinion in regards to evolution. Dharmic traditions are so often overshadowed by the Big Three Abrahamic traditions that finding someone who actually cares about the Dharmic opinion is a refreshing change. This survey forces Hindus to examine the relationship between their faith and science, and this kind of self inquiry can create a deeper understanding of Hindu philosophy, spirituality, and cosmology among Hindus, something that is sorely lacking in the Hindu mind today but more necessary than ever.

Needless to say, it was precisely my interest in seeking "a Dharmic opinion" that motivated me to undertake the survey. I hope that the earlier chapters of this book have sufficiently contextualized the questions in the survey that they will not be seen as somehow equating "Hinduism" with the Abrahamic traditions. Rather, my intention was to bring out what is distinctive about Hindu responses to Darwinism among Hindus today. These responses reflect the ancient Vedic traditions but as refracted through the writings of the nineteenth- through twenty-first-century Hindu thinkers grappling with the challenges of colonialism, Christian evangelism, and modern science.

The survey contained twenty-three demographic questions and twenty-seven substantive questions on evolution and related issues dealing with Hinduism and modern science.[8] The substantive questions used a five-level Likert-scale introduced with the instruction, "Do you agree or disagree with the following?" Five responses were available: "Agree strongly," "Agree," "Neither agree nor disagree," "Disagree," and "Disagree strongly." In summarizing the Likert-scale data throughout this analysis, I focus on the positive responses, combining "Strongly agree" and "Agree" unless otherwise indicated. All the substantive questions are listed in Appendix A.

The survey was placed on a SurveyMonkey website. Notices of the survey were sent by e-mail to the academic list-serv RISA-L (Religion in South Asia list), to various Hindu temples and organizations, Hindu student groups, and Hindu faculty and staff members of various university and research institutions around the world. By far the most productive strategy was the use of Facebook to contact the administrators of dozens of Hindu Facebook groups. Most administrators contacted were willing to send notice of the survey to their members.

The respondents represent a convenience rather than a random sample, being limited to select groups of English-speaking Hindus with access to the Internet.[9] Any summary claims or conclusions I make about various groups (e.g., physicians, or members of ISKCON) thus refer only to the groups of respondents in this survey who may not be representative of their groups as a whole. At the same time, I obtained a fairly wide range of respondents from diverse cultural, ethnic, and educational backgrounds. Selected background data is given in Appendix B.

The survey generated 1,690 responses, of which 1,301 were complete. However, among the completed surveys were a number of respondents who indicated that they were not Hindu per se but Buddhist, Jain, Sikh, or Roman Catholic. Others were not sure what they were. I eliminated all who were not of Hindu background or who indicated they were not converts. In this manner I ended up, quite by accident, with an even 1,000.

The substantive, Likert-scale questions may be divided into three broad and somewhat overlapping categories concerning: (1) epistemological beliefs, specifically regarding reliable sources of knowledge about the physical world; (2) general beliefs relevant to modern science; and (3) beliefs explicitly related to evolution and Darwinism. I will consider these in order, discussing those patterns of responses and cross-tabulations that I found to be most illuminating of the issues dealt with in this book.

Hindu beliefs regarding reliable sources of knowledge about the physical world

Five questions in the survey ask about possible sources of knowledge concerning the physical world and the relative reliability of the different sources. These sources include scripture (the Vedas), spiritually enlightened persons, superconscious states of mind, and modern science.

The Vedas, whether restricted to the ancient four Vedas or extended to include more recent texts like the *Bhāgavata Purāṇa*, are generally viewed as divine revelations perceived or received by the ancient Vedic seers, and as authoritative regarding all spiritual truths. But are they also authoritative regarding empirical knowledge? One survey question asks whether "[t]he Vedas contain all knowledge concerning both spiritual and scientific truths." That is, do the Vedas have a dual competence in the realms of both spirit and matter? A majority of all respondents (64 percent) accepts the Vedas as both a spiritual and scientific authority (Table 14.1, question 28). Acceptance of Vedic scientific and spiritual competence is strong for most levels of educational attainment, ranging from 63 to 72 percent for high school, associate's, bachelor's, master's, and M.D.s (Table 14.2). The one notable exception is the group of Ph.D.s, of whom only 45 percent recognize Vedic competence in both realms. Regarding alleged familiarity with modern science, I note that a very few of those who claim "thorough and extensive" familiarity indicate in comments that they are experts in Vedic or spiritual science. In any case, overall, those who claim either "thorough and extensive" familiarity or "thorough in some areas, not in others," are less likely to accept both Vedic scientific and spiritual competence, at 60 and 59 percent, respectively, than were those who indicate they are only "somewhat familiar and knowledgeable" or "only vaguely familiar or knowledgeable," both at 72 percent (Table 14.3). Greater familiarity with modern science thus generally correlates with less acceptance of Vedic dual competence, almost certainly due to increased skepticism regarding the scientific competence of the Vedas. However, it should be noted that solid majorities of all respondents, regardless of degree of familiarity with modern science, accept the Vedas as a valid scientific authority. Women are somewhat more accepting of Vedic dual competence (67 percent) than men (62 percent) (Table 14.4). Greater degree of religiosity/spirituality correlates well, unsurprisingly, with acceptance of Vedic dual authority, the "quite religious/spiritual" at 73 percent, the "somewhat religious/spiritual" at 58 percent, and the "neutral or indifferent" at 47 percent (Table 14.5). Of the very few who are "somewhat religious/spiritual" or "quite non-religious/spiritual," only 25 and 17 percent, respectively, accept Vedic dual authority.

A second survey question asks whether "[t]he Vedas and modern science are in complete harmony with each other." A majority of respondents (57 percent) affirm their complete harmony (Table 14.1, question 29). With regard to educational attainment, there are small affirming majorities at most levels (54 to 59 percent), with Ph.D.s somewhat lower at 50 percent (Table 14.6). Those respondents indicating "thorough and extensive" familiarity with modern science,

Table 14.1 Questions relating to Hindu beliefs regarding reliable sources of knowledge about the physical world

Do you agree or disagree with the following?	*Agree strongly*	*Agree*	*Neither agree nor disagree*	*Disagree*	*Disagree strongly*
28. The Vedas contain all knowledge concerning both spiritual and scientific truths. (985)	40%	24%	22%	11%	4%
	64%			15%	
29. The Vedas and modern science are in complete harmony with each other. (990)	27%	30%	28%	12%	3%
	57%			15%	
31. Concerning factual knowledge about the physical universe, the statements of spiritually enlightened persons are more trustworthy than those of modern scientists. (987)	24 %	20%	31%	17%	8%
	44%			25%	
32. Concerning factual knowledge about the physical universe, the statements of scripture are more trustworthy than those of modern science. (981)	20%	19%	35%	18%	9%
	38%			27%	
33. Superconscious or yogic states of mind are capable of discovering most or all truths about the physical world. (982)	35%	33%	19%	9%	4%
	69%			13%	

Note: For all tables in this chapter, the n (in parentheses) is provided for each question and/or each group. Percentages may not add up to 100 due to rounding. For tables 14.1, 14.15, and 14.16, the sums of the "agree strongly" and "agree" responses and of the "strongly disagree" and "disagree" responses are included on the second row of percentages for each question.

"thorough in some areas, not in others," or "somewhat familiar and knowledgeable" all tend to accept Vedic-scientific harmony (57 to 60 percent), but not quite half of the few respondents "only vaguely familiar or knowledgeable" did so (49 percent) (Table 14.7). Men and women generally affirm the harmony (both at 57 percent), although women were a bit less skeptical (only 12 percent) than men (16 percent) (Table 14.8). A greater degree of religiosity/spirituality correlates well for the most part with affirmation of Vedic-scientific harmony: the "quite religious/spiritual" at 62 percent, the "somewhat religious/spiritual" at 58 percent, and the "neutral or indifferent" at 41 percent (Table 14.9). Of the very few who are "somewhat non-religious/spiritual" or "quite non-religious/spiritual," 13 and 24 percent, respectively, accept such harmony.

Table 14.2 Cross-tabulations between educational attainment and belief in the Vedas as sources for all knowledge, spiritual and scientific

28. The Vedas contain all knowledge concerning both spiritual and scientific truths.	*H.S. (120)*	*Assoc. (46)*	*Bachelor's (318)*	*Master's (323)*	*Ph.D. (82)*	*M.D. (59)*
Strongly agree or agree	63%	72%	66%	65%	45%	63%
Neither agree nor disagree	24%	26%	20%	21%	28%	22%
Strongly disagree or disagree	13%	2%	15%	14%	27%	15%

Table 14.3 Cross-tabulations between alleged familiarity with modern science and belief in the Vedas as sources for all knowledge, spiritual and scientific

28. The Vedas contain all knowledge concerning both spiritual and scientific truths.	*Thorough and extensive (242)*	*Thorough in some areas, not in others (392)*	*Somewhat familiar and knowledgeable (306)*	*Only vaguely familiar or knowledgeable (35)*
Strongly agree or agree	60%	59%	72%	72%
Neither agree nor disagree	21%	26%	16%	20%
Strongly disagree or disagree	19%	15%	12%	9%

Table 14.4 Cross-tabulations between sex and belief in the Vedas as sources for all knowledge, spiritual and scientific

28. The Vedas contain all knowledge concerning both spiritual and scientific truths.	*Male (690)*	*Female (290)*
Strongly agree or agree	62%	67%
Neither agree nor disagree	21%	23%
Strongly disagree or disagree	17%	10%

Table 14.5 Cross-tabulations between religiosity and belief in the Vedas as sources for all knowledge, spiritual and scientific

28. The Vedas contain all knowledge concerning both spiritual and scientific truths.	*Quite religious/ spiritual (549)*	*Somewhat religious/ spiritual (312)*	*Neutral or indifferent (71)*	*Somewhat non-religious/ spiritual (8)*	*Quite non-religious/ spiritual (29)*
Strongly agree or agree	73%	58%	47%	25%	17%
Neither agree nor disagree	17%	27%	35%	38%	10%
Strongly disagree or disagree	11%	14%	18%	38%	72%

Table 14.6 Cross-tabulations between educational attainment and belief in harmony of the Vedas with modern science

29. The Vedas and modern science are in complete harmony with each other.	*H.S. (120)*	*Assoc. (47)*	*Bachelor's (322)*	*Master's (323)*	*Ph.D. (82)*	*M.D. (59)*
Strongly agree or agree	54%	55%	58%	59%	50%	56%
Neither agree nor disagree	31%	28%	27%	27%	38%	25%
Strongly disagree or disagree	15%	17%	15%	14%	12%	19%

Table 14.7 Cross-tabulations between alleged familiarity with science and belief in harmony of the Vedas with modern science

29. The Vedas and modern science are in complete harmony with each other.	*Thorough and extensive (244)*	*Thorough in some areas, not in others (395)*	*Somewhat familiar and knowledgeable (307)*	*Only vaguely familiar or knowledgeable (35)*
Strongly agree or agree	57%	57%	60%	49%
Neither agree nor disagree	30%	27%	27%	34%
Strongly disagree or disagree	14%	16%	13%	17%

Table 14.8 Cross-tabulations between sex and belief in harmony of the Vedas with modern science

29. The Vedas and modern science are in complete harmony with each other.	*Male (695)*	*Female (290)*
Strongly agree or agree	57%	57%
Neither agree nor disagree	26%	32%
Strongly disagree or disagree	16%	12%

Table 14.9 Cross-tabulations between religiosity and belief in harmony of the Vedas with modern science

29. The Vedas and modern science are in complete harmony with each other.	*Quite religious/ spiritual (553)*	*Somewhat religious/ spiritual (314)*	*Neutral or indifferent (71)*	*Somewhat non-religious/ spiritual (8)*	*Quite non-religious/ spiritual (29)*
Strongly agree or agree	62%	58%	41%	13%	24%
Neither agree nor disagree	24%	31%	44%	25%	31%
Strongly disagree or disagree	15%	12%	16%	63%	45%

The Vedas are not generally regarded as revealed books like the Bible, but rather as sacred sounds, syllables, or words reverberating throughout the cosmos.[10] They were heard or seen, that is, directly cognized, by the ancient seers through their yogic attunement to the sounds. Can such yogic or supernatural cognition also have direct and immediate access to the truths about the material universe? A third epistemological question therefore asks whether "[s]uperconscious or yogic states of mind are capable of discovering most or all truths about the physical world." We find a great majority of all respondents, 69 percent, accepts the scientific authority of superconscious or yogic states of mind, a bit more than the 64 percent who accept Vedic scientific authority (Table 14.1, question 33). Acceptance of such authority is strong for most levels of education except the very highest, ranging from 70 to 81 percent for high school, associate's, bachelor's, and master's degrees (Table 14.10). Ph.D.s and M.D.s are considerably less accepting at 53 percent and 49 percent, respectively. As for familiarity with modern science, those who claim thorough and extensive familiarity are least likely to accept the scientific authority of superconscious-yogic states, at 61 percent, compared to those with thorough knowledge only in some areas, at 68 percent, those who are only somewhat familiar, at 75 percent, or those only vaguely familiar, at 83 percent (Table 14.11). Women are somewhat more accepting of superconscious-yogic scientific authority, at 72 percent, than men, at 67 percent (Table 14.12). Greater degree of religiosity/spirituality correlates especially well with acceptance of superconscious scientific authority, the quite religious/spiritual at 78 percent, the somewhat religious/spiritual at 66 percent, and the neutral or indifferent at 37 percent (Table 14.13). Of the very few who are somewhat or quite non-religious/spiritual, 0 and 24 percent, respectively, accept superconscious-yogic scientific authority.

Turning to relative reliability of sacred versus secular-scientific sources for knowledge about the physical universe, let me emphasize that the two questions involved are not concerned with statements about ethical ideals, moral values, or the meaning of life. Regarding the first question, whether "[c]oncerning factual knowledge about the physical universe, the statements of spiritually enlightened persons are more trustworthy than those of modern scientists," 44 percent of

Table 14.10 Cross-tabulations between educational attainment and belief in superconscious or yogic states of mind as sources for most or all knowledge about the physical world

33. Superconscious or yogic states of mind are capable of discovering most or all truths about the physical world.	*H.S. (118)*	*Assoc. (46)*	*Bachelor's (321)*	*Master's (321)*	*Ph.D. (81)*	*M.D. (59)*
Strongly agree or agree	70%	81%	73%	70%	53%	49%
Neither agree nor disagree	20%	17%	19%	18%	22%	24%
Strongly disagree or disagree	11%	2%	8%	13%	25%	27%

Table 14.11 Cross-tabulations between alleged familiarity with science and belief in superconscious or yogic states of mind as sources for most or all knowledge about the physical world

33. Superconscious or yogic states of mind are capable of discovering most or all truths about the physical world.	*Thorough and extensive (242)*	*Thorough in some areas, not in others (393)*	*Somewhat familiar and knowledgeable (302)*	*Only vaguely familiar or knowledgeable (35)*
Strongly agree or agree	61%	68%	75%	83%
Neither agree nor disagree	22%	18%	18%	17%
Strongly disagree or disagree	17%	15%	8%	0%

Table 14.12 Cross-tabulations between sex and belief in superconscious or yogic states of mind as sources for most or all knowledge about the physical world

33. Superconscious or yogic states of mind are capable of discovering most or all truths about the physical world.	*Male (691)*	*Female (286)*
Strongly agree or agree	67%	72%
Neither agree nor disagree	19%	20%
Strongly disagree or disagree	14%	8%

Table 14.13 Cross-tabulations between religiosity and belief in superconscious or yogic states of mind as sources for most or all knowledge about the physical world

33. Superconscious or yogic states of mind are capable of discovering most or all truths about the physical world.	*Quite religious/ spiritual (550)*	*Somewhat religious/ spiritual (310)*	*Neutral or indifferent (70)*	*Somewhat non-religious/ spiritual (7)*	*Quite non-religious/ spiritual (29)*
Strongly agree or agree	78%	66%	37%	0%	24%
Neither agree nor disagree	14%	24%	26%	57%	21%
Strongly disagree or disagree	8%	10%	37%	43%	55%

respondents prefer the statements of spiritually enlightened persons (Table 14.1, question 31). Interestingly, almost as large a minority (40 percent) of scientists likewise favors enlightened persons.[11] As for the second question, whether "[c]oncerning factual knowledge about the physical universe, the statements of scripture are more trustworthy than those of modern science," 38 percent of respondents place greater trust in scripture (Table 14.1, question 32). A similar percentage of scientists (36 percent) also prefer the statements of scripture.[12]

Finally, let us look at cross-tabulations between epistemological views and religious affiliation. I was able to identify four groups of at least forty members on the basis of their own self-descriptions. These are: (1) members of ISKCON/ followers of Swami Prabhupada, (2) members of the Self-Realization Fellowship/ followers of Yogananda, (3) followers of Sathya Sai Baba, and (4) general followers of Advaita Vedānta. The last three are somewhat overlapping, since followers of Yogananda and Sai Baba often have Advaitic leanings, but some self-identify as following another (usually Vedāntic) school, or do not list any philosophical affiliation. Some respondents specifically identified themselves as followers of the major Neo-Vedāntic figures Aurobindo and Vivekananda, but their numbers were small, and in any case are likely to be included among the Advaitins. There were also a few respondents indicating that they were disciples of a "Dayananda Saraswati," but unfortunately they did not clarify which of the two modern gurus by that name they followed (and very few identified themselves unequivocally as members of the Arya Samaj). As for the three religious figures whose followers clearly identified themselves, we have previously met Swami Prabhupada, whose anti-evolutionism was discussed in Chapter 13. Paramahamsa Yogananda (1893–1952) was an important Neo-Vedānta missionary to America who founded the Self-Realization Fellowship in 1935, in Southern California. Sathya Sai Baba (1926–2011), a popular guru especially among wealthy urban and Westernized Hindu Indians, was noted for his charitable works and materializations of small physical objects. Table 14.14 provides the data for the four groups and includes for comparative purposes the figures for all respondents.

Substantial majorities (65 to 78 percent) of the four groups accept the idea that the Vedas are scientifically authoritative, with Advaitins at the low end (Table 14.14, question 28). Somewhat smaller majorities (55 to 69 percent), with one major exception, affirm the harmony of the Vedas with modern science (Table 14.14, question 29). The exception is ISKCON, at only 11 percent. Very large majorities (67 to 97 percent) accept the scientific authority of superconscious-yogic states of mind (Table 14.14, question 33). Regarding the scientific authority of spiritually enlightened persons (Table 14.14, question 31), however, there is more of a split with only substantial minorities of followers of Sai Baba (43 percent) and Advaitins (45 percent) accepting such authority, compared to strong majorities of ISKCON members (70 percent) and Yogananda followers (74 percent). Regarding the scientific authority of scripture (Table 14.14, question 32), ISKCON again stands clearly apart from the other groups with 70 percent accepting it, while only 37 to 41 percent of all other groups are so inclined. In general the data for all respondents is quite similar to the responses of the Advaitins.

In summary, the data regarding epistemological issues reveal that a substantial majority of all Hindus surveyed accept both the scientific authority of the Vedas (64 percent) and especially of yogic, superconscious states of mind (69 percent), with relatively few dissenters. Clearly, Vivekananda's idea of the scientific authority of yogic vision or intuition has become a broadly accepted notion. As for Hindu acceptance of the authority of religious or spiritual sources versus that of modern science regarding knowledge about the physical universe, substantial

Table 14.14 Cross-tabulations between Hindu beliefs about scientific authority and religious affiliation

Strongly agree or agree [A] *Neither agree nor disagree [N]* *Strongly disagree or disagree [D]*	*ISKCON (47)*	*Yogananda (58)*	*Sai Baba (43)*	*Advaita (128)*	*All (1000)*
28. The Vedas contain all knowledge concerning both spiritual and scientific truths.	70% A 21% N 9% D (47)	72% A 19% N 9% D (57)	78% A 20% N 2% D (41)	65% A 18% N 17% D (127)	64% A 22% N 15% D (985)
29. The Vedas and modern science are in complete harmony with each other.	11% A 21% N 68% D (47)	69% A 26% N 5% D (58)	55% A 33% N 12% D (42)	60% A 24% N 17% D (127)	57% A 28% N 15% D (990)
31. Concerning factual knowledge about the physical universe, the statements of spiritually enlightened persons are more trustworthy than those of modern scientists.	70% A 17% N 13% D (47)	74% A 19% N 7% D (58)	43% A 41% N 17% D (42)	45% A 32% N 23% D (127)	44% A 31% N 25% D (987)
32. Concerning factual knowledge about the physical universe, the statements of scripture are more trustworthy than those of modern science.	70% A 19% N 11% D (47)	41% A 45% N 14% D (58)	41% A 52% N 7% D (42)	37% A 36% N 27% D (126)	38% A 35% N 27% D (981)
33. Superconscious or yogic states of mind are capable of discovering most or all truths about the physical world.	67% A 22% N 11% D (46)	97% A 2% N 2% D (58)	79% A 19% N 2% D (42)	72% A 14% N 13% D (127)	69% A 19% N 13% D (982)

minorities of respondents privilege the scientific authority of scripture (38 percent) and of spiritually enlightened persons (44 percent). Much of this acceptance is due to the scientizing of tradition by the "scientific" sadhus like Vivekananda.[13] Interestingly, while all groups prefer to some degree the statements of spiritually enlightened persons and of scripture to those of modern science, ISKCON is the only group with a large majority (70 percent) preferring the scientific authority of both enlightened persons and scriptures.[14] ISKCON's distinctive perspective is clearly related to its inerrantist and literalist approach to its scriptures encouraged by its founder, a perspective that is especially reflected in its attitude towards Darwinism, as we shall see.

Hindu beliefs relevant to science

There are eight relevant questions here. Two concern the relationship of Hinduism and India to science, and six pertain to Hindu beliefs about animals, karma, and the history of life on earth (Table 14.15).

Table 14.15 Questions relating to Hindu beliefs relevant to modern science

Do you agree or disagree with the following?	*Agree strongly*	*Agree*	*Neither agree nor disagree*	*Disagree*	*Disagree strongly*
24. Animals have the same moral rights that human beings do. (991)	41%	34%	15%	8%	1%
	76%			9%	
25. Nature is predominantly a fierce struggle for survival of the fittest. (988)	25%	34%	22%	14%	5%
	59%			20%	
26. Both human beings and animals possess some sort of non-material, individual soul that survives the death of the body. (987)	56%	29%	9%	3%	3%
	85%			6%	
27. Karma, like the law of gravity, is a fundamental scientific law of nature. (986)	54%	29%	10%	5%	3%
	82%			8%	
30. Hinduism is the most scientific religion in the world. (988)	40%	24%	29%	5%	2%
	64%			7%	
34. Human beings have existed on earth for hundreds of millions of years or longer. (982)	28%	28%	22%	14%	8%
	56%			22%	
35. The birthplace or cradle of humanity was ancient Africa. (982)	12%	24%	44%	13%	7%
	36%			20%	
36. India was the fountainhead of all sciences and civilization. (981)	32%	28%	28%	11%	2%
	59%			13%	

The notion that "Hinduism is the most scientific religion in the world," much promoted by Vivekananda and later Hindu cultural nationalists, is espoused by a solid majority (64 percent) of respondents (Table 14.15, question 30). In similar fashion, the belief that "India was the fountainhead of all sciences and civilization," an idea first broached by Rammohan Roy and aggressively asserted by Dayananda Saraswati, is accepted by 59 percent (Table 14.15, question 36). Those Hindus affirming these views reflect a thoroughly scientized perspective on the Hindu Dharmic tradition, with significant implications for attitudes towards Darwinism. For instance, those who believe that all sciences originated in India may be inclined to affirm that the ancient Vedic seers discovered evolution centuries before Darwin.

A large majority (85 percent) of respondents, unsurprisingly, accepts the traditional Hindu belief that "[b]oth human beings and animals possess some sort of non-material, individual soul that survives the death of the body" (Table 14.15, question 26). A similar number (82 percent) believes that "[k]arma, like the law of gravity, is a fundamental scientific law of nature" (Table14.15, question 27). The explicit identification of karma with scientific laws like gravity, of course, is a modern, scientizing twist on an ancient theme. A slightly smaller majority (76 percent) affirms that "[a]nimals have the same moral rights that human beings do" (Table 14.15, question 24). While animals may have full moral rights, a significant majority (59 percent) believes that "[n]ature is predominantly a fierce struggle for survival of the fittest" (Table 14.15, question 25), a rather bleak outlook that may well reflect some familiarity with popular Darwinian ideas. A somewhat smaller majority (56 percent) accepts the traditional belief that "[h]uman beings have existed on earth for hundreds of millions of years or longer" (Table 14.15, question 34). One can speculate that modern paleontological and other scientific evidences have had a somewhat corrosive effect on this particular traditional belief concerning a matter of empirical record, while leaving the more philosophical issues of souls and karma largely untouched. Such evidences may also be responsible for the 36 percent who accept that "[t]he birthplace or cradle of humanity was ancient Africa" (Table 14.15, question 35), rather than India as promoted by some Hindu cultural nationalists.

The last six questions above have direct relevance for Hindu perspectives on evolution. Beliefs about animal souls and karma, on one hand, affirm the ideals of spiritual evolution but without necessarily affirming or denying Darwinian theory. For many Hindus, belief in rebirth leaves open the option of accommodating Darwinian evolution within an overarching metaphysical framework of karmic spiritual progress—despite Darwinian explanations of behavioral and character propensities that conflict with, or at least are in considerable tension with, karmic explanations. On the other hand, the belief in the ancient antiquity of human beings on earth (for hundreds of millions of years or longer) clearly and unambiguously conflicts with modern scientific evolutionary reconstructions of the history of terrestrial life. Such considerations bring us to the survey questions dealing explicitly with Darwinian evolution.

Hindu beliefs related to evolution and Darwinism

There are fourteen survey questions relating directly to evolution and Darwinism. One introductory question concerns whether respondents see a conflict between Hinduism and evolution—the term *evolution* being left undefined. Seven questions deal with the mechanisms and explanations of "evolution" and are highly relevant for interpreting responses to the introductory question. Two historical questions deal with beliefs about the discovery of evolution, and four other questions with philosophical issues regarding the general nature of Darwinian theory. The results for these fourteen questions are given in Table 14.16.

The introductory question, that "[t]here is no conflict between Hinduism and evolution," receives a large affirming majority (65 percent) (Table 14.16, question 37). But as noted earlier in relation to the Pew survey, without defining more carefully the term *evolution*, we have little way of knowing exactly what Hindus are affirming. This becomes apparent in looking at the questions relating to the mechanisms and explanations of evolution. For instance, only 51 percent accept that "[h]uman beings evolved from lower animal species by processes of random mutation and natural selection" (Table 14.16, question 40). Yet 59 percent accept that the evolution of species is preceded by an involution of spirit (Table 14.16, question 44)—in accord with the evolutionism of neo-Vedāntins like Vivekananda and Aurobindo. In similar fashion, 54 percent believe that "[k]arma and rebirth are better explanations for the origin and diversity of life than Darwinian evolution" (Table 14.16, question 43). Yet only 45 percent affirm that "[h]umans evolved from lower animal species by processes of spiritual evolution" (Table 14.16, question 41). A major reason that Hindu respondents refrain from endorsing a robust Darwinism appears to be the issue of consciousness. A very large majority (75 percent) agree that "[c]onsciousness cannot be explained by Darwinian evolution" (Table 14.16, question 45), even though a significant minority (43 percent) accept that "[c]onsciousness is an emergent property of matter" (Table 14.16, question 46). But as we have seen, this idea may reflect the Neo-Vedāntic idea that consciousness is a re-emergent property of matter, rather than a materialistic conception of consciousness.

Not as many respondents, 40 percent, believe that "[i]ntelligent design is a better explanation for the origin and diversity of life than Darwinian evolution" (Table 14.16, question 42), but that is still a significant number who reject basic Darwinian theory. Interestingly, relatively few Hindus, 27 percent, believe that "Darwinian evolution insists that complex organisms are the products of mere blind chance," while 36 percent reject such a view of Darwinism (Table 14.16, question 49). The rejection of the idea that Darwinism is the equivalent of mere-chance theory on the part of many Hindus may well be due to their belief that Darwinian theory at least allows for a complementary and non-random evolutionary process guided by karmic mechanisms.

Regarding historical issues, exactly half of respondents espouse the view that "Hindus discovered evolution centuries before Darwin" (Table 14.16, question 38). Respondents who affirmed this were asked to indicate who they thought responsible. A very few mentioned specific sages, such as Patañjali (Vivekananda's favorite) and Śaṅkara. A far more common response was the generic "Vedic rishis," who were said to be responsible for the discovery of evolution in the sense of karma and rebirth. Another common response was the Purāṇic sages and their discovery of the avataric sequence of Viṣṇu's incarnations. One question specifically asks whether "[t]he ancient Hindu concept of divine incarnations (avataras) directly anticipated Darwinian evolutionary theory," and 39 percent of respondents answer affirmatively (Table 14.16, question 39).

As for the general nature of Darwinism, a majority (60 percent) affirm that "Darwinian evolution is true, but very limited," a view that readily allows for

Table 14.16 Questions relating to evolution and Darwinism

Do you agree or disagree with the following?	*Agree strongly*	*Agree*	*Neither agree nor disagree*	*Disagree*	*Disagree strongly*
37. There is no conflict between Hinduism and evolution. (984)	33%	33%	22%	8%	5%
	65%			13%	
38. Hindus discovered evolution centuries before Darwin. (975)	29%	21%	40%	7%	3%
	50%			10%	
39. The ancient Hindu concept of divine incarnations (avataras) directly anticipated Darwinian evolutionary theory. (977)	16%	23%	41%	14%	7%
	39%			20%	
40. Human beings evolved from lower animal species by processes of random mutation and natural selection. (970)	21%	31%	24%	13%	12%
	51%			25%	
41. Humans evolved from lower animal species by processes of spiritual evolution. (979)	18%	27%	30%	15%	10%
	45%			25%	
42. Intelligent design is a better explanation for the origin and diversity of life than Darwinian evolution. (972)	18%	22%	36%	13%	11%
	40%			24%	
43. Karma and rebirth are better explanations for the origin and diversity of life than Darwinian evolution. (976)	27%	26%	27%	14%	6%
	54%			20%	

44. The evolution of species is preceded by an involution of spirit (Brahman/God). (968)	25%		34%	28%	9%		4%
		59%				13%	
45. Consciousness cannot be explained by Darwinian evolution. (975)	38%		36%	18%	6%		1%
		75%				7%	
46. Consciousness is an emergent property of matter. (963)	14%		29%	30%	13%		15%
		43%				27%	
47. Darwinian evolution is true, but very limited. (971)	11%		48%	24%	10%		6%
		60%				16%	
48. Darwinian evolutionary theory is not scientific. (969)	6%		10%	33%	37%		14%
		16%				51%	
49. Darwinian evolution insists that complex organisms are the products of mere blind chance. (964)	6%		21%	37%	26%		10%
		27%				36%	
50. Darwinism is atheism. (967)	7%		13%	37%	30%		13%
		20%				42%	

accommodation of modern evolution within a hierarchical framework of material and spiritual sciences (Table 14.16, question 47). Very few respondents (16 percent) believe that "Darwinian evolutionary theory is not scientific" (Table 14.16, question 48), and a similarly small minority (20 percent) accepts that "Darwinism is atheism" (question 50). It is telling that those who think that Darwinism is atheistic tend to view it as not scientific. For instance, of those who "strongly agree" that Darwinian theory is atheistic (n = 69), 65 percent "strongly agree" or "agree" that the theory is not scientific. As one such respondent comments, Darwinian evolution is "[a] misunderstanding by modern science, based on speculation and atheistic wishful thinking."

Let us now turn to two noteworthy demographic cross-tabulations, beginning with respondents who indicate that they were, are, or will be employed in a scientific field. I restrict my analysis to five fields: (1) biological sciences, (2) health care and medicine, (3) physical sciences, (4) engineering, and (5) computer science. I also compare the average of responses for all five fields with the average for all non-scientists. The results are given in Table 14.17.

Majorities in all fields agree that there is no conflict between Hinduism and evolution (Table 14.17, question 37). However, the biologists are rather more skeptical, with only 55 percent affirming, compared to the others, ranging from 66 to 72 percent. The biologists also stand out, sometimes along with those in health care and medicine (henceforth called health care providers), with regards to the mechanisms of evolution, being considerably more accepting of naturalistic mechanisms and more skeptical of spiritual or supernatural explanations for diversity of life. With reference, for instance, to whether "[h]uman beings evolved from lower animal species by processes of random mutation and natural selection," biologists, at 74 percent, and health care providers at 66 percent are the most accepting, far more than the physical scientists at 55 percent, engineers at 52 percent, and computer scientists at 52 percent (Table 14.17, question 40). Respondents not involved in scientific careers (non-scientists) are even less accepting of random mutations and natural selection, at 45 percent.

Regarding supernatural explanations, biologists, with only 30 percent accepting the idea that spiritual processes were involved in the evolution of humans from animals, are more skeptical than other scientists (at 41 to 45 percent) (Table 14.17, question 41). The idea that intelligent design is a better explanation than Darwinian evolution is affirmed by only 31 percent of biologists and 30 percent of health care providers, over against 56 percent of physical scientists, 38 percent of engineers, and 49 percent of computer scientists (Table 14.17, question 42). And only 42 percent of biologists and 34 percent of health care providers accept that karma and rebirth are better explanations, over against other scientists (at 53 to 58 percent) (Table 14.17, question 43). Approximately half of biologists and health care providers accept the idea of an involution of spirit preceding the evolution of species, but still are somewhat more skeptical than the other groups (Table 14.17, question 44). And over two-thirds of biologists and health care providers believe that Darwinism cannot explain consciousness, again just somewhat more skeptical than other scientists (Table 14.17, question 45). Just over half of biologists and physical scientists (at 53 percent) accept consciousness as an emergent property of matter, a few to several percentage points more than the others (Table 14.17, question 46).

In general, the physical scientists, engineers, and computer scientists, in their greater acceptance of supernatural explanations, are closer to the views of non-scientists than were the biologists and health providers.[15] Better familiarity with biology and pathology clearly correlates with greater acceptance of modern evolutionary theory and greater skepticism regarding traditional or supernatural explanations of evolution like karma and intelligent design.

Regarding historical issues, 51 percent of scientists overall and 50 percent of non-scientists agree that "Hindus discovered evolution centuries before Darwin" (Table 14.17, question 38). Once again, among the scientists, biologists are the most skeptical, with only 41 percent agreeing. Scientists overall are somewhat more inclined (42 percent) than non-scientists (37 percent) to agree that "[t]he ancient Hindu concept of divine incarnations (avataras) directly anticipated Darwinian evolutionary theory" (Table 14.17, question 39).

As one would expect, a majority of scientists overall reject the statement that "Darwinian evolutionary theory is not scientific" (Table 14.17, question 48). Biologists and health care providers are the most skeptical, with only 6 and 10 percent, respectively, agreeing with the statement, and 69 and 68 percent, respectively, disagreeing. At the other extreme, 23 percent of computer scientists agree, while only 45 percent disagree. Overall, only 15 percent of scientists affirm that Darwinism is not scientific, with non-scientists only slightly more affirming, at 18 percent. At the same time, 62 percent of scientists overall agree that "Darwinian evolution is true, but very limited" (Table 14.17, question 47). Only about a third of scientists overall see Darwinian evolution as simply a process of blind chance (Table 14.17, question 49). Regarding the statement that "Darwinism is atheism," only 20 percent of scientists overall agree, with biologists (16 percent) and health care providers (15 percent) the least affirming (Table 14.17, question 50). Non-scientists, at 23 percent agreeing, are close to the views of the scientists.

In conclusion, let us look at selected cross-tabulations according to religious affiliation (Table 14.18). Members of ISKCON stand out for their opposition to Darwinian evolution and espousal of spiritual or supernatural explanations for the origin and diversity of life. For instance, only 4 percent accept the idea that humans evolved from lower animal species as a result of random mutation and natural selection (Table 14.18, question 40). Yogananda's group is a distant second, with 30 percent accepting this naturalistic explanation of human origins, followed by followers of Advaita at 51 percent, and of Sai Baba at 60 percent. We should keep in mind, however, that affirmation of naturalistic explanations for many Hindus does not exclude more supernaturalistic explanations such as karma and rebirth, the two sorts of explanation often being viewed as complementary.

ISKCON members are also more inclined, at 42 percent, to see Darwinism as a process of blind chance than the other groups, which range from 15 to 35 percent (Table 14.18, question 49). As indicated earlier, such a view is a misperception or

Table 14.17 Cross-tabulations between Hindu beliefs about evolution and field of science

Strongly agree or agree [A] Neither agree nor disagree [N] Strongly disagree or disagree [D]	Biological sciences (66)	Health care and medicine (112)	Physical sciences (33)	Engi-neering (138)	Computer science (117)	Ave. for all five sciences (466)	Non-scientists (455)
37. There is no conflict between Hinduism and evolution.	55% A 32% N 14% D (66)	70% A 18% N 12% D (111)	72% A 16% N 13% D (32)	71% A 20% N 9% D (137)	66 % A 23% N 11% D (115)	67% A 22% N 11% D (461)	65% A 20% N 15% D (448)
38. Hindus discovered evolution centuries before Darwin.	41% A 45% N 14% D (64)	52% A 36% N 12% D (109)	59% A 31% N 9% D (32)	50% A 42% N 8% D (135)	54% A 36% N 10% D (115)	51% A 39% N 11% D (455)	50% A 41% N 9% D (446)
39. The ancient Hindu concept of divine incarnations (avataras) directly anticipated Darwinian evolutionary theory.	38% A 40% N 22% D (65)	42% A 40% N 18% D (110)	41% A 44% N 16% D (32)	45% A 41% N 14% D (136)	40% A 40% N 21% D (114)	42% A 41% N 18% D (457)	37% A 40% N 23% D (444)
40. Human beings evolved from lower animal species by processes of random mutation and natural selection.	74% A 16% N 11% D (64)	66% A 16% N 17% D (110)	55% A 24% N 21% D (29)	52% A 26% N 23% D (136)	52% A 24% N 25% D (114)	59% A 21% N 20% D (453)	45% A 24% N 31% D (441)
41. Humans evolved from lower animal species by processes of spiritual evolution.	30% A 41% N 30% D (64)	45% A 30% N 25% D (111)	41% A 31% N 28% D (32)	45% A 30% N 26% D (137)	44% A 32% N 25% D (114)	42% A 32% N 26% D (458)	49% A 27% N 24% D (445)
42. Intelligent design is a better explanation for the origin and diversity of life than Darwinian evolution.	31% A 28% N 41% D (64)	30% A 33% N 37% D (109)	56% A 16% N 28% D (32)	38% A 38% N 24% D (138)	49% A 33% N 18% D (112)	39% A 33% N 28% D (455)	43% A 38% N 19% D (439)

43. Karma and rebirth are better explanations for the origin and diversity of life than Darwinian evolution.	42% A 27% N 30% D (66)	34% A 35% N 31% D (111)	56% A 19% N 25% D (32)	53% A 27% N 21% D (137)	58% A 25% N 17% D (111)	48% A 28% N 24% D (457)	59% A 24% N 17% D (440)
44. The evolution of species is preceded by an involution of spirit (Brahman/God).	48% A 26% N 26% D (65)	50% A 34% N 17% D (109)	53% A 34% N 13% D (32)	58% A 27% N 16% D (134)	63% A 22% N 15% D (113)	55% A 28% N 17% (453)	66% A 24% N 11% D (441)
45. Consciousness cannot be explained by Darwinian evolution.	68% A 15% N 17% D (65)	71% A 16% N 13% D (111)	75% A 6% N 19% D (32)	77% A 20% N 4% D (137)	77% A 20% N 4% D (114)	73% A 17% N 10% D (459)	77% A 18% N 5% D (441)
46. Consciousness is an emergent property of matter.	53% A 29% N 18% D (62)	37% A 36% N 27% D (110)	53% A 16% N 31% D (32)	43% A 29% N 28% D (134)	47% A 30% N 23% D (114)	45% A 30% N 25% D (452)	42% A 31% N 27% D (438)
47. Darwinian evolution is true, but very limited.	71% A 11% N 19% D (65)	55% A 24% N 21% D (111)	66% A 19% N 16% D (32)	64% A 21% N 15% D (136)	59% A 30% N 11% D (112)	62% A 22% N 16% D (456)	58% A 26% N 16% D (439)
48. Darwinian evolutionary theory is not scientific.	6% A 25% N 69% D (64)	10% A 22% N 68% D (110)	19% A 25% N 56% D (32)	15% A 36% N 49% D (134)	23% A 32% N 45% D (113)	15% A 29% N 56% D (453)	18% A 38% N 45% D (440)
49. Darwinian evolution insists that complex organisms are the products of mere blind chance.	37% A 19% N 45% D (65)	29% A 28% N 44% D (109)	29% A 36% N 36% D (31)	31% A 33% N 36% D (135)	31% A 41% N 28% D (112)	31% A 32% N 37% D (452)	25% A 41% N 34% D (437)
50. Darwinism is atheism.	16% A 35% N 49% D (63)	15% A 32% N 53% D (111)	26% A 26% N 48% D (31)	22% A 44% N 34% D (136)	22% A 42% N 36% D (112)	20% A 38% N 42% D (453)	23% A 35% N 42% D (438)

Table 14.18 Cross-tabulations between Hindu beliefs about evolution and religious affiliation

Strongly agree or agree [A] *Neither agree nor disagree [N]* *Strongly disagree or disagree [D]*	*ISKCON* *(47)*	*Yogananda* *(58)*	*Sai Baba* *(43)*	*Advaita* *(128)*	*All* *(1000)*
37. There is no conflict between Hinduism and evolution.	24% A 13% N 63% D (46)	75% A 19% N 5% D (57)	83% A 14% N 2% D (42)	75% A 18% N 7% D (126)	65% A 22% N 13% D (984)
38. Hindus discovered evolution centuries before Darwin.	36% A 39% N 25% D (44)	52% A 47% N 2% D (58)	62% A 36% N 2% D (42)	58% A 34% N 8% D (127)	50% A 40% N 10% D (975)
39. The ancient Hindu concept of divine incarnations (avataras) directly anticipated Darwinian evolutionary theory.	13% A 29% N 58% D (45)	33% A 55% N 12% D (58)	57% A 41% N 2% D (42)	44% A 43% N 13% D (127)	39% A 41% N 20% D (977)
40. Human beings evolved from lower animal species by processes of random mutation and natural selection.	4% A 7% N 89% D (46)	30% A 32% N 39% D (57)	60% A 21% N 19% D (42)	51% A 21% N 29% D (126)	51% A 24% N 25% D (970)
41. Humans evolved from lower animal species by processes of spiritual evolution.	66% A 9% N 26% D (47)	67% A 21% N 12% D (58)	61% A 15% N 24% D (41)	54% A 30% N 17% D (127)	45% A 30% N 25% D (979)
42. Intelligent design is a better explanation for the origin and diversity of life than Darwinian evolution.	87% A 7% N 7% D (45)	55% A 33% N 12% D (58)	42% A 44% N 15% D (41)	38% A 34% N 29% D (125)	40% A 36% N 24% D (972)

43. Karma and rebirth are better explanations for the origin and diversity of life than Darwinian evolution.	94% A 2% N 4% D (47)	72% A 21% N 7% D (58)	56% A 32% N 12% D (41)	50% A 29% N 21% D (125)	54% A 27% N 20% D (976)
44. The evolution of species is preceded by an involution of spirit (Brahman/God).	59% A 30% N 11% D (44)	68% A 28% N 4% D (57)	68% A 29% N 2% D (41)	69% A 24% N 6% D (124)	59% A 28% N 13% D (968)
45. Consciousness cannot be explained by Darwinian evolution.	96% A 2% N 2% D (47)	88% A 9% N 4% D (57)	68% A 20% N 12% D (41)	76% A 17% N 7% D (126)	75% A 18% N 7% D (975)
46. Consciousness is an emergent property of matter.	11% A 4% N 85% D (45)	43% A 33% N 24% D (58)	54% A 39% N 7% D (41)	49% A 21% N 30% D (124)	43% A 30% N 27% D (963)
47. Darwinian evolution is true, but very limited.	17% A 15% N 68% D (47)	70% A 25% N 5% D (57)	64% A 27% N 10% D (41)	63% A 26% N 11% D (125)	60% A 24% N 16% D (971)
48. Darwinian evolutionary theory is not scientific.	74% A 15% N 11% D (46)	14% A 38% N 48% D (58)	13% A 39% N 49% D (39)	10% A 34% N 56% D (124)	16% A 33% N 51% D (969)
49. Darwinian evolution insists that complex organisms are the products of mere blind chance.	42% A 20% N 38% D (45)	35% A 26% N 40% D (58)	15% A 49% N 36% D (39)	28% A 39% N 33% D (123)	27% A 37% N 36% D (964)
50. Darwinism is atheism.	57% A 19% N 23% D (47)	14% A 37% N 49% D (57)	13% A 48% N 40% D (40)	15% A 35% N 50% D (121)	20% A 37% N 42% D (967)

mischaracterization of Darwinian evolution. Given the *apparent* design in much of nature and the patent absurdity of purely chance processes accounting for the integrated complexity of organisms, the ISKCON view of Darwinian evolution as a purely random process helps to explain not only their opposition to Darwinism in general, but also their strong belief that Darwinism is not scientific. While the other groups range between only 10 and 14 percent in affirming the non-scientific nature of Darwinism, 74 percent of ISKCON members maintain this view (Table 14.18, question 48). Given these results, it is not surprising that only 24 percent of ISKCON members see no conflict between Hinduism and evolution, while solid majorities in the other groups see no conflict, ranging between 75 and 83 percent (Table 14.18, question 37).

Followers of Advaita generally have little difficulty in reconciling evolution with Hinduism, 75 percent seeing no conflict between the two. And as noted above, just over half of Advaitins (51 percent) accept evolution of humans by random mutation and natural selection, although 38 percent still see intelligent design as a better explanation for the origin and diversity of life than Darwinism, but still far below the 87 percent of ISKCON members (Table 14.18, question 42). Further, 50 percent of Advaitin respondents see karma and rebirth as a better explanation than Darwinism, once again well below ISKCON's 94 percent (Table 14.18, question 43). Sixty-nine percent of Advaitins believe that the evolution of species is preceded by an involution of the spirit—the hallmark of Vivekananda's Advaitic Evolutionism (Table 14.18, question 44).

The followers of Sai Baba regard Hinduism as thoroughly scientific, despite the role of alleged miracles in their guru's ministry. Typical of the group's outlook is the following statement by a follower in an essay posted on the Internet: "As a human being is developed through biological evolution—mutations and genetic recombination where behaviour [is] determined partly by genes, so the further development would proceed through spiritual evolution determined by Atmic (Cosmic) consciousness" (Anon. n.d.). Such a perspective accounts for the strong agreement (83 percent) with the idea that there is no conflict between Hinduism and evolution, as well as the relatively high acceptance (60 percent) of naturalistic mechanisms (random mutations and natural selection) for evolution. Karma and genetic change are seen simply as complementary processes.

Yogananda's group tends to see evolution primarily as a spiritual process. As Yogananda argues in his *Autobiography*, quoting his own master, "God created the human species by materializing the bodies of man and woman through the force of His will" (1955: 177). God had formerly manifested himself only in animal bodies bound by instinct. But for the sake of spiritual or "upward evolution," he transferred animal souls into the human bodies he had created, with reason predominating in the male body and feeling in the female body. Yogananda concludes in the words of his master:

> The human body was therefore not solely a result of evolution from beasts, but was produced by an act of special creation by God. The animal forms were too crude to express full divinity; the human being was uniquely given

> a tremendous mental capacity—the "thousand-petaled lotus" of the brain—as well as acutely awakened occult centers in the spine.
>
> (ibid.: 178)

The "thousand-petaled lotus" mentioned in the passage is the highest energy center at the crown of the head, representing the supreme transcendental consciousness in Kundalini Yoga. Yogananda and his master thus dismiss Darwinism, clearly rejecting a naturalistic explanation of human evolution—though the condescending qualification that human bodies were "not solely a result of evolution from beasts" perhaps allows for a minor role for naturalistic evolution. Human mental and spiritual capacities, in any case, are attributed to a divine bequest for the purposes of spiritual evolution. In this manner, followers of Yogananda can heartily endorse (at 75 percent) the idea of no conflict between Hinduism and evolution, while approaching ISKCON's general rejection of naturalistic mechanisms and endorsement of supernatural explanations for the evolution of life and humankind.

In summary, the strongly theistic respondents (ISKCON members) are far more likely to reject Darwinian evolution than the monistic idealists. Yet the Advaitins do not accept a robust Darwinism, radically subordinating it to spiritual evolutionism, or dismissing it as irrelevant to the spiritual quest for self-realization. The results of the survey thus generally confirm the trends discussed earlier in the book.

15 Conclusion

What role for Brahmā?

The wide spectrum of views we have covered in Parts II and III of this book constitutes an important aspect of the Hindu Renaissance. By the term *Hindu Renaissance*, following Agehananda Bharati, I mean the general movement among Hindu intellectual, political, and spiritual leaders beginning in the early nineteenth century to negotiate the complex interactions between traditional and emerging transitional ideologies in modern India. This Renaissance is a cultural-religious revival that has redefined religious practice but without challenging fundamental religious ideals and values (Bharati 1970: 268). Bharati characterizes the basic tenets of the Hindu Renaissance as follows:

> India has forgotten her marvelous past: this past contained not only material and cultural wealth, it also offered a complete solution of all problems of the individual and society. There is nothing—material, spiritual, or cultural—which ancient India has not brought forth.
>
> (ibid.: 276)

As we have seen, a marked feature of these tenets, especially as propounded by the major, Neo-Vedāntic wing of the Renaissance, is that the Vedic tradition, while inclusive of all religions, is nonetheless the best among the religions of the world. Among the reasons proffered is that the Vedic tradition is the most scientific.

Such a perspective has resulted in the scientizing of tradition and the spiritualizing of science. It helps to account for the frequent use of scientific metaphors and ideas in the Hindu apologetic, manifest in the phenomenon of what Bharati calls the "'scientific' sadhu" or holy man (ibid.: 277). Prominent among these "scientific" sadhus are men like Dayananda, Vivekananda, and Aurobindo who, often dressed in the venerable attire of monks and wandering saints, represent themselves as "expert in salvation-giving meditation," updating the ancient traditions, while generally eschewing traditional ritual practices (ibid.). In the last decades of the twentieth century, as a complement to the "scientific" sadhus, we see the emergence of Hindu scientific and technology professionals in such fields as medicine, engineering, and physics, who have joined the Hindu apologetic as sadhu-inspired and sadhu-citing scientists. Of note among these are thinkers like

Subhash Kak, Nrusingh Charan Panda, Gopala Rao, Varadaraja V. Raman, and Amit Goswami.

These scientists employ their expertise to confirm the teachings of the sadhus and, not infrequently, to deplore the naturalistic theory of Darwinian evolution. But as is often the case with Western anti-Darwinian scientists, the major Hindu scientists critiquing modern evolutionary theory are trained in fields outside evolutionary biology. Their expertise is marginal and hardly relevant. The predictable result has been a revival of the design argument: dressing out the latest teleological versions with the most recent discoveries of quantum mechanics, genetics, and information science. In the process, the classical Dharmic notions of evolution (*pariṇāma*) and manifestation (*vivarta*) are revalorized in the service of Modern Vedic Evolutionism and Creationism. Darwin and Dharma are increasingly discordant.

As we reach the end of this journey from the ancient Vedic seers to the "scientific" sadhus and sadhu-inspired scientists of the post-colonial period, we may well wonder if the Darwin–Dharma conflict is inevitable. Certainly, such conflict is much more likely in any world view wherein science and religion (or metaphysics) are regarded as continuous or inseparable. Regarding the various classical Dharmic traditions, the philosopher Jitendra Nath Mohanty summarizes: "We can say, without taking much risk, that for most Indian thinkers science and metaphysics were continuous" (1992: 222). He views Advaita as an exception, since from its perspective, "the world, being unreal … is left to the empirical sciences, and metaphysics—if that is what *parāvidyā* [higher knowledge] is—is the knowledge of the one Being underlying the many beings" (ibid.). But clearly modern Neo-Advaitins, whether they affirm the reality of this world or not, have re-enfolded science into their metaphysical embrace.

Mohanty's view of traditional Indian thinkers seems to apply to many Hindus today, according to the data presented in Chapter 14. We saw that a large majority of Hindus surveyed (64 percent) see Hinduism as the most scientific religion in the world, and the same percentage believe that the Vedas contain all knowledge concerning both spiritual and scientific truths. A somewhat greater majority (69 percent) accepts that superconscious or yogic states of mind can discover most or all truths about the physical world. And another substantial majority of respondents (65 percent) see no conflict between Hinduism and evolution (although the term *evolution* was left undefined so that this cannot be read as saying that 65 percent see no conflict between Hinduism and *Darwinian* evolution). But why is there this general sense of concordance between religion and science/evolution on the part of Hindus? One insight into this question is provided by looking into the models for the religion and science relationship promoted by Hindus.

Hindu models of the religion and science relationship

One possibility for the perceived concordance is that Hindus regard religion and science as belonging to two separate domains, or in Stephen Jay Gould's phrase, two non-overlapping magisteria. Incidentally in this regard, while judiciously acknowledging his ignorance, Gould suspected that "the standard attitude of all major Western religions," which he characterized as asserting that science and religion do not conflict, also generally prevails among "Eastern religions" (Gould 1997: 18). He went on to assert that the lack of conflict arises because the domains of religion and science are different. However, this does not seem to be the case for large numbers of Hindus, as Mohanty's views above already suggest.

Recall that for the Brahmo leader Keshab Chandra Sen, "All science is religion, and all religion is science" (1940: 356), a perspective elaborated upon by Vivekananda whose ideas have greatly impacted contemporary Hindu perceptions of science and religion. As Swami Ranganathananda Maharaj, a former president of the Ramakrishna Math and Mission, declared in a message to the magazine *Hinduism Today* in 1999, "The Sanatana Dharma is mainly the product of an extraordinary literature, the ten or twelve *Upanishads* … in which you find a scientific and experiential approach to religion" (Ranganathananda 1999: 52).

Nor is it just the religious leaders who espouse the unity of science and religion. One of the greatest of colonial-period Hindu scientists, Jagadish Chandra Bose, whose ideas on the all-pervasive nature of consciousness we have already referred to, proclaimed that "India is the only country where so far from there being a conflict between science and religion, knowledge is regarded as religion itself" (quoted in Anon. 1921: 29).[1] In confirmation of this assessment, he invoked the ancient seers, for whom, he says, "knowledge and religion are one" (J.C. Bose 1921: 39).

The claims regarding the unity and even identity of science and religion are based on two different sorts of assumptions. The first, already touched upon, is epistemological in nature: superconscious or yogic states of mind are valid sources of empirical knowledge. The second is the ontological idea that nature encompasses both spirit and matter, and thus there is really nothing supernatural in the universe. As Swami Ranganathananda elaborates upon his comment above, "The supernatural has no place in Vedanta or Sanatana Dharma, because the concept of nature is very wide, including the Self or subject of knowledge along with sensory objects of knowledge" (ibid.). The term *Self* for Ranganathananda, of course, does not refer to the empirical self of an individual, but to the universal Self or Ātman, identical with Brahman. The epistemological assumption, among other things, calls into question the exclusive right of "Western" or "materialist" scientists to set forth and expound empirical knowledge, while the ontological assumption justifies the epistemological. That is, materialist scientists neglect, or rather cannot even comprehend, the supersensory reality of Brahman, but spiritual scientists, by understanding Ātman/Brahman, understand the empirical world and its workings. The laboratories of modern science are not necessary; only meditation is.

The combining of the epistemological and ontological assumptions allows for the ready identification of scientific and experiential modes of verification. Such a move results in the absorption of modern science into the traditional Hindu notion of *vidyā*, sacred knowledge, frequently translated by Hindus today as "science" and encompassing both spiritual and natural sciences.[2] But as noted earlier, such a traditional view is in considerable tension with modern ideas of science, as the former views "knowledge" as eternal and already known by the ancient sages. This view of knowledge means that experience can only result in the rediscovery of ancient truths, in the meditational/experiential confirmation of knowledge already known. Such experiential science is thus not the open-ended empirical quest that characterizes modern science.[3]

In summary, those Hindus who accept the epistemological and ontological assumptions outlined above clearly reject the independence model of the science–religion relationship. Rather than the ideal of two non-overlapping magisteria, these Hindus promote a model of hierarchical integration, with material science either subordinate to and/or subsumed into spiritual science. But other Hindus do entertain the Gouldian model, a model favored, incidentally, by late-nineteenth-century Hindu scientists who "partition[ed] the world into an inner moral realm and an outer physical realm, each having its own methodological privileges" (Raina and Habib 1996: 32). As my survey reveals, there are substantial minorities of Hindus who reject the epistemological and ontological assumptions underlying the integrative and hierarchical models. And some of the respondents in their open-ended comments make this point decisively. For instance, a "quite spiritual/religious" Advaitin follower of Ramana Maharshi notes:

> The domains of spirituality and science are entirely different. Science can have nothing to say about meaning, sense of self, etc ... On the other hand, science is a pure and unbiased quest for objective Truth and any religious "scripture" that makes unproven declarations that run contrary to this should be abandoned.

A member of the Arya Samaj similarly observes: "Intelligent design is barking up the wrong tree. Science seeks to explain the external, the physical world. Religion seeks to explain the internal, the spiritual world. Don't confuse the two."

I may note in passing that, while such clear demarcation of the two realms is relatively easy with regards to inanimate physical entities and processes, the boundary gets considerably blurred with regard to the internal world of consciousness, a subject of intense scientific investigation with potentially major implications for traditional beliefs regarding an immaterial soul and notions of some sort of universal consciousness. In this connection, we may recall that, while 43 percent of Hindus in my survey agreed that consciousness is an emergent property of matter, 75 percent seemed to think that consciousness could not be explained by Darwinian evolution. Such views seem consonant with Aurobindo's teaching that the consciousness manifest in humans is simply a re-emergence of a more fundamental consciousness.

Another revealing comment regarding the science–religion relationship was offered by a respondent who did not identify his specific religious affiliation but who strongly agreed that the Vedas and modern science are in complete harmony with each other, while strongly disagreeing that the Vedas contain all knowledge concerning spiritual and scientific truths. According to this respondent, one has to understand the Vedānta views of God (Īśvara) and of Nature (Prakṛti) in order

> to adequately understand how deeply separated the concerns of Hinduism are from the practical considerations of science. Indeed, Stephen Jay Gould['s] principle of NOMA is best demonstrated in how many of the acharyas approach science and the practical world in general. The putative conflict between evolution and Hinduism, as well as the "modern" apologetic rationalizing of Hinduism prefiguring evolution, is largely a mass, ill-informed response …

The contemporary Indian philosopher Daya Krishna in like manner speaks of various human quests, including the religious and the scientific, with each having "a fundamental autonomy" of its own (1994: 13). Specifically, he argues that knowledge resulting from the scientific quest "has little to do with the existential-experiential concerns of man" (ibid.). He concludes:

> There are, of course, claims to the contrary on behalf of science just as there are those on behalf of religion, but basically these are illusions generated by successes achieved in any field. The converse claims of the vindication of religion by science or of science by religion are equally mistaken as they usually arise not only by ignoring the fact of plurality of science and religions and the provisional tentativeness of their contentions, but also because they ignore the radical disparateness of the quests which make them fundamentally incommensurabe [sic].
>
> (ibid.)

Another contemporary Indian philosopher, Sangeetha Menon, makes a similar point:

> A great danger in the golden age of dialogues based on spiritual meeting points is the hasty and immature appropriations of different domains of which science and religion are both sometimes guilty. When religion tries to claim that the current trends, theories, and findings of science existed already in ancient thought, forgetting the historicity and cultural specificity of religion, the way is unfortunately paved for primitive competitions rather than for the healthy pursuit of knowledge. Likewise, when science tries to dispossess religion of its central role in defining new meanings for human identity and spiritual well-being, it critically impinges on our ability to conceptualize the very meaning of existence and survival—that is, human imagination and the pursuit of something still beyond.
>
> (Menon 2006: 22)

What is unclear to me is how seriously the more ardent of the Neo-Hindus—who tend to see completeness and finality in their religious claims—will take notice of such ideas and heed their warnings. The problem becomes acute, for instance, when "scientific" claims based on mystical or yogic experience conflict with claims based on empirical science—that is, when the experiential and the empirical collide. In such collisions, it can no longer be just a case in which empirical science is merely incomplete but still consilient with experiential, spiritual, or higher dimensional science. As J.C. Bose warned, despite his insistence on the unity of science and religion, "A true inquirer has … to guard against being self-deceived; he has at every step, to compare his own thought with the external fact; he has remorselessly to abandon all in which these are not agreed" (quoted in Anon. 1921: 27).

Non-consilience is perhaps nowhere more apparent than in the controversy surrounding the creative capacity of "blind" Darwinian mechanisms to generate radically new and different forms of life. This issue highlights the problem of the God of the gaps, or in more contextualized Hindu terms, the problem of the role of the creator god Brahmā—the Brahmā of the gaps.

The problem of Brahmā of the gaps

Shortly after the publication of *On the Origin of Species*, Darwin's mentor, the geologist Charles Lyell, made the following entry in one of his scientific journals, for March of 1860:

> If we take the three attributes of the deity of the Hindoo Triad, the Creator, Brahma, the preserver or sustainer, Vishnu, & the destroyer, Siva, Natural Selection will be a combination of the last two but without the first, or the creative power, we cannot conceive the others having any function.
>
> (Lyell 1970: 369)

For Lyell, the naturalistic mechanism of natural selection readily accounts for the preservation of species, as well as for the annihilation of unfit variations—thus fitting in with his idea of the fixity of species.[4] But only a supernatural creative force could account for the creation of life; only such a divine force was endowed with the "variety-making" power responsible for speciation and the gradual appearance of sentience and mind.[5] Thus, while Viṣṇu and Śiva could be dispensed with as their preservative and destructive roles have a ready naturalistic replacement, Brahmā alone retained in Lyell's thinking—however metaphorically—a cosmic role: a supernatural deity must be responsible for the origin and development of life and its countless species.[6]

Lyell shared his reflections above with Darwin, for in April Darwin wrote him a letter questioning such theological speculations. Darwin writes:

> I must say one more word about our quasi-theological controversy about natural Selection … Do you consider that the successive variations in the size of the crop of the Pouter Pigeon, which man has accumulated to please his caprice, have been due to "the creative & sustaining powers of Brahma."
>
> (Darwin 2008: 9)

Darwin goes on to protest that he finds it hard to believe that "a maker of the Universe sh.[d] care about the crop of a Pigeon solely to please men's silly fancies" (ibid.). He concludes:

> But if you agree with me in thinking such an interposition of the Deity uncalled for, I can see no reason whatever for believing in such interposition in the case of natural beings, in which strange & admirable peculiarities have been naturally selected for the creature's own benefit.
>
> (ibid.)

Darwin's response emphasized the theological problem of God's seemingly fanatical attention to the fanciful details of variation found both in artificial and natural selection, a non-scientific response appropriate enough given the theological nature of Lyell's challenge. In other venues, Darwin provided responses of a more scientific nature in defending the creative role of natural selection. Of course his defense was quite incomplete and in some ways mistaken, as he had no idea as to the chemical and molecular mechanisms underlying inheritance and Lyell's variety-making power, and thus no way of explaining in any detail the interaction between inheritance and natural selection. He simply stressed the importance of the guiding role of natural selection relative to that of chance variation (Beatty 2010: 22–4). In any case, two months later Lyell accused Darwin of deifying natural selection, giving it creative powers beyond its actual capacities. In leveling this charge, Lyell was making "what has since become the familiar case [of anti-Darwinian creationists] that natural selection can only preserve or eliminate, it cannot create" (ibid.: 23). We may recall that this is not an uncommon objection among Hindu anti-Darwinians today, including theistic creationists like Michael Cremo and Advaitic evolutionists like Amit Goswami. In the former case, of course, Brahmā gives way to a Kṛṣṇa of the gaps; in the latter, Brahmā becomes the Brahman of the gaps.

The persistence of Brahmā's (or Kṛṣṇa's or Brahman's) role in speciation among Neo-Hindu proponents of Vedic Creationism and Evolutionism (or Devolutionism in Cremo's terminology) symbolizes their antagonism to Darwinism. In certain ways this persistence is quite understandable. One major key to explaining the natural mechanisms underlying innovation in animal forms is the close relationship between embryological development and evolutionary change (Evo-Devo). As Sean B. Carroll points out, Darwin (and Thomas Huxley) understood and appreciated the "intimate connection between development and evolution" (2005: 5). But as Carroll further indicates, for more than a century after Darwin's and Huxley's major writings, "virtually no progress was made in understanding the mysteries of development" (ibid.: 6). Thus, embryology "played no part in the so-called Modern Synthesis of evolutionary thought that emerged in the 1930s and 1940s" (ibid.).

It seems that many of the anti-Darwinian arguments currently made by Hindu creationists and Vedāntic evolutionists harken back to the state of knowledge that prevailed in the heyday of the Modern Synthesis, when the claim that biologists cannot explain this or that detail of development was largely true. But it seems disingenuous for an Amit Goswami today simply to dismiss the modern discoveries of Evo-Devo by likening them to a Ptolemaic epicycle that attempts to fix a fundamentally flawed model of the solar system. It is a claim, a preference, but not a reasoned argument. The nature of the scientific enterprise is to update continuously its data and theories. In the case of Ptolemy, the discoveries of Kepler and Galileo seriously eroded Ptolemaic geocentrism. The discoveries of Evo-Devo and other emerging fields in the Extended Synthesis of evolutionary thought have to date only further confirmed many of the most basic insights of Darwin. At the very least, it would seem that the evolutionary details discovered in these emerging fields must be grappled with on a more serious level than simply dismissing them as unsatisfying. Evo-Devo may not satisfy everyone emotionally, but then one inevitably suspects this dissatisfaction stems from the fact that the new discoveries do not fit readily into an *a priori* metaphysical framework. But if the Hinduism and science or Hinduism and evolution discourse is to remain compelling, it seems that serious and thoughtful consideration must be given to these discoveries. Otherwise, for those not already "a true believer" in Vedic metaphysics, the Brahmā/Brahman of the gaps will seem increasingly irrelevant.

Halbfass has remarked, "For the Hindu tradition, the encounter with modern science and technology was the most momentous part of its encounter with the West" (1988: 399). Reviled or revered, modern science has inevitably been engaged. Halbfass also points out that, for Neo-Hindus, Dharma is universal, universalizable, historically flexible, and open "to the discoveries of science and contemporary conditions in the modern world" (ibid.: 338). But as he further notes, it is rare for Neo-Hindu authors to attempt "to fully secularize the *dharma* concept" (ibid.). It thus seems unlikely that Neo-Hindus will accept the independence of science from religion, nor be willing to engage in serious dialogue either with a robust Darwinism or, within its own tradition, to engage the Cārvāka—the ancient Indian materialist view so reviled and repeatedly denounced. Yet the Cārvāka view does not go away: as Madhava said, it is a hard theory to drive out.[7] In Madhava's view, it was hard to expunge because it rationalized the people's pursuit of sensual pleasures. But today, it may be harder to drive out because it has the unanticipated support of Darwinian evolutionary theory.

Appendix A
The substantive survey questions regarding attitudes towards evolution and related issues

The substantive questions below followed the twenty-three demographic questions. Each of the substantive questions was preceded by the instruction, "Do you agree or disagree with the following?"

24 Animals have the same moral rights that human beings do.
25 Nature is predominantly a fierce struggle for survival of the fittest.
26 Both human beings and animals possess some sort of non-material, individual soul that survives the death of the body.
27 Karma, like the law of gravity, is a fundamental scientific law of nature.
28 The Vedas contain all knowledge concerning both spiritual and scientific truths.
29 The Vedas and modern science are in complete harmony with each other.
30 Hinduism is the most scientific religion in the world.
31 Concerning factual knowledge about the physical universe, the statements of spiritually enlightened persons are more trustworthy than those of modern scientists.
32 Concerning factual knowledge about the physical universe, the statements of scripture are more trustworthy than those of modern science.
33 Superconscious or yogic states of mind are capable of discovering most or all truths about the physical world.
34 Human beings have existed on Earth for hundreds of millions of years or longer.
35 The birthplace or cradle of humanity was ancient Africa.
36 India was the fountainhead of all sciences and civilization.
37 There is no conflict between Hinduism and evolution.
38 Hindus discovered evolution centuries before Darwin.
39 The ancient Hindu concept of divine incarnations (avataras) directly anticipated Darwinian evolutionary theory.
40 Human beings evolved from lower animal species by processes of random mutation and natural selection.
41 Humans evolved from lower animal species by processes of spiritual evolution.
42 Intelligent design is a better explanation for the origin and diversity of life than Darwinian evolution.

43 Karma and rebirth are better explanations for the origin and diversity of life than Darwinian evolution.
44 The evolution of species is preceded by an involution of spirit (Brahman/God).
45 Consciousness cannot be explained by Darwinian evolution.
46 Consciousness is an emergent property of matter.
47 Darwinian evolution is true, but very limited.
48 Darwinian evolutionary theory is not scientific.
49 Darwinian evolution insists that complex organisms are the products of mere blind chance.
50 Darwinism is atheism.

Appendix B

Selected demographic data for respondents from completed surveys

Note: percentages may not add up to 100 due to rounding; the numbers of respondents given in parentheses may not add up to the total of 1000 respondents who completed the survey, due to those who skipped a particular question.

Table 1 Sex

Male	*Female*
71% (703)	29% (292)

Table 2 Age

18–22 yrs	*23–29 yrs*	*30–45 yrs*	*46–60 yrs*	*61 yrs +*
17% (168)	27% (273)	33% (325)	18% (181)	5% (51)

Table 3 Ethnic background

Indian & Hindu	*Indian, Non-Hindu*	*Hindu, Non-Indian*	*Non-Indian, Non-Hindu*
75% (745)	0% (1)	19% (194)	6% (60)

Table 4 Born in

India	*U.S.*	*Europe*	*U.K.*	*Africa*	*Cen./So. Am.*	*Carib.*	*Canada*	*Austral.*	*Mid. East*
66% (564)	17% (141)	4% (35)	3% (25)	3% (24)	3% (22)	2% (16)	1% (11)	1% (9)	0% (3)

There were others not counted in the percentages who were born primarily in Nepal, Burma, and Southeast Asia.

Table 5 Currently living in

India	*U.S.*	*Europe*	*U.K.*	*Africa*	*Cen./So. Am.*	*Carib.*	*Canada*	*Austral.*	*Mid. East*
32% (285)	43% (382)	5% (46)	5% (44)	2% (16)	2% (16)	1% (7)	5% (45)	4% (31)	1% (9)

There were others living mainly in parts of South and Southeast Asia.

Table 6 English as mother tongue

Yes	*No*
21% (214)	79% (784)

Of those respondents whose mother tongue was not English, 75 percent indicated an Indian language.

Table 7 Convert to Hindu faith

Yes	*No*	*Not applicable, not sure*
18% (177)	78% (771)	5% (45)

Those who responded "Not applicable, not sure" also indicated that they were of Hindu ethnic background and that they still followed one or more Hindu teachers or schools; thus they were included as Hindus.

Table 8 Highest year of school or degree completed

< H.S.	*H.S.*	*Assoc.*	*Bachelor's*	*Master's*	*Ph.D.*	*M.D.*	*LL.D.*
0% (3)	12% (120)	5% (48)	34% (327)	34% (324)	9% (83)	6% (59)	0% (3)

Table 9 Familiarity with and knowledge of modern science

Thorough and extensive	*Thorough in some areas, not in others*	*Somewhat familiar and knowledgeable*	*Only vaguely familiar or knowledgeable*	*Not familiar or knowledgeable at all*
25% (244)	40% (395)	32% (314)	4% (35)	1% (7)

Table 10 Past, present, or future career as a scientist

Yes	*No*	*Not sure*
49% (488)	46% (455)	5% (53)

Table 11 Of those answering yes for a career in science, specific fields named

Biological sciences (excluding health care and medicine)	*Health care and medicine*	*Physical sciences (astronomy, chemistry, geology, physics)*	*Psychology*	*Sociology and anthropology*	*Engineering*	*Computer science*	*Mathematics*
12% (66)	21% (112)	6% (33)	6% (30)	5% (26)	26% (138)	22% (117)	3% (14)

There were others not counted in the percentages who mentioned educational science, library science, political science, management science, environmental science, astrology, parapsychology, metaphysical sciences, and yoga.

Table 12 Religious affiliation

Advaita	*Self-Realization Fellowship (followers of Yogananda)*	*ISKCON (followers of Prabhupada)*	*Sai Baba*
128	58	47	43

Religious affiliation was determined from five questions asking whether or not a respondent: (1) follows a particular saint or spiritual teacher, (2) worships a specific deity, (3) follows a general school of thought such as Advaita, Vedānta, or Dvaita, (4) identifies with a particular social class, community, or group, and (5) attends or affiliates with a particular temple or religious institution. Respondents who answered yes to any of the questions were asked to specify the deity, school, and so forth. By analyzing the results, I identified four groups with at least forty members. In some cases, these involve the name of a school or movement; in others, the name of a spiritual teacher. Due to possible overlap and other factors, only numbers, not percentages, are given. Needless to say, there were many other groups with much smaller numbers.

Table 13 Religiosity/spirituality

Quite religious/ spiritual	*Somewhat religious/ spiritual*	*Neutral or indifferent*	*Somewhat non-religious/ spiritual*	*Quite non-religious/ spiritual*	*Not applicable or not sure*
56% (553)	32% (319)	7% (71)	1% (8)	3% (29)	1% (13)

Table 14 Frequency of religious/spiritual practices

Frequency	*Prayer*	*Attendance at religious services*	*Recitation of scripture*	*Meditation or yoga*	*Fasting or ritual vows/practices*
Once a week or more	78% (769)	30% (295)	54% (535)	53% (518)	27% (264)
A few times a year or more	13% (128)	54% (529)	26% (259)	21% (202)	41% (401)
Seldom/ never	10% (95)	17% (163)	20% (193)	27% (264)	33% (326)

Notes

1 Introduction

1 Ironically, as Wilhelm Halbfass points out, the original use of *dharma* as an analogue for *religion* derives from the Baptist missionaries in Bengal at the beginning of the nineteenth century, who referred to their own religion as *satya-dharma*, the "true dharma" (1988: 340). For discussion of the traditional meanings of *dharma* and its modern reinterpretations, see Halbfass (ibid.: 310–48). Despite the innovations, Halbfass concludes: "Even in its modern reinterpretation and universalization, the concept of *dharma* has remained the expression of the Hindu sense of identity and continuity … no radical break with tradition has taken place" (ibid.: 347–8). Specifically, Halbfass notes that the notion of *dharma* in Hindu self-understanding excludes the "historicism, relativism, and radical secularization" that form so prominent a part of Western cultural experience.

2 For discussion of the evolutionary naturalism of the Materialists, see pp. 21–5 below.

3 I have supplied in brackets the Hindi words from the Hindi text (Dayananda 2002: 150).

4 On the aesthetic version of the design argument, see Michael Ruse, *The Darwinian Revolution* (1999: 72). Ruse also notes briefly Augustine's appeal to beauty in *Darwin and Design* (2003: 21).

5 For discussion of this passage, see pp. 121–2 below.

6 See Śaṇkara, *BSB* 1.4.23, and my discussion of Śaṇkara on p. 27 below.

7 Cf. Ruse (2003: 43–4).

8 This term, of course, comes from Daniel C. Dennett's book *Darwin's Dangerous Idea: Evolution and the Meanings of Life*.

9 I have discussed some of these assumptions in an article for *Zygon* (Brown 2007a: 425). When I wrote that article, I was more inclined to think that the Hindu tradition as a whole was more favorably disposed in general to Darwinism than the Abrahamic traditions. I have changed my mind since then as a result of my inquiry into traditional Hindu design arguments, and prompted by critiques of that article by Meera Nanda (personal conversations).

10 Cf. Dermot H. Killingley (1990: 153), who highlights the unique spiritual position of humankind in the Hindu perspective. Cf. also Nanda (2003: 86). For an ancient version of the Hindu Great Chain of Being, see *Bṛhadārāṇyaka Upaniṣad* 1.4.

11 See Singh (1971: pt. I, p. 132) and Dayananda (1958: 15–16; 2003: 262–3, 267).

12 Cf. Killingley (1995: 186–8).

13 For example, Lawrence A. Babb describes the contemporary Radhasoami cosmology in the following words: "The Supreme Being is here no watchmaker, nor is the cosmos a watch. The cosmos originated in, and is sustained by, the surging, fluid, wavelike current of the Supreme Being's mauj, a flowing forth of divine substance" (1986: 41).

14 Lynette E.L. Thistlethwayte attributes this passage directly to Dayananda. Her source for the quotation, however, is Kenneth W. Jones (1976: 45), where the passage is apparently attributed simply to one of "those [early Arya reformers] constructing Arya ideology." In any case, the quotation accurately reflects the views of Dayananda himself. Jones indicates that the quotation is from an article, entitled "The Theory of Evolution from an Aryan Point of View," appearing in the *Arya* (March 1882).

15 While there have been various Buddhist commentators on Darwinism, it seems not to have been as major a concern as in other religious traditions. According to Piet Hut, Buddhists have been much less concerned with Darwinism than Christians (2003: 410). The same seems to be true vis-à-vis Hindus. At the same time, a panel at the 2010 national meeting of the American Academy of Religion was devoted to "The Impact of Biological Theories of Evolution on East Asian Buddhism Around the Early Twentieth Century," suggesting there are resources on the Buddhism and evolution discourse not yet well examined. And recently, the Fourteenth Dalai Lama has commented, if rather briefly, on Darwinism in his *The Universe in a Single Atom*, devoting a chapter to the topic of "Evolution, Karma, and the World of Sentience" (2005: 95–115).

2 Creationism and evolutionism in the ancient and classical traditions

1 Although the verbal form used here, from *vivṛt*, may suggest the Advaitic notion of manifestation (*vivarta*), thereby implying the work of creation is merely an apparent evolution, in this early text the meaning is rather that the creation consisting of the five elements begins to turn round or move about, impelled by God. Cf. *Śvetāśvatara Upaniṣad* 6.6.

2 This interpretation is apparently based on those manuscripts or editions that have *kale* ("at a certain time"), rather than *kala* (time in the nominative case), and which supply an understood "Lord" as the subject in place of time.

3 There has been considerable argument as to whether Sāṃkhya is truly atheistic. Certainly theistic forms of Sāṃkhya-like teachings are found both early and late. But the earliest source belonging to the school proper, the *Sāṃkhya-Kārikā* of Īśvarakṛṣṇa (ca. fifth century CE), gives little indication of a theistic bent. That is, the Vedic gods, while not denied, are simply irrelevant, and in any case belong to the side of Prakṛti, not Puruṣa. For an extended discussion of this problem, see Gerald James Larson 1969: 132–9. In any case, Śaṅkara's argument is clearly directed against an atheistic form of the Sāṃkhya, while he was aware of more theistic forms (see, for example, *BSB* 2.2.37–8).

4 Cf. Ninian Smart (1964: 81).

5 Immediately preceding this account, Bhṛgu provides a different cosmogonic scheme, in which Cosmic Mind (Mānasa), by mental will (*manas*), created all things, beginning with water, apparently substituting Cosmic Mind for the more ancient original source, space (*Mahābhārata* 12.176.2–4).

6 See p. 166 below.

7 The first fundamental dogma is the acceptance of matter as the sole reality.

8 The relation of accidentalism and evolutionary naturalism to the Cārvāka is disputed. Kiwal Krishan Mittal rejects the interpretations of scholars like Mysore Hiriyanna who regard Cārvāka as a combination of *svabhāva-vāda* and *yadṛcchā-vāda*. Hiriyanna claims that "Indian Materialism as set forth in works like the *Sarva-darśana-Saṁgraha* is a curious mixture of *Svabhāva-* and *Yadṛcchā-vādas*" (1957: 73, n. 2; see also Hiriyanna 1932: 104, 187). Cf. Gopinath Kaviraj, who, unlike both Mittal and Hiriyanna, argues that *svabhāva* and *yadṛcchā* are nearly synonymous (1990: 442–8).

9 A similar account of the materialist view of the emergence of consciousness is given by Śaṅkara in *BSB* 3.3.53. The analogy of fermenting liquids to illustrate the emergence

of consciousness in the body may be quite ancient. According to one summary of Ajita Keśakambalī's teachings, this contemporary of the Buddha argued: "Just as, certain ingredients coming together produces [sic] the power of intoxication, so the elements came together and produced what is life" (Shastri 1982: 30). K.M. Ganguli's translation of *Mahābhārata* 12.218.29 (vol. 2, p. 171), which incorporates explanatory material from a traditional commentator, Nīlakaṇṭha, in describing the philosophy of skeptics, refers not only to the analogy of fermenting liquids, but also to seeds producing mature plants with flowers and fruit, to cows producing milk and butter from grass and water, to two pieces of wood when rubbed together producing fire, and so on. In all these cases, entities emerge which are different from the producing causes, just as consciousness emerges from material bodies.

10 I use the term *Neo-Darwinian* in its generally accepted sense today, referring to contemporary evolutionary theory, not to its more narrow application, originating with George Romanes, to describe August Weismann's theory of natural selection devoid of any Lamarckian ideas.

3 Manifestation and apparent design in the Advaita Vedānta of Śaṅkara

1 For discussion of the post-Śaṅkara use of the term *vivarta* to distinguish apparent manifestation from real transformation (*pariṇāma*), see Richard King (1999: 220). While Śaṅkara may not have used *vivarta* in this technical sense, his general cosmological model is clearly one of apparent manifestation.

2 Cf. *BSB* 2.1.4–11 and Francis X. Clooney (1993: 103–6).

3 In his introductory comment to *Brahma-Sūtras* 2.1.1, Śaṅkara does not actually refer to the spider analogy, but likens the retraction of the emitted world back into Brahman at the end of a cosmic cycle to the reabsorption of all beings into the earth—certainly evoking the spider image.

4 For useful summaries of Śaṅkara's arguments in his commentary on the *Brahma-Sūtras*, see Karl H. Potter (1981: 119–80) and George Thibaut (1890: xxxii–lxxxv).

5 Śaṅkara in his explanation cites the scriptural verse that Bādarāyaṇa alludes to: "That from which all these beings are born, by which they live once born, and into which they enter on dying, … That is Brahman" (*Taittirīya Upaniṣad* 3.1.1).

6 In *BSB* 1.1.1, Śaṅkara rejects a number of similar views, but with regard to the origin of consciousness (*caitanya*). Specifically, he refers to the common view, promulgated by the Lokāyatikas, that the Self (Ātman) or consciousness is merely a special quality of the body alone (*deha-mātra*).

7 As Clooney points out, however, these rational refutations are not meant to shift "the analysis to a forum where reason would be the primary standard," but rather just to show the implausibility, on rational grounds, of opposing views (1993: 107). Needless to say, the various opponents, including the Sāṃkhyas, are not presented in "a 'real-life' debate," as the whole is orchestrated to produce a clear victory for the Advaitin side.

8 The idea that the effect pre-exists in its cause is usually referred to as *satkārya*. This latter term appears, for instance, in *Sāṃkhya-Kārikā* 9.

9 For discussion of Śaṅkara's rational refutations of the Sāṃkhya, see Clooney (1993: 110–13) and Chanddradhar Sharma (1960: 254–6).

10 The three qualities refer to the three *guṇas* of Primordial Matter, *sattva* (goodness, lightness, pleasure), *rajas* (activity, passion, pain), and *tamas* (darkness, inertia, delusion).

11 Regarding the possible innate purposeful creativity of Brahman, see Anantanand Rambachan (2006: 91–6).

12 See p. 15.

13 Chakravarthi Ram-Prasad sees the apparent contamination of God by the doctrine of the non-difference between individual self and God as one of the major critiques of the Advaita by other schools (2001: 215). Śaṅkara's resolution, he points out, is ultimately to reduce God to "an expression of the linguistic imagination." Cf. K. Satchidananda Murty (1959: 3–5).

14 The cosmological and teleological arguments frequently form an easy complement. In the West, we find such a juxtaposition as early as Plato (*Laws*, book X). In a theistic context, Robert Boyle affirms "that the origin of motion in matter is from God; and not only so, but that thinking it very unfit to be believed that matter, barely put into motion and left to itself, should casually constitute this beautiful and orderly world, I think also further that the wise Author of things did, by establishing the laws of motion among bodies and by guiding the first motions of the small parts of matter, bring them to convene after the manner requisite to compose the world" (1979: 19).

15 See note 10 above.

16 Cf. Īśvarakṛṣṇa's *Sāṃkhya-Kārikā* (1969: vv. 56–7): "This evolution initiated by Prakṛti, from the intellect (*mahat*) to the specific gross elements, is for the sake of another, the liberation of each soul, as if for its own sake. Just as the flowing of insentient milk is for the sake of nourishing the calf, so the activity of Pradhāna is for the sake of liberating souls."

17 The problem of the unseen karmic force as a creative power will be addressed in the next chapter.

4 Theism and atomistic design in the Nyāya-Vaiśeṣika of Udayana

1 The Nyāya and Vaiśeṣika share similar realist and pluralist outlooks (Hiriyanna 1932: 225–6). Thus, the two schools "from the first considered themselves as mutually supportive, Nyāya specializing in epistemology and methodology, Vaiśeṣika in metaphysics" (Potter 1977: 12).

2 Karl H. Potter (1977: 100) notes that Kaṇāda does not even mention God, while the reference to God in the *Nyāya-Sūtras* (1913: 4.1.19) likely belongs to an objector's argument.

3 Kewal Krishan Mittal similarly notes that these schools had "no 'craze for liberation' as the starting point" that characterizes the other orthodox philosophical schools (1974: 167). See also Dale Riepe (1961: 246) and Debiprasad Chattopadhyaya (1964: 163–5).

4 See Potter (1977: 36, 95–6) and Surendranath Dasgupta (1961: 363).

5 Cf. Gaṅganātha Jhā's English translation of Uddyotakara (1984: 1466).

6 Cf. ibid.: 1463.

7 For brief discussions of the problem of motion in Kaṇāda, see Gopikamohan Bhattacharyya (1961: 162) and Arthur Berriedale Keith (1921: 221–4).

8 Cf. Uddyotakara (1984: 1473).

9 See Bhattacharyya (1961: 62–4, 142–3) and Ninian Smart (1964: 153–4).

10 Potter sees three general varieties of arguments in the *Kusumāñjali*: cosmoteleological arguments, arguments from the existence of language and thought, and negative ontological arguments (1977: 101–2). The last are not particularly pertinent for our purposes.

11 Cf. N.S. Dravid's commentary in Udayana (1996: 378).

12 See Bhattacharyya (1961: 51–2, 56–7).

13 My translation from the Sanskrit text in George Chemparathy (1972: 88). Bimal Krishna Matilal (1977: 594) provides a useful summary of the *Kiraṇāvalī*'s arguments for the existence of God.

14 Chemparathy (1972: 184) sees the maintenance and destruction arguments for the existence of God as original with Udayana.

15 Mādhava (1981: 240–3) discusses the problem of a disembodied creator with specific reference to Udayana's *Kusumāñjali*.

16 For a brief summary of Udayana's arguments in his *Ātmatattvaviveka* for God's disembodied agency, see V. Varadachari (1977: 555).

17 Cf. the summary of these arguments in Potter and Sibajiban Bhattacharya (1977: 572–4).

18 Cf. Śaṅkara's denial (*BSB* 2.1.1) of the ability of sages like Kapila to have supersensuous perception of transcendental realities without recourse to scripture.

19 Udayana, incidentally, largely ignored the issue of multiple creators, relying on the views of earlier theistic Atomists like Praśastapāda and Uddyotakara. Regarding the major Nyāya-Vaiśeṣika arguments for a single creator, see Aruna Goel (1984: 159–60), and K. Sadchidananda Murty (1959: 204).

20 Goel seems unfamiliar with the epistemological and methodological underpinnings of modern science, but at least candidly admits that her training is in Sanskrit and Indian philosophy, with little expertise in the basic sciences (1984: xviii).

21 Cf. Halbfass (1988: 271–2, 362).

5 Design discounted in the Viśiṣṭādvaita Vedānta of Rāmānuja

1 See S.S. Antarkar (1999: 144) and John Braisted Carman (1974: 262).

2 For a clear and succinct summary of the various Vedāntic perspectives, see Eric Lott (1980: 191–3).

3 Viśiṣṭādvaita is often translated as "qualified non-dualism," but this is somewhat inaccurate. Most literally, the term translates as "non-dualism of what is differentiated" (cf. Van Buitenen 1968: 1n).

4 For an extended discussion of Īśvara's attributes in Rāmānuja, see Carman (1974: 65–113).

5 On the Self–body model of Brahman's relation to the world, see Anne Hunt Overzee 1992: 63–83.

6 For discussion of the efficient and material causality in Rāmānuja's doctrine of God, see Carman 1974: 114–17.

7 See, for example, *ŚB* 1.1.13 (Thibaut 1966: 226).

8 For Śaṅkara's parallel argument against the Sāṃkhya, see p. 30 above.

9 Rāmānuja (n.d.) makes the same arguments in his *Gītā Bhāṣya* 7.12–14 (regarding play) and 9.9 (regarding the role of karma in creating inequality). In the latter instance, Rāmānuja cites the *Brahma-Sūtras* (2.1.34–5). For a nuanced discussion of Rāmānuja's conception of divine sport, see Carman (1974: 117–22).

10 Ninian Smart (1964: 154–5) has a brief summary of Rāmānuja's objections to the design argument, along with possible Rational-Atomist rejoinders. See also Surendranath Dasgupta (1963: 189–92). For a comparison of the Western tradition of natural theology and Rāmānuja's rejection of its Hindu equivalent, including that of the Nyāya-Vaiśeṣika, see Carman (1974: 258–64).

11 Śaṅkara in commenting on this aphorism also claims that scripture rather than inference is the source of our knowledge about Brahman, but then makes no further comment at this point regarding the design argument. Madhva, however, in his comment on the aphorism in his *Anu-Vyākyāna*, draws attention like Rāmānuja to the inadequacy of unaided reason to provide any knowledge about God (B.N.K. Sharma 1986: vol. 1, p.83).

12 A similar line of argument was put forward by the Śaivites (Smart 1964: 156).

13 See, for instance, *ŚB* 2.1.32–3 (Thibaut 1966: 476–7).

14 Cf. Overzee (1992: 67).

6 Design and Darwin in the colonial period

1 Accounts of Darwin's theory appeared in Kolkata as early as 1860, when an article by the geologist Joseph G. Medlicott, entitled "Darwin on the Origin of Species," appeared in the *Calcutta Review* (Thistlethwayte 1998: 82, 89n). Interestingly, Darwin himself wrote to Medlicott praising the article as the best essay on the *Origin* yet to appear (Roday and Purohit 2010: 1137). From early on, then, English-educated Hindus in Bengal had access to reasonably accurate reports of Darwin's book.

2 For an insightful discussion of Sircar's views on evolution and design or purpose in nature, see Dhruv Raina and S. Isfan Habib (1996).

3 For early British reticence to introducing theoretical scientific education to India and reluctance to encourage indigenous scientific enterprise up to 1857, see Satpal Sangwan (1991a: 146–52).

4 For discussion of this passage and a general overview of the obstacles Sircar faced in establishing the IACS, see Chittabrata Palit (1991).

5 Pratik Chakrabarti (2004: 182–212) gives an excellent summary of Bose's scientific work on electric waves and plant sensitivity, and his integration of his discoveries with his Vedic monism. See also Ashis Nandy (1995: 17–87) and Subrata Dasgupta (1999: 105–44).

6 P. Chakrabarti provides an insightful analysis of Sircar's religious, scientific, and nationalist ideals that underlay the mission of the IACS (2004: 150–79).

7 For a succinct overview of the transmission of modern science into colonial India and of Indian responses, see ibid.: 10–15.

8 Regarding Brahminical views of modern experimental science, see Kapil Raj (1991).

9 See p. 87 below.

10 See, for instance, the comments of the Indian Nationalist Lajpat Rai (1916: 114–18).

11 For discussion of Mill's "scale of civilization," see Javed Majeed (1992: 135–6).

12 For a critique of Mill's denial of Indian scientific originality, see Amartya Sen (2005: 78–9).

13 See Zaheer Baber (1998).

14 George Chemparathy notes, in partial defense of Cowell, that Cowell had only the verse portions of the *Kusumāñjali* before him and thus "was not in a position to make an adequate study of Udayana or sufficiently appreciate his contribution" (1972: 44).

15 Regarding the dark view, Jesuit missionaries beginning with Francis Xavier in the sixteenth century complained about the heat of the tropics with its excessive fecundity, which they saw as inducing a fertile but immoderate imagination in the inhabitants, manifesting in the "sin of idolatry" (Županov 2005: 8). The Jesuit Jacome Fenicio at the beginning of the seventeenth century emphasized the debauchery manifest in the Hindu cosmogonic tales, noting especially the sexual relations of Śiva and Śakti (ibid.: 178–80). For Fenicio, as for Mill, the fundamental theological and cosmological error of the Hindus was seeing copulation as the means for creating animate beings. Both also feared that unconstrained flights of imagination and the glorification of Eros would undermine the supremacy of reason (cf. Majeed 1992: 183, 185).

16 For early Baptist missionary attitudes regarding Hindus and their inability to advance towards higher civilization, see E. Daniel Potts (1967: 1–3).

17 Mill came upon Ward's translation in preparing the final draft of his *History*, but dismissed it as likely due to recent, foreign influence (Mill 1820: 330n).

18 For similar assessments made by Protestant missionaries, see Wilhelm Halbfass (1988: 46–8).

19 As Rai observes, "The Theosophical Society began to praise and justify every Hindu institution and to find science in every custom" (1916: 117).

20 Gosling notes a similar absence of controversy regarding Darwinism in the science-oriented editorials of the popular and conservative daily *Sambad Prabhakar*, from 1860 until 1900, as well as in other Indian sources dealing with science in general and

occasionally touching upon Darwinism. Raina and Habib point out that, at the turn of the century in India, the theory of biological evolution, unlike that of social evolution with its political and cultural implications, "was hardly considered to be problematic" (1996: 31).

21 For discussion of Bankim's essay, his invoking of science in defense of the Hindu trinity, and the general role of Positivism in late nineteenth-century Bengal, see Gyan Prakash (1999: 57–63).

22 From an 1896 lecture, "Philosophy of Indian Theism." For a summary of theology in general with references to this lecture, see James Kellock (1926: 165–80).

23 Ramana Maharshi (1968a: 75).

24 Ramana Maharshi (1968b: 218).

7 The Vedāntic Deism of Rammohan Roy

1 The title of this work has been variously translated: "A Present to the Believers in One God," "Gift to Theists," "Gift to Monotheists," and "[A] Gift to [the] Deists." This last, in various versions, is proffered, for instance, by Hiren Mukerjee (1975: 16), Radharaman Chakraborti (1979: 23), and Willhelm Halbfass (1988: 202). Given the thoroughly deistic character of this text, which portrays God as little more than creator of the universe and warrantor of morality, I have followed the latter.

2 The Mughal administration of Bengal only began to give way to the British around the time of Rammohan's birth, and Persian continued to be the official language of government until English was adopted in the late 1830s.

3 See also Tapan Raychaudhuri (1999: 50–1).

4 Mukerjee sees the *Gift to Deists* as showing that Rammohan derived his "new rationalism from native sources," but without specifics (1975: 16). Ajit Kumar Ray claims that the *Dabistān* was the probable source of inspiration (1976: 21–2). Copies of the *Dabistān* were available in Bengal in the late eighteenth century, for Sir William Jones first read it in 1787. Orientalist scholars in the Persian Department at Fort William College at the beginning of the nineteenth century were quite familiar with it. The identity of the author remains unknown.

5 For the state of Sanskrit and Vedāntic studies at this time, see Dermot Killingley (1977: 98–102). He notes William Ward's estimate, apparently for Bengal, that while 40–50 percent of Brahmins learned Sanskrit poetry (*kāvya*), and another 30 percent learned Nyāya, only 0.6 percent studied Veda and the *darśanas* other than the Nyāya.

6 Rammohan's condemnation of these European apologists for Hinduism is similar to that of the Baptist missionary William Ward (1985: vol. 3, xcvi–ciii).

7 Also known as Śaṅkara's *Brahmasutra Bhashya* (*BSB*).

8 Rammohan in his Bengali summary of the *Vedānta Sūtras* refers to Śaṅkara as facilitating the study of the Vedānta (Killingley 1982: 14).

9 *Vedānta Sūtras* 1.1.2 reads *janmādyasya yataḥ*, which literally means "from whence the birth etc. of this," where Brahman is assumed to be that from which "this," the universe, originates. The "etc." is taken to include preservation and destruction.

10 See p. 30 above.

11 The title pages of the four English translations of the Upaniṣads, for instance, all indicate that they conform to the glosses of Śaṅkara.

12 Cf. Rammohan's 1817 comment: "I have found the doctrines of Christ more conducive to moral principles, and better adapted for the use of rational beings, than any other which have come to my knowledge" (quoted in N.S. Bose 1976: 38).

13 For a detailed account of Rammohan's interest in Christian Unitarianism, see Killingley (1993: 135–47).

14 Cf. Killingley (1993: 57–8).

15 Halbfass notes that the German historian of philosophy C.J.H. Windischmann "remarked that Rammohun had wanted to find 'pure deism' in the Vedas," and

criticized Rammohan's Unitarianism as an indication that he "was not ready for the Christian trinity" (1988: 199). Similarly, the French Sanskritist Eugene Bournouf observed of Rammohan's writings: "The purpose of his numerous publications was to establish the existence of one sole, eternal, infinite God, who requires no cult from his worshippers except the practice of a strict morality. He published extracts from the Vedas, in Bengali and English successively, to show that these ancient books taught only the purest deism" (quoted in Killingley 1977: 338).

16 Cf. Nalin C. Ganguly (1934: 150).

17 Peter Gay notes that, despite the naiveté and eventual eclipse of the European Deists, their rejection of priestly superstition and of humankind's fallen state, their love of ethical teachings and optimism about human nature, and their faith in the capacity of human reason to find solutions to the problems of society and to define and circumscribe the mysteries of the universe through science made them into "powerful agents of modernity" (1968: 13).

8 Design and the intuitive theism of Debendranath Tagore

1 See p. 37 above for the *Nyāya-Sūtras* and p. 142 below for Vivekananda.

2 The goddess Kali is worshiped at the temple of Kalighat. The *shalgram* is a sacred stone symbolizing Viṣṇu.

3 I have relied on Debendranath's *Autobiography* for several of the details of his life, as well as David Kopf's detailed study of the Brahmo Samaj (1979).

4 Debendranath Tagore tells us that his family's accountant, Rajendranath Sircar, came to him in 1845 with a sorrowful tale regarding the conversion and baptism of his youngest brother and his wife by Alexander Duff (1994: 29). Outraged over the conversion of one of the ladies of the *zenana*, Debendranath immediately set A.K. Dutt the task of responding to this challenge by Duff, who did so in the pages of the *Tattvabodhini Patrika*. Incidentally, Dutt used the *Patrika* to popularize various scientific subjects of the day (Lourdusamy 2004: 45).

5 For discussions of this decision, see Kopf (1979: 170); Meredith Borthwick (1977: 20–1); and Anantanand Rambachan (1994: 260–3).

6 The second part of the *Brahmo Dharma*, composed later, provides broad moral precepts derived from *Manu*, the *Mahābhārata*, and other ancient scriptures, but ignoring their caste-based moral codes, subscribing instead to a general humanistic ethic of helping one's fellow human beings (cf. Kopf 1979: 163–4).

7 This is the verse frequently cited by commentators like Śaṅkara in interpreting the second aphorism of the *Vedānta Sūtras*. Cf. Ranade's use of the verse, p. 77 above.

8 As we shall see in the next chapter, Keshab Chandra Sen cites the European writers who inspired his views on the role of intuition vis-à-vis reason. Debendranath, while familiar with many of these writers, provides no references.

9 Selections from these sermons are appended to Debendranath's *Autobiography*.

10 Debendranath's son, Satyendranath, argues that the Hebrew Scriptures were foreign in spirit to his father, whom he says never quoted the Bible (S. Tagore 1994: vii). But clearly Debendranath was familiar with biblical themes as evidenced by his cosmogony.

9 Design and the religious evolutionism of Keshab Chandra Sen

1 For extended treatments of Keshab's life, see Meredith Borthwick (1977) and Frans L. Damen (1983). For his role in the Brahmo Samaj, see David Kopf (1979).

2 For an excellent discussion of the increasing role of intuition vis-à-vis reason and Vedic authority in the Brahmo Samaj, see Anantanand Rambachan (1994). Rambachan sees Keshab as the "most articulate" of the Brahmo leaders in giving intuition the leading role.
3 The contrast between Debendranath and Keshab on this issue is nicely illustrated by the latter's *Ślokasaṃgraha*, a collection of scriptural citations from several religious traditions of the world, as opposed to Debendranath's *Brahmo Dharma* that focused on the Upaniṣads (Wilhelm Halbfass 1988: 225). As for the development of cultural nationalism and the rise of militant nationalism in the Brahmo Samaj, see Kopf (1979: 176–214).
4 C. Mackenzie Brown (2007a: 432). See also Damen (1983: 192–3) and Kopf (1979: 268–9).
5 Cf. William Hamilton's remarks in an 1836 lecture on Metaphysics (1874: 25).
6 Cf. Hamilton (1874: 29).
7 Cf. Ronald Numbers on the traditional generic meaning of science in the West and its changing definition in the nineteenth century (2009: 16).
8 I discuss Spencer's Unknown and its role in Vivekananda's thought in Chapter 11.
9 See p. 97 above.
10 Keshab frequently emphasizes three sources of knowledge about God: the external world, inner intuition, and history. This threefold source of divine knowledge echoes that of Unitarians like Theodore Parker's account of methods for constructing a "theory of God" (Grodzins 2002: 274; Rambachand 1994: 270–1).
11 See for instance, J.D. Morell (1848: 658).
12 For an extended discussion of the roots of Keshab's Avataric Evolutionism, see Brown (2007a: 431–44).

10 The Modern Vedic Creationism of Dayananda Saraswati

1 See, for example, Dayananda's *Introduction to the Commentary on the Four Vedas*, the sections on "The Art of Building Ships and Aerial Cars" and "The Science of Telegraphy" (1981: 254–66). Cf. Vedānta Prakash Garg (1984).
2 Detailed notes of the debate were taken and translated into English by a follower of Dayananda, a revised version of which is provided by Dayananda's biographer, Chajju Singh (1971: pt. 1, pp. 124–43).
3 For a comparison of the two editions of the *Satyārth Prakāś*, see Jordens (1997: 99–102, 249–56).
4 I have added the Hindi words in brackets from the Hindi text (Dayananda 2002: 118).
5 For the Hindi text, see ibid.: 139.
6 For the Hindi text, see ibid.: 142.
7 For the Hindi text, see ibid.: 143.
8 For earlier references to this verse, see pp. 17 and 27 above.
9 Dayananda's account of creation here is very similar to the one he gave at the Chandapur fair. See Singh (1971: pt. 1, pp. 129–30).
10 A similar argument is presented by Dayananda in Chapter 12 of the *Satyārth Prakāś*, against the Cārvākas.
11 For the Hindi text, see Dayananda (2002: 145–7).
12 These refer to the first three states of consciousness in classical Hindu philosophy. Dayananda thus apparently envisions different compartments in the body where the soul can disport itself or rest in these three modes.
13 The same argument appears in Chapter 12 of the *Satyārth Prakāś*, in arguing against atheists.
14 The same quotation appears in V.P. Garg, "Dayananda and Science" (1984: 501).
15 For details of Gurudatta's life, I have relied on two main sources, both hagiographic in tone, but apparently sound on basic factual matters: Anon. (1902) and Chamupati (n.d.).

16 Gurudatta's early death was due to tuberculosis.

17 Singh (1971: pt. 2, pp. 4–5n) quotes the same passage in his discussion of Dayananda's anti-evolutionary lecture at Roorkee.

18 For a critique of this finding, see Glen J. Kuban (2006–2008). Kuban notes that the "print" was actually discovered in 1917 or before, by Albert E. Knapp, rather than John T. Reid. The two surviving, available photographs of the "print" strongly suggest that it is of inorganic origin and appears to be an ironstone concretion.

11 The Modern Advaitic Evolutionism of Swami Vivekananda

1 Cf. *CW* 1:371–2.

2 Tapan Raychaudhuri notes that Vivekananda similarly dismissed Hindu attempts to justify various religious practices—superstitions in the Swami's eyes—as scientifically efficacious or necessary (1999: 123).

3 Cf. *CW* 4:348–9.

4 For a brief summary of Vivekananda's life, see George M. Williams (1981).

5 These influences have been discussed in more detail in C. Mackenzie Brown (2011).

6 Blavatsky (1888) expounds these ideas in *The Secret Doctrine*, especially its Proem, available online at: http://www.theosociety.org/pasadena/sd/sd1-0-co.htm#contents. See E.B. Sellon and R. Weber (1992) for discussion of Blavatsky's assimilation of Vedāntic themes.

7 Blavatsky's influence on Vivekananda's evolutionary thinking is nicely illuminated by Meera Nanda in her insightful essay "Madame Blavatsky's Children: Modern Hindu Encounters with Darwinism" (2011).

8 Almost fifty years earlier, the English critic of all things Hindu, Thomas Babington Macaulay, had noted the incredible conceptions of ancient Hindu astronomy that "would move laughter in girls at an English boarding school" (1968: 46). Narendranath likely imbibed such assessments during his college years. For discussion of early English criticism of Indian sciences, including astronomy, see David Arnold (2000: 35–9).

9 For Vivekananda's conflating of Darwinian and Sāṃkhyan evolution, see Mark Singleton (2007: 131).

10 Williams notes the tension between the Advaitic view of the world as illusion and the more realistic orientation of the ethics of service espoused by Vivekananda (1974: 74).

11 See Brown (2011: 237–8).

12 See pp. 31, 42, and 55 above.

13 See p. 54 above.

14 Cf. the pronouncement of Vivekananda's follower, Ramatirtha, that the world (especially the West with its idea of a God struggling with Satan as found in Genesis) should give up its outmoded ideas of God: "He [God] should be above the profession of an architect, a world-builder or maker. It is high time for the whole world to take up Vedanta" (1930–1932: vol. 4, p. 141).

15 See p. 32 above.

16 Cf. Vivekananda's summary of the early history of religion in India (*CW* 6:157).

17 The Upaniṣadic reference is to *Chāndogya Upaniṣad* 6.1.4.

18 Thomas Huxley himself also gave some support to such a view, claiming that evolutionary conceptions of the universe were extant among the philosophers of India and Greece in the sixth century BCE (1894: 53–4). I have not found any direct citation of this claim by Vivekananda, but there is a good probability that he was familiar with Huxley's essay "Evolution and Ethics," in which it occurs.

19 Cf. Ramatirtha (1930–1932: vol. 6, pp. 186–8), who argues that the Law of Karma and the Law of Heredity are complementary, and that the Law of Karma gives the simplest,

most parsimonious explanation for the different propensities in children. The Law of Karma, in fact, he claims, explains the Law of Heredity and is thus the more scientific.

20 See p. 128 above.

21 Swami Ramakrishnananda, in a lecture entitled "Evolution and Revolution," utilizes this passage in explaining Darwin's ideas (1993: 34). Like Vivekananda, Ramakrishnananda criticizes all material theories of evolution for neglecting the spiritual evolution of consciousness. Cf. Ramatirtha (1930–1932: vol. 3, p. 350).

22 Cf. *CW* 5:255, where he states, "Every evolution is preceded by an involution. The whole of the tree is present in the seed, its cause."

23 See also *CW* 2:209, 2:427.

24 The use of the law of energy conservation to refute the idea of the creation of the universe out of nothing has become part of the standard rhetoric of Hindu apologists in the Hinduism and science discourse. See, for instance, Nem Kumar Jain (1982: 1).

25 Killingley (1998: 151), in commenting on a similar passage (*CW* 1:156–7), notes the affinity with Spencer's evolutionary views. Killingley adds: "But whereas Spencer was interested in the observable organisms of increasing complexity, Vivekananda passes beyond these, and beyond humankind, to what he here calls 'God,' and a few lines later 'perfection.' It is this part of the evolutionary scale which chiefly concerns him." Cf. Vivienne Baumfield (1998: 205).

26 The use of the amoeba-and-perfected-man example to illustrate the involution-evolution process is often repeated by later followers of Vivekananda. See, for instance, Nikhilananda 1958, 58.

27 Cf. *CW* 4:484.

28 Cf. *CW* 3:335.

29 Cf. *CW* 5:520.

30 See p. 57 above.

31 See Killingley (1990: 173–4) for a brief discussion of this claim.

32 Cf. *CW* 5:278.

33 For discussion of "eugenic evolutionism" in Aurobindo and its role in Hindu revisioning of yoga as a means of "genetic and spiritual perfection" in the early twentieth century, see Singleton (2007: 126, 133).

34 Cf. Swami Mukhyananda: "It [Vedānta] is Spiritual Science, encompassing all sciences (*Brahma-vidyā Sarva-vidyā Pratiṣhṭhā—Muṇḍaka Upaniṣad I.i.1*). It is metaphysics, philosophy, axiology, epistemology, cosmology, religion, ethics, science, and psychology—all in one" (1998: 19).

35 Cf. Satpal Sangwan (1991a: 8–11; 1991b: 32–4).

36 See p. 110 above. Cf. Mary Louise Burke (1983–1987: vol. 3, pp. 433–44).

37 Cf. Brown (2010: 715).

12 The Integrative Evolutionism of Sri Aurobindo Ghose

1 For details of Aurobindo's biography, I have largely relied on Peter Heehs (1989).

2 The transformation does not mean a radical break in Aurobindo's ideals and values. As Robert Neil Minor points out, Aurobindo's ultimate concern during his nationalist days was to spread the Vedāntic message of the oneness of Ātman and Brahman to the world (1978: 89–90). After his withdrawal to Pondicherry, Aurobindo assumed that the work of Indian independence would continue apace, while he would devote himself to yogic practice.

3 Heehs notes that the appellation Sri Aurobindo only came into use in 1926 (1989: 75n).

4 Rama Shanker Srivastava, an ardent exponent of Aurobindo's evolutionary philosophy, finds the roots of the involution-evolution idea in the Vedic hymn to Puruṣa (10.90) (1968: 8). According to Srivastava, involution is seen in the self-involvement of the Absolute through the immolation of the Puruṣa, and since only one-quarter of the

Puruṣa has so far become involved, there will be further, emergent evolution of the world. Srivastava here follows his teacher S. K. Maitra, for whom the Puruṣa Sūkta "is the foundation for the emergent view of evolution in our country [India]" (1956: 262).

5 For a discussion of the evolution in meaning of the term *evolution*, see Robert J. Richards (1992: 5–16).

6 For a discussion of these categories and their development, see Minor (1978: 106–13).

7 Srivastava notes that in traditional Hindu theories of evolution, as in the Sāṃkhya, "evolution is always followed by dissolution, and this takes away all its significance," and argues that Aurobindo is the first philosopher, East or West, to give evolution "its rightful place" (1968: viii). This follows Maitra (1956: 52–3).

8 Aurobindo's ideal of transformation is often seen by his followers as not only key to his theory, but also as what sets his evolutionary philosophy apart from all preceding evolutionary thinkers, both eastern and Western, and fulfills them (e.g., Srivastava 1968: viii–ix; Maitra 1956: 326–7).

9 Srivastava uses the term *integral evolution*, which he defines as the process whereby "when the higher principle emerges the lower principles are also transformed" (1968: 88). Cf. Maitra (1956: 39–40, 99).

10 The book's dedication reads: "Dedicated to H.P.B. [Helena Petrovna Blavatsky] by Her Grateful Student."

11 Another renowned Neo-Hindu, Sarvepalli Radhakrishnan, also cites Bergson's example of the independent evolution of the eye in mollusks and in vertebrates in refuting the idea that chance variations could thus have produced the same result twice (1937: 255).

12 For a brief overview of Aurobindo's ideas of involution and evolution, see Robert A. McDermott (1974). A detailed treatment is provided by Srivastava (1968).

13 For superman in Nietzsche and Aurobindo, see K.D. Sethna (1981: 266) and Mark Singleton (2007: 133–4).

14 In similar vein, Radhakrishnan notes: "When we pass from animal to man, we find not a gradual development but a sudden break, a leap into a new form of experience … man is not simply the animal gone up any more than an animal is a man gone down. Between the two there is a gulf. No amount of scientific observation can help us to explain the astonishing change" (1937: 262–3).

15 For discussion of the term *svabhāva*, see p. 22 above.

16 Despite Bergson's rejection of mechanism, Maitra (1956: 37–9), followed by Srivastava (1968: vii, 423–4), still consider Bergson, in the end, to be a mechanist.

17 Maitra (1956: 41) and Srivastava (1968: 321) consider Alexander's philosophy of evolution to have the closest resemblance to Aurobindo's of all Western evolutionary theories.

18 Cf. Lloyd Morgan's statement: "I acknowledge God as the Nisus through whose Activity emergents emerge, and the whole course of emergent evolution is directed" (1931: 36). For critiques of Alexander's views by advocates of Aurobindo's philosophy, see Maitra (1956: 41–9) and Srivastava (1968: vii–viii, 322–31, 424), who assess Alexander's theory as ultimately purely materialistic, with mere Space-Time as a material ultimate reality and his nisus as an extraneous principle unconnected with the Space-Time matrix.

19 Cf. Radhakrishnan, who argues that emergent evolution, properly understood in its creativity, indicates that the irreversible processes of matter "mean plans and ends, however unconscious matter may be of them" (1937: 241). And like Aurobindo, he acknowledges that evolutionary creativity, with its process of trial and error, provides no evidence for any teleological plan or purpose. He concludes: "In spite of signs of lack of design, there is a general trend in evolution towards specific forms not yet realised" (ibid.: 332).

13 Design and Darwin in the post-colonial period

1 For a brief overview of the rise of Hindu revivalism and of Hindutva nationalism, see T.N. Madan (1997: 203–27). For a more extended discussion, see Chetan Bhatt (2001).
2 See Shishya (2006–2009a).
3 For critiques of the notion that evolution is "just a theory," see Richard Dawkins (2009: 3–18); Jerry A. Coyne (2009: 1–19); and Eugenie C. Scott (2004: 11–20). See also Kenneth R. Miller's 2008 book *Only a Theory*.
4 See Shishya (2006–2009b).
5 The discussion of Prabhupada's views is based on my 2002 and 2010 articles.
6 For a scientific answer to such arguments, see Dawkins (2009: 25–6).
7 See p. 171 above.
8 Personal communication, November 2009.
9 See pp. 37 and 142 above.
10 Panda frequently attacks the idea of creation out of nothing explicitly in the text, citing among other textual authorities *Bhagavad-Gītā* 2.16 and *Vedānta Sūtras* 2.2.26 in support (2002: 648, 734).
11 See p. 29 above.
12 Panda cites here not Śaṅkara's comment on the *Vedānta Sūtras*, but Śaṅkara's *Vivekacudamaṅi* (v. 110 [108 in the 1983b edition]), where the Advaitin that: "Maya, by whom this whole world is produced, is to be inferred (*anumeya*) by her effects" (my translation).
13 The article is written under the pseudonym Paul Utukuru. In a version of the paper sent to me by the author, the article is subtitled: "A Step towards Bridging the Gap between Science and Human Religiosity."
14 The first edition was published in 1997, the third in 2006. The authorship of the book is given as "Amit Goswami with Maggie Goswami," the latter being his wife. Given that the essential content of the book is Amit Goswami's, I attribute the book's arguments to him.
15 Cf. Goswami (2008: 73–4, 80).
16 See Stephen J. Gould (2002: 986).
17 Goswami briefly mentions hopeful monsters in *Science and Spirituality* (2006: 59), and elaborates upon the theme in *Creative Evolution* (2008: 163–5).
18 I have relied on Gould (2002: 451–66) for my summary of Goldschmidt's views.
19 I have relied on Gould (2002: 528–31) for my summary of Simpson's views.
20 The first edition was published in 1996, the third in 2007.

14 Design and Darwin in the contemporary Hindu world

1 Some Hindus, especially Neo-Vedāntins, insist that there is nothing supernatural in Hinduism, since nature for them includes both spiritual realities and sensory objects. Thus, there are only supersensory, but not supernatural, realities. I will discuss the implications of this ontological perspective in Chapter 15, simply noting here that I am using the term *supernatural* to refer to entities or mechanisms that transcend the realm of what Hindus call *prakṛti*, which is usually translated into English as *nature*, or what the *Bhagavad-Gītā* (6.7) refers to as Kṛṣṇa's lower nature (*apara prakṛti*).
2 The term *evolution*, for instance, was once used in the West to describe the theory of preformationism, the idea that the adult form of an organism was simply the unfolding (evolution) of a miniature but complete version of the adult in the egg. See Clara Pinto-Correia (1997).
3 British Council (2009), "Results of British Council Darwin International Survey." A copy of the report was sent to me upon request. A very general summary is available online: http://www.britishcouncil.org/darwin_now_survey_global.pdf. Last accessed 1/29/2011.

4 See also the *India Science Report* (Shukla 2005), which polled 30,255 Indians in 2004. The survey was primarily concerned with issues of science education and public attitudes towards science and technology, but included one question relevant to Darwinism: "How did human beings/man evolve?" Unfortunately, the question itself assumes evolution is true, and the response choices ("Biological theory," "Monkey," and "Adam/Brahma") were oddly worded, making conclusions difficult to draw.

5 David Gosling's survey was first published in 1976 in his *Science and Religion in India*. A fairly complete summary is available in Gosling (2007: 102–29).

6 Some 800 questionnaires were distributed, and just under 700 were completed.

7 Ronald Numbers, personal communication (1/8/2011).

8 The survey also included three open-ended questions asking respondents for their definitions of *science*, *Hinduism*, and *evolution*, respectively, and there was also the opportunity for general comments.

9 For a similar survey on Buddhist practice and health, see Sullivan, Wiist, and Wayment (2010).

10 There are theistic Hindus, like Dayananda Saraswati, who regard the eternal Vedas as "revealed," in the sense of being retaught to humankind at the beginning of each cosmic cycle.

11 To avoid more tables, I summarize the results here for scientists overall. The 40 percent figure is based on 460 respondents.

12 The number of respondents is 458.

13 For discussion of "scientific" sadhus, see Chapter 15 below.

14 The Yogananda group is a partial exception, being like ISKCON with a large majority (74 percent) preferring the statements of spiritually enlightened persons to those of modern science. But it is similar to non-ISKCON groups with only a substantial minority (41 percent) preferring the statements of scripture.

15 With regards to traditional religious beliefs, a similar divide between those in biological fields and more applied areas of science occurs among Western scientists, with the latter more closely resembling the views of the general public (Brown 2003).

15 Conclusion

1 The text cited seems to be a summary of Bose's comments, but certainly reflects his views. David Arnold (2000: 174–5) attributes these words to Bose.

2 Cf. Dhruv Raina and S. Irfan Habib (1996: 29–30).

3 See Wilhelm Halbfass (1988: 186) and my discussion of Vivekananda's use of the term *vidyā*, pp. 151–2 above.

4 For further discussion of Lyell's use of the Hindu triad and Darwin's theory regarding natural selection as a creative force, see Stephen Jay Gould 2002: 137–41.

5 In Chapter 6 we noted that Bankim Chandra Chatterji used the trinity of Brahmā, Viṣṇu, and Śiva to argue that Hinduism was more in accord with Darwinian theory than Christian theology (1986: 70). And like Lyell, he equated natural selection with the destructive force of Śiva, complementing the creative and preserving powers responsible for the evolution of species.

6 Charles Lyell's attitude towards Hindu cosmogony in general is well expressed in the first volume of his *Principles of Geology*, published in 1830. He argues that the cosmogony, as he found it in William Jones' translation of *Manu*, contained passages of "much grandeur of conception and sublimity of diction," as well as "puerile conceits and monstrous absurdities" (Lyell 1990: 6).

7 For discussion of Madhava's summary of the Cārvāka view, see pp. 23–4 above.

Bibliography

References to Bādarāyaṇa's *Brahma-Sūtras* are to the texts contained in Śaṅkara (1983a) *Brahmasutra Bhashya* and in Rāmānuja (1985–1991) *Śrībhāṣyam*. References to the *Bhagavad-Gītā* are to the text published in Śaṅkara (n.d.) *Srīmad Bhagavad Gītā Bhāṣya of Sri Sāṃkarācārya*. References to the *Mahābhārata* are to the critical edition (1927–1966), unless otherwise noted. References to Upaniṣadic texts are to S. Radhakrishnan (1953) *The Principal Upaniṣads*.

Abbreviations

BSB Śaṅkara (1983a) *Brahmasutra Bhashya*, rev. edn, vol. 7 in *Complete Works of Sri Sankaracharya in the Original Sanskrit*, Madras: Samata Books.

CW Vivekananda (2003) *The Complete Works of Swami Vivekananda*, 9 vols, in *Swami Vivekananda: Life, Works & Research* (multimedia CD), Calcutta: Advaita Ashrama.

EW Rammohan Roy (1978) *The English Works of Raja Rammohun Roy, with an English Translation of "Tuhfatul Muwahhiddin,"* New York: AMS Press.

LD Aurobindo Ghose (1982) *The Life Divine*, Pondicherry: Sri Aurobindo Ashram.

ŚB Rāmānuja (1985–1991) *Śrībhāṣyam: Śārīrakamīmāmsābhāṣyam*, 4 vols, Melkote: Academy of Sanskrit Research.

References

Aiyer, K.N. (1914) *The Purāṇas in the Light of Modern Science*, Adyar, Madras: Theosophical Society.

Alexander, S. (1950) *Space, Time, and Deity: The Gifford Lectures at Glasgow 1916–1918*, 2 vols, New York: Humanities Press.

Anantharaman, T.R. (2007) *Ancient Yoga and Modern Science*, New Delhi: Centre for Studies in Civilization.

Anon. (n.d.) *Seeking the Divine Perfection*. Available at http://www.saibaba.ws/articles1/seekingdivineperfection.htm (accessed 18 May 2011).

Anon. (1902) "Biographical Sketch," in Guru Datta Vidyarthi, *The Works of Late Pandit Guru Datta Vidyarthi, M.A., with Biographical Sketch*, Lahore: The Aryan Printing, Publishing & Trading Co.

Anon. (1921) "Sir Jagadish Chander Bose: A Sketch," in J.C. Bose, *Sir Jagadish Chander Bose: His Life, Discoveries and Writing*, Madras: G.A. Natesan.

Anon. (1948) *Life of Sri Ramakrishna: Compiled from Various Authentic Sources*, Mayavati, Almora: Advaita Ashrama.

Antarkar, S.S. (1999) "Hinduism and Some of Its Major Cosmologies," in K. Acharya, N. Manca, and L. Namjoshi (eds) *A Dialogue: Hindu-Christian Cosmology and Religion*, Mumbai: Somaiya Publications Pvt. Ltd.

Arnold, D. (2000) *Science, Technology and Medicine in Colonial India*, Cambridge: Cambridge University Press.

Aurobindo Ghose (1944) *Evolution*, Calcutta: Arya.

— (1971) "The Purpose of Avatarhood," in Sri Aurobindo, *Letters on Yoga* (vol. 22 of the Sri Aurobindo Birth Centenary Library), Part One, Sec. VII, Pondicherry: Sri Aurobindo Ashram.

— (1972) *The Upanishads: Texts, Translation and Commentaries*, Pondicherry: Sri Aurobindo Ashram.

— (1982) *The Life Divine*, Pondicherry: Sri Aurobindo Ashram.

Babb, L.A. (1986) *Redemptive Encounters: Three Modern Styles in the Hindu Tradition*, Berkeley: University of California Press.

Baber, Z. (1998) *The Science of Empire: Scientific Knowledge, Civilization, and Colonial Rule in India*, Delhi: Oxford University Press.

Bagemihl, B. (1999) *Biological Exuberance: Animal Homosexuality and Natural Diversity*, New York: St. Martin's Press.

Barbour, I.G. (1997) *Religion and Science: Historical and Contemporary Issues*, New York: HarperCollins.

Barker, E. (1986) "World Congress for the Synthesis of Science and Religion: A Personal Account," *ISKCON Review*, 2: 133–47.

Baumfield, V. (1998) "Science and Sanskrit: Vivekananda's Views on Education," in W. Radice (ed.) *Swami Vivekananda and the Modernization of Hinduism*, Delhi: Oxford University Press.

Beatty, J. (2010) "Reconsidering the Importance of Chance Variation," in M. Pigliucci and G.B. Müller (eds) *Evolution—The Extended Synthesis*, Cambridge, Massachusetts: MIT Press.

Belludi, C. (2009) "Vedanta and the Theory of Evolution," *Vedanta Kesari*, 33–6. Available at http://www.scribd.com/doc/11795254/Vedanta-and-Evolution-Vedanta-Kesari-Jan-2009 (accessed 14 March 2011).

Bergson, H. (1911) *Creative Evolution*, trans. A. Mitchell, New York: Random House.

Bharati, A. (1970) "The Hindu Renaissance and Its Apologetic Patterns," *Journal of Asian Studies*, 29, no. 2: 267–87.

Bhatt, C. (2001) *Hindu Nationalism: Origins, Ideologies and Modern Myths*, Oxford: Berg.

Bhattacharyya, G. (1961) *Studies in Nyāya-Vaiśeṣika Theism*, Calcutta: Sanskrit College.

Blavatsky, H.P. (1877) *Isis Unveiled*. Available at http://www.sacred-texts.com/the/iu/ (accessed 16 October 2010).

— (1888) *The Secret Doctrine: The Synthesis of Science, Religion, and Philosophy*. Available at http://www.theosociety.org/pasadena/sd/sd1-0-co.htm#content (accessed 8 February 2011).

Borthwick, M. (1977) *Keshub Chunder Sen: A Search for Cultural Synthesis*, Calcutta: Minerva Associates.

Bose, J.C. (1921) *Sir Jagadish Chander Bose: His Life, Discoveries and Writing* (paginated separately from Anon., "Sir Jagadish Chander Bose: A Sketch," in the same volume), Madras: G.A. Natesan.

Bose, N.S. (1976) *Indian Awakening and Bengal*, 3rd edn, Calcutta: Firma K.L. Mukhopadhyay.

Bowler, P.J. (2001) *Reconciling Science and Religion: The Debate in Early-Twentieth-Century Britain*, Chicago: University of Chicago Press.

Boyle, R. (1979) "The Origin of Forms and Qualities According to the Corpuscular Philosophy," in M.A. Stewart (ed.) *Selected Philosophical Papers of Robert Boyle*, Manchester: Manchester University Press.

Brown, C.M. (2002) "Hindu and Christian Creationism: 'Transposed Passages' in the Geological Book of Life," *Zygon*, 37, no. 1: 95–114.

— (2003) "The Conflict between Religion and Science in Light of the Patterns of Religious Belief among Scientists," *Zygon*, 38, no. 3: 603–32.

— (2007a) "The Western Roots of Avataric Evolutionism in Colonial India," *Zygon*, 42, no. 2: 425–49.

— (2007b) "Colonial and Post-Colonial Elaborations of Avataric Evolutionism," Zygon, 42, no. 3: 715–47.

— (2008) "The Design Argument in Classical Hindu Thought," *International Journal of Hindu Studies*, 12, no. 2: 103–51.

— (2010) "Hindu Responses to Darwinism: Assimilation and Rejection in a Colonial and Post-Colonial Context," *Science & Education*, 19: 705–38.

— (2011) "Vivekananda and the Scientific Legitimation of Advaita Vedānta," in J.R. Lewis and O. Hammer (eds) *Handbook on Religion and the Authority of Science*, Leiden: Brill.

Burke, M.L. (1983–1987) *Swami Vivekananda in the West: New Discoveries*, 3rd edn, 6 vols, Mayavati, India: Advaita Ashrama.

Carman, J.B. (1974) *The Theology of Rāmānuja: An Essay in Interreligious Understanding*, New Haven: Yale University Press.

Carroll, S.B. (2005) *Endless Forms Most Beautiful: The New Science of Evo Devo*, New York: W.W. Norton & Company.

Chakrabarti, A. (1989) "From the Fabric to the Weaver," in R.W. Perrett (ed.) *Indian Philosophy of Religion*, Dordrecht: Kluwer Academic Publishers.

Chakrabarti, K.K. (1999) *Classical Indian Philosophy of Mind: The Nyāya Dualist Tradition*, Albany: State University of New York Press.

Chakrabarti, P. (2004) *Western Science in Modern India: Metropolitan Methods, Colonial Practices*, Delhi: Permanent Black.

Chakraborti, R. (1979) "Rammohun Ray: His Vision of Social Change," in A.K. Mukhopadhyay (ed.) *The Bengali Intellectual Tradition: From Rammohun Ray to Dhirendranath Sen*, Calcutta: K.P. Bagchi & Co.

Chakraborty, R. (1979) "Aksayakumar Datta: The First Social Scientist," in A.K. Mukhopadhyay (ed.) *The Bengali Intellectual Tradition: From Rammohun Ray to Dhirendranath Sen*, Calcutta: K.P. Bagchi & Co.

Chamupati (n.d.) *A Biographical Sketch of Pandit Gurudatta*. Available at http://aryamissionaries.blogspot.com/2009/11/biographical-sketch-of-pandit-gurudatta.html (accessed 31 December 2010).

— (ca. 1930) *The Ten Commandments of Dayananda*, Lahore: Arya Pustakalaya.

— (1937) *Glimpses of Dayananda*, 2nd edn, Delhi: Sharada Mandir Ltd.

Chatterji, B.C. (1986) "Mill, Darwin, and Hinduism," in S.N. Mukerjee and M. Maddern (eds) *Bankimchandra Chatterjee: Sociological Essays: Utilitarianism and Positivism in Bengal*, trans. S.N. Mukerjee and M. Maddern, Calcutta: Ṛddhi.

Chattopadhyaya, D. (1964) *Indian Philosophy: A Popular Introduction*, New Delhi: People's Publishing House.

Chaudhuri, H. (1960) *Sri Aurobindo: The Prophet of Life Divine*, 2nd edn, Pondicherry: Sri Aurobindo Ashram.

Chemparathy, G. (1972) *An Indian Rational Theology: Introduction to Udayana's Nyāyakusumāñjali*, Vienna: Indologisches Institut der Universität Wien.

Chopra, D. (2005a) "Intelligent Design without the Bible," *Intent Blog*. Available at http://www.intentblog.com/archives/2005/08/intelligent_des_1.html#more (accessed 14 March 2011).

— (2005b) "Rescuing Intelligent Design—But from Whom?" *The Huffington Post*. Available at http://www.huffingtonpost.com/deepak-chopra/rescuing-intelligent-desi_b_6164.html (accessed 14 March 2011).

Clooney, F.X. (1993) *Theology after Vedānta: An Experiment in Comparative Theology*, Albany: State University of New York Press.
Collet, S.B. (1914) *The Life and Letters of Raja Rammohun Roy: Compiled and Edited by the Late Sophia Dobson Collet and Completed by a Friend*, 2nd edn, Calcutta. Available at http://ia700404.us.archive.org/12/items/lifelettersofraj00collrich/lifelettersofraj00collrich.pdf (accessed 31 December 2010).
Cowell, E.B. (ed.) (1983) *The Kusumanjali, or Hindu Proof of the Existence of a Supreme Being, by Udayana Acharya, with the Commentary of Hari Dasa Bhattacharya*, trans. E.B. Cowell, Delhi: Caxston.
Coyne, J.A. (2009) *Why Evolution is True*, New York: Viking.
Cremo, M. (2003) *Human Devolution: A Vedic Alternative to Darwin's Theory*, Los Angeles: Bhaktivedanta Book Publishing, Inc.
Cremo, M. and Thompson, R.L. (1993) *Forbidden Archaeology: The Hidden History of the Human Race*, San Diego: Bhaktivedanta Institute.
Dabistān, The (1973) trans. D. Shea and A. Troyer, Lahore: W. Pakistan.
Dalai Lama (2005) *The Universe in a Single Atom: The Convergence of Science and Spirituality*, New York: Morgan Road Books.
Damen, F.L. (1983) *Crisis and Religious Renewal in the Brahmo Samaj (1860–1884): A Documentary Study of the Emergence of the "New Dispensation" under Keshab Chandra Sen*, Leuven, Belgium: Katholieke Universiteit.
Darling, D. (n.d.) "Frogs in Stone," *The Internet Encyclopedia of Science*. Available at http://www.daviddarling.info/encyclopedia/F/frogs_in_stones.html (accessed 6 September 2010).
Darwin, C. (2008) *Evolution: Selected Letters of Charles Darwin, 1860–1870*, ed. F. Burkhardt, S. Evans and A. Pearn, Cambridge: Cambridge University Press.
Dasgupta, Subrata (1999) *Jagadis Chandra Bose and the Indian Response to Western Science*, Oxford: Oxford University Press.
— (2007) *The Bengal Renaissance: Identity and Creativity from Rammohun Roy to Rabindranath Tagore*, Delhi: Permanent Black.
Dasgupta, Surendranath (1961) *A History of Indian Philosophy*, vol. 1, Cambridge: Cambridge University Press.
— (1963) *A History of Indian Philosophy*, vol. 3, Cambridge: Cambridge University Press.
Davids, T.W.R. (intro.) (1990) *Sāmañña-phala-sutta*, trans. T.W.R. Davids, in D. Chattopadhyaya and M.K. Gangopadhyaya (eds) *Cārvāka/Lokāyata: An Anthology of Source Materials and Some Recent Studies*, New Delhi: Indian Council of Philosophical Research.
Dawkins, R. (1997) *Climbing Mount Improbable*, New York: W.W. Norton & Company.
— (2006) *The God Delusion*, Boston and New York: Houghton Mifflin Company.
— (2007) *The Blind Watchmaker*, London: Folio Society.
— (2009) *The Greatest Show on Earth: The Evidence for Evolution*, New York: Free Press.
Dayananda Saraswati (n.d.) *The Light of Truth: English Translation of Svami Dayananda's Book Satyartha Prakahsa*, rev. edn, trans. G.P. Upadhyaya, Allahabad: Kala Press.
— (1958) *Introduction to the Commentary on the Vedas*, trans. G.Ram, Delhi: Sarvadeshik Arya Pratinidhi Sabha.
— (1970) *An English Translation of the Satyarth Prakash, Literally: Exposé of Right Sense (of Vedic Religion) of Maharshi Swami Dayananda Saraswati, "The Luther of India," Being a Guide to Vedic Hermeneutics*, 2nd edn, trans. D. Prasad, New Delhi: Jan Gyan Prakashan.
— (1981) *Svāmī Dayānanda Sarasvatī's Ṛgvedādi-Bhāṣya-Bhūmikā, Being an Introduction to the Commentary on the Four Vedas*, trans. Parmanand, New Delhi: Meharchand Lachhmandas.
— (2002) [*Satyārthaprakāśaḥ*] [2nd edn] [Hindi text] [Delhi: Ārṣa Sāhitya Pracāra Drasṭa].

— (2003) *Light of Truth, or An English Translation of the Satyarth Prakash*, 2nd edn, trans. C. Bharadwaya, New Delhi: Arvadeshik Arya Pratinidhi Sabha.
Dennett, D.C. (1995) *Darwin's Dangerous Idea: Evolution and the Meanings of Life*, New York: Simon and Schuster.
Doniger, W. (trans.) (1981) *The Rig Veda: An Anthology*, New York: Penguin Books.
Drayton, H.S. and McNeill, J. (1882) *Brain and Mind; or, Mental Science Considered in Accordance with the Principles of Phrenology, and in Relation to Modern Physiology*, New York: Fowler and Wells.
Duff, A. (1988) *India and India Missions: Including Sketches of the Gigantic System of Hinduism Both in Theory and Practice*, Delhi: Swati Publications.
Edis, T. (2007) *An Illusion of Harmony: Science and Religion in Islam*, Amherst, New York: Prometheus Books.
Feuerstein, G., Kak, S., and Frawley, D. (1995) *In Search of the Cradle of Civilization: New Light on Ancient India*, Wheaton, Illinois: Quest Books, Theosophical Publishing House.
Flood, G. (1996) *An Introduction to Hinduism*, Cambridge: Cambridge University Press.
Frauwallner, E. (1973) *History of Indian Philosophy*, 2 vols, trans. V.M. Bedekar, Delhi: Motilal Banarsidass.
Ganguly, N.C. (1934) *Raja Ram Mohun Roy*, Calcutta: Y.M.C.A Publishing House.
Garg, G.R. (1984) *World Perspectives on Swami Dayananda Saraswati*, New Delhi: Concept Publishing Company.
Garg, V.P. (1984) "Dayananda and Science," in G.R. Garg, *World Perspectives on Swami Dayananda Saraswati*, New Delhi: Concept Publishing Company.
Gay, P. (1968) "Toward an Interpretation," in P. Gay (ed.) *Deism: An Anthology*, Princeton: D. Van Nostrand Co.
Goel, A. (1984) *Indian Philosophy: Nyāya-Vaiśeṣika and Modern Science*, New Delhi: Sterling Publishers.
Gosling, D.L. (1976) *Science and Religion in India*, Madras: The Christian Literature Society.
— (2007) *Science and the Indian Tradition: When Einstein Met Tagore*, London: Routledge.
Goswami, A. (2006) *Science and Spirituality: A Quantum Integration*, 3rd edn, New Delhi: Centre for Studies in Civilization.
— (2008) *Creative Evolution: A Physicist's Resolution between Darwinism and Intelligent Design*, Wheaton, Illinois: Quest Books, Theosophical Publishing House.
Gould, S.J. (1993) *Eight Little Piggies: Reflections in Natural History*, New York: W.W. Norton & Co.
— (1997) "Nonoverlapping Magisteria," *Natural History*, 106, no. 2: 16–22.
— (2002) *The Structure of Evolutionary Theory*, Cambridge, Massachusetts: Harvard University Press.
Grodzins, D. (2002) *American Heretic: Theodore Parker and Transcendentalism*, Chapel Hill: University of North Carolina Press.
Gurudatta, V. (1902a) "Evidences of the Human Spirit," in Guru Datta Vidyarthi, *The Works of Late Pandit Guru Datta Vidyarthi, M. A., with Biographical Sketch*, Lahore: The Aryan Printing, Publishing & Trading Co.
— (1902b) "Mandukyopanishat," in Guru Datta Vidyarthi, *The Works of Late Pandit Guru Datta Vidyarthi, M. A., with Biographical Sketch*, Lahore: The Aryan Printing, Publishing & Trading Co.
Halbfass, W. (1988) *India and Europe: An Essay in Understanding*, Albany: State University of New York Press.
— (1992) *On Being and What There Is: Classical Vaiśeṣika and the History of Indian Ontology*, Albany: State University of New York Press.
Hamilton, W. (1874) *Lectures on Metaphysics and Logic*, vol. 1, ed. H.L. Mansel and J. Vetch, Boston: Gould and Lincoln.

Hay, S. (1996) "Review of Bruce Carlisle Robertson, *Raja Rammohan Roy: The Father of Modern India*," *Journal of the American Oriental Society*, 116, no. 4: 753–4.

Heehs, P. (1989) *Sri Aurobindo: A Brief Autobiography*, Oxford: Oxford University Press.

Heimsath, C.H. (1964) *Indian Nationalism and Hindu Social Reform*, Princeton: Princeton University Press.

Hiriyanna, M. (1932) *Outlines of Indian Philosophy*, London: George Allen & Unwin.

— (1949) *The Essentials of Indian Philosophy*, London: George Allen & Unwin.

— (1957) *Indian Philosophical Studies*, Mysore: Kavyalaya.

Hoodbhoy, P. (1991) *Islam and Science: Religious Orthodoxy and the Battle for Rationality*, London: Zed Books.

Hume, D. (1964) *A Treatise of Human Nature: Being an Attempt to Introduce the Experimental Method of Reasoning into Moral Subjects; and Dialogues Concerning Natural Religion*, vol. 2, ed. T.H. Green and T.H. Grose, Aalen, Germany: Scientia Verlag.

Hut, P. (2003) "Conclusion: Life as a Laboratory," in B.A. Wallace (ed.) *Buddhism & Science: Breaking New Ground*, New York: Columbia University Press.

Hutchison, J.C. (1998) "The Design Argument in Scientific Discourse: Historical-Theological Perspective from the Seventeenth Century," *Journal of the Evangelical Theological Society*, 41, no. 1: 85–105.

Huxley, T.H. (1869) *An Introduction to the Classification of Animals*, London: John Churchill and Sons.

— (1894) *Evolution and Ethics and Other Essays*. New York: D. Appleton and Co.

Īśvarakṛṣṇa (1969) *The Sāṁkhyakārikā of Īśvarakṛṣṇa* [with Sanskrit text], trans. G.J. Larson, in G.J. Larson, *Classical Sāṃkhya: An Interpretation of Its History and Meaning*, Delhi: Motilal Banarsidass.

Iyengar, T.C.R. (1908) *The Hindu-Aryan Theory on Evolution and Involution or the Science of Raja-Yoga*, New York: Funk & Wagnalls Company.

Jain, N.M. (1982) *Science and Scientists in India (Vedic to Modern)*, Delhi: Indian Book Gallery.

Jones, K.W. (1976) *Arya Dharm: Hindu Consciousness in 19th-Century Punjab*, Berkeley: University of California Press.

Jordens, J.T.F. (1997) *Dayananda Sarasvati: His Life and Ideas*, Delhi: Oxford University Press.

— (1998) *Dayananda Sarasvati: Essays on His Life and Ideas*, New Delhi: Manohar.

Joshi, P. (1997) "The Relationship between Religion and Science: The Hindu Perspective," *Scriptura*, 61: 109–20.

Kak, S. (2002) "A Most Subversive Idea," *Sulekha*. Available at http://subhash-kak.sulekha.com/blog/post/2002/06/a-most-subversive-idea.htm (accessed 16 June 2006).

Kaviraj, G. (1990) "Lokāyata and the Doctrine of Svabhāva," in D. Chattopadhyaya and M.K. Gangopadhyaya (eds) *Cārvāka/Lokāyata: An Anthology of Source Materials and Some Recent Studies*, New Delhi: Indian Council of Philosophical Research.

Keith, A.B. (1921) *Indian Logic and Atomism: An Exposition of the Nyāya and Vaiśeṣika Systems*, Oxford: Oxford University Press.

Kellock, J. (1926) *Mahadev Govind Ranade: Patriot and Social Servant*, Calcutta: Association Press.

Keshab Chandra Sen (n.d.) "Scientists—A Doubtful Title," in S. Sen (ed.) *The Place of Science in Religion*, Calcutta: Sujata Sen.

— (1901) *Keshub Chunder Sen's Lectures in India*, London: Cassell and Company, Ltd.

— (1940) *Life and Works of Brahmananda Keshav*, ed. P.S. Basu, Calcutta: Navavidhan Publication Committee.

— (1979) *Keshub Chunder Sen*, ed. D.C. Scott, Madras: The Christian Literature Society.

Ketelaar, J.E. (1991) "Strategic Occidentalism: Meiji Buddhists at the World's Parliament of Religions," *Buddhist-Christian Studies*, 11: 37–56.

Killingley, D. (1977) "Rammohun Roy's Interpretation of the Vedānta," unpublished thesis, University of London.
— (1982) *The Only True God: Works on Religion by Rammohun Roy, Selected and Translated from Bengali and Sanskrit, with an Introduction and Notes*, Newcastle upon Tyne: Grevatt & Grevatt.
— (1990) "*Yoga-Sūtra* IV, 2–3 and Vivekānanda's Interpretation of Evolution," *Journal of Indian Philosophy*, 18: 151–79.
— (1993) *Rammohun Roy in Hindu and Christian Tradition: The Teape Lectures 1990*, Newcastle upon Tyne: Grevatt & Grevatt.
— (1995) "Hinduism, Darwinism and Evolution in Late Nineteenth-Century India," in D. Amigoni and J. Wallace (eds) *Charles Darwin's The Origin of Species: New Interdisciplinary Essays*, Manchester: Manchester University Press.
— (1998) "Vivekananda's Western Message from the East," in W. Radice (ed.) *Swami Vivekananda and the Modernization of Hinduism*, Delhi: Oxford University Press.
King, R. (1999) *Indian Philosophy: An Introduction to Hindu and Buddhist Thought*, Washington, D.C.: Georgetown University Press.
Kopf, D. (1969) *British Orientalism and the Bengal Renaissance: The Dynamics of Indian Modernization 1773–1835*, Berkeley: University of California Press.
— (1979) *The Brahmo Samaj and the Shaping of the Modern Indian Mind*, Princeton: Princeton University Press.
Krishna, D. (1994) "Religion and Science—Reflections from the Indian Perspectives," *Journal of Religious Studies*, 25, no. 2: 3–13.
Kuban, G.J. (2006–2008) *Nevada Shoe Print*. Available at http://paleo.cc/paluxy/nevada.htm (accessed 29 March 2011).
Kumar, D. (2010) "Reason, Science and Religion: Gleanings from the Colonial Past," *Current Science*, 99, no. 5: 671–8.
Kumar, F.L. (1991) *The Philosophies of India: A New Approach*, Lewiston, New York: The Edwin Mellon Press.
Lafont, E. (2003) "Father Lafont's Comments on Hypothesis, Induction and Religious Belief," in A.K. Biswas (ed.) *Collected Works of Mahendralal Sircar, Eugene Lafont and the Science Movement (1860–1910)*, Kolkata: The Asiatic Society.
Larson, G.J. (1969) *Classical Sāṃkhya: An Interpretation of Its History and Meaning*, Delhi: Motilal Banarsidass.
Lightman, B. (2007) *Victorian Popularizers of Science: Designing Nature for New Audiences*, Chicago: University of Chicago Press.
Lipner, J. (1986) *The Face of Truth: A Study of Meaning and Metaphysics in the Vedāntic Theology of Rāmānuja*, London: Macmillan Press Ltd.
Llewellyn, J.E. (1993) *The Arya Samaj as a Fundamentalist Movement: A Study in Comparative Fundamentalism*, New Delhi: Manohar.
Lopez, D. (2008) *Buddhism and Science: A Guide for the Perplexed*, Chicago: University of Chicago Press.
Lott, E. (1980) *Vedāntic Approaches to God*, London: Macmillan Press Ltd.
Lourdusamy, J. (2004) *Science and National Consciousness in Bengal (1870–1930)*, New Delhi: Orient Longman.
Lurquin, P. and Stone, L. (2007) *Evolution and Religious Creation Myths: How Scientists Respond*, Oxford: Oxford University Press.
Lyell, C. (1970) *Sir Charles Lyell's Scientific Journals on the Species Question*, ed. L.G. Wilson, New Haven: Yale University Press.
— (1990) *Principles of Geology*, 3 vols, Chicago: University of Chicago Press.
Macaulay, T.B. (1968) "Minute on Education," in W.T. de Bary (ed.) *Sources of Indian Tradition*, vol. 2, New York: Columbia University Press.

Madan, T.N. (1997) *Modern Myths, Locked Minds: Secularism and Fundamentalism in India*, Delhi: Oxford University Press.
Mādhava [Mādhavācārya] (1981) *Sarva-Darśana-Saṅgraha of Mādhavācārya: Sanskrit Text, English Translation, Notes & Appendix*, ed. K.L. Joshi, trans. E.B. Cowell and A.E. Gough, Ahmedabad: Parimal Publications.
Mahabharata of Krishna-Dwaipayana Vyasa, The (1884–1894) 18 vols, trans. K.M. Ganguli, Calcutta: Bharata Press.
Mahābhārata, The (1927–1966) 19 vols, ed. V.S. Sukthankar et al., Poona: Bhandarkar Oriental Research Institute.
Maitra, S.K. (1956) *The Meeting of the East and the West in Sri Aurobindo's Philosophy*, Pondicherry: Sri Aurobindo Ashram.
Majeed, J. (1992) *Ungoverned Imaginings: James Mill's* The History of British India *and Orientalism*, Oxford: Oxford University Press.
Mathur, H.C. (2002) *A Text Book of Hindu Psyche (Scientific Rationale of Hindu Faith): To Bridge the Gap between Rational Scientific Thinking and Dogmatic Hindu Faith*, New Delhi: JDB Book Distributor.
Matilal, B.K. (1977) "Kiraṇāvalī on Praśastapāda's Padārthadharma-Samgraha," in K.H. Potter (ed.) *Encyclopedia of Indian Philosophies: Indian Metaphysics and Epistemology: The Tradition of Nyāya-Vaiśeṣika up to Gaṅgeśa*, Princeton: Princeton University Press.
McDermott, R.A. (1974) "*The Life Divine*: Sri Aurobindo's Philosophy of Evolution and Transformation," in R.A. McDermott (ed.) *Six Pillars: Introductions to the Major Works of Sri Aurobindo*, Chambersburg, Pennsylvania: Wilson Books.
Menon, S. (2006) "Hinduism and Science," in P. Clayton and Z. Simpson (eds) *The Oxford Handbook of Religion and Science*, Oxford: Oxford University Press.
Mill, J. (1820) *The History of British India*, 2nd edn, 6 vols, London: Baldwin, Cradock, and Joy.
Miller, K.R. (2008) *Only a Theory: Evolution and the Battle for America's Soul*, New York: Viking.
Minor, R.N. (1978) *Sri Aurobindo: The Perfect and the Good*, Columbia, Missouri: South Asia Books.
Mittal, K.K. (1974) *Materialism in Indian Thought*, New Delhi: Munshiram Manoharlal.
Mohanty, J.N. (1992) *Reason and Tradition in Indian Thought: An Essay on the Nature of Indian Philosophical Thinking*, Oxford: Clarendon Press/Oxford University Press.
Monod, J. (1971) *Chance and Necessity: An Essay on the Natural Philosophy of Modern Biology*, New York: Alfred A. Knopf.
Morell, J.D. (1848) *Historical and Critical View of the Speculative Philosophy of Europe in the Nineteenth Century*, New York: Robert Carter.
Morgan, L. (1931) *Emergent Evolution: The Gifford Lectures, Delivered in the University of St. Andrews in the Year 1922*, New York: Henry Holt and Company.
Mukerjee, H. (1975) *Indian Renaissance and Raja Rammohun Roy*, Poona: University of Poona.
Mukhyananda (1998) *Vedānta: The Science of Integral Reality vis-a-vis Modern Science*, Calcutta: Centre for Reshaping Our World-View.
Müller, F.M. (1873) *Introduction to the Science of Religion, Four Lectures Delivered at the Royal Institution, with Two Essays on False Analogies, and the Philosophy of Mythology*, Oxford: Longmans, Green, and Co.
Murty, K.S. (1959) *Revelation and Reason in Advaita Vedānta*, Waltair: Andhra University.
Nanda, M. (2003) *Prophets Facing Backward: Postmodern Critiques of Science and Hindu Nationalism in India*, New Brunswick: Rutgers University Press.
— (2006) "Vedic Creationism in America," *The Global Spiral.* Available at http://www.metanexus.net/magazine/tabid/68/id/9470/Default.aspx (accessed 28 July 2010).
— (2011) "Madame Blavatsky's Children: Modern Hindu Encounters with Darwinism," in J.R. Lewis and O. Hammer (eds) *Handbook of Religion and the Authority of Science*, Leiden: Brill.

Nandy, A. (1995) *Alternative Sciences: Creativity and Authenticity in Two Indian Scientists*, Delhi: Oxford University Press.

Narayana, H. (1895) *The Vedic Philosophy, or an Exposition of the Sacred and Mysterious Monosyllable (Aum): The Māndukya Upanishad (Text) with English Translation and Commentary and an Introduction*, Bombay: Tatva-Vivechaka Press.

Nikhilananda (1958) "Hinduism and the Idea of Evolution," in R. Buchsbaum (ed.) *A Book that Shook the World: Anniversary Essays on Charles Darwin's* Origin of Species, Pittsburgh: University of Pittsburgh Press.

Numbers, R. (2009) "Aggressors, Victims, and Peacemakers: Historical Actors in the Drama of Science and Religion," in H.W. Attridge (ed.) *The Religion and Science Debate: Why Does It Continue?*, New Haven: Yale University Press.

Nyāya Sūtras (1913) *The Nyâya Sutrâs of Gotama* [with Sanskrit text], trans. M.S.C. Vidyâbhuṣana, Allahabad: Panini Office.

Overzee, A.H. (1992) *The Body Divine: The Symbol of the Body in the Works of Teilhard de Chardin and Ramanuja*, Cambridge: Cambridge University Press.

Paley, W. (2006) *Natural Theology, or Evidence of the Existence and Attributes of the Deity, Collected from the Appearances of Nature*, ed. M.D. Eddy and D. Knight, Oxford: Oxford University Press.

Palit, C. (1991) "Mahendra Lal Sircar, 1833–1904: The Quest for National Science," in D. Kumar (ed.) *Science and Empire: Essays in Indian Context (1700–1947)*, Delhi: Anamika Prakashan.

Panda, N.C. (2002) *Cyclic Universe: Cycles of the Creation, Evolution, Involution and Dissolution of the Universe*, 2 vols [continuous pagination], New Delhi: D.K. Printworld.

Pew Forum on Religion & Public Life (2008) *U.S. Religious Landscape Survey: Religious Beliefs and Practices: Diverse and Politically Relevant.* Available at http://religions.pewforum.org/pdf/report2-religious-landscape-study-full.pdf#page=99 (accessed 29 January 2011).

Pinto-Correia, C. (1997) *The Ovary of Eve: Egg and Sperm and Preformation*, Chicago: University of Chicago Press.

Plato (1961) *Laws*, trans. A.E. Taylor, in E. Hamilton and H. Cairns (eds) *The Collected Dialogues of Plato, Including the Letters*, New York: Bollingen Foundation.

Potter, K.H. (1977) "Introduction to the Philosophy of the Nyāya-Vaiśeṣika," in K.H. Potter (ed.) *Encyclopedia of Indian Philosophies: Indian Metaphysics and Epistemology: The Tradition of Nyāya-Vaiśeṣika up to Gaṅgeśa*, Princeton: Princeton University Press.

— (ed.) (1981) *Encyclopedia of Indian Philosophies: Advaita Vedānta up to Śaṃkara and His Pupils*, Princeton: Princeton University Press.

Potter, K.H. and Bhattacharya, S. (1977) "Nyāyākusumāñjali," in K.H. Potter (ed.) *Encyclopedia of Indian Philosophies: Indian Metaphysics and Epistemology: The Tradition of Nyāya-Vaiśeṣika up to Gaṅgeśa*, Princeton: Princeton University Press.

Potts, E.D. (1967) *British Baptist Missionaries in India 1793–1837: The History of Serampore and Its Missions*, Cambridge: Cambridge University Press.

Prabhupada, A.C.B. (1972) *The Bhagavad Gita As It Is*, New York: Collier-Macmillan, Collier Books.

— (1979) *Life Comes from Life: Morning Walks with His Divine Grace A. C. Bhaktivedanta Swami Prabhupada*, Los Angeles: Bhaktivedanta Book Trust.

Prakash, G. (1999) *Another Reason: Science and the Imagination of Modern India*, Princeton: Princeton University Press.

Prakashanand Saraswati (1999–2001a) "The Unbroken Continuity of Indian Civilization and Its History," *Encyclopedia of Authentic Hinduism.* Available at http://www.encyclopediaofauthentichinduism.org/articles/1_the_unbroken_continuity.htm (accessed 14 March 2011).

— (1999–2001b) "A New Trend of Anti-Hinduism that has Developed in the Name of Hinduism," *Encyclopedia of Authentic Hinduism*. Available at http://www.encyclopediaofauthentichinduism.org/articles/42_a_new_trend.htm (accessed 14 March 2011).

—(1999–2001c) "A Review of the Most Popular Scientific Theories of the World," *Encyclopedia of Authentic Hinduism*. Available at http://www.encyclopediaofauthentichinduism.org/articles/44_a_review_of.htm (accessed 14 March 2011).

— (1999–2001d) "Actual Age of the Universe," *Encyclopedia of Authentic Hinduism*. Available at http://www.encyclopediaofauthentichinduism.org/articles/49_actual_age.htm (accessed 14 March 2011).

— (1999–2001e) "Age of the Vedas and the Puranas," *Encyclopedia of Authentic Hinduism*. Available at http://www.encyclopediaofauthentichinduism.org/articles/58_age_of.htm (accessed 14 March 2011).

Radhakrishnan, S. (1937) *An Idealist View of Life*, London: George Allen & Unwin.

— (ed.) (1953) *The Principal Upaniṣads, Edited with Introduction, Text, Translation and Notes*, trans. S. Radhakrishnan, London: George Allen & Unwin.

Rai, L. (1916) *Young India: An Interpretation and a History of the Nationalist Movement from Within*, New York: B.W. Huebsch.

Raina, D. and Habib, S.I. (1996) "The Moral Legitimation of Modern Science: Bhadralok Reflections on Theories of Evolution," *Social Studies of Science*, 26: 9–42.

Raj, K. (1991) "Knowledge, Power and Modern Science: The Brahmins Strike Back," in D. Kumar (ed.) *Science and Empire: Essays in Indian Context (1700–1947)*, Delhi: Anamika Prakashan.

Raju, P.T. (1985) *Structural Depths of Indian Thought*, Albany: State University of New York Press.

Ram-Prasad, C. (2001) *Knowledge and Liberation in Classical Indian Thought*, New York: Palgrave.

Ramakrishnananda (1993) "Evolution and Revolution," in G. Northe (ed.) *God is Evolution, Evolution is God: Essays and Lectures from Message of the East, 1912–1961*, Cohasset, Massachusetts: Vedanta Centre.

Raman, V.V. (2003) "Traditional Hinduism and Modern Science," in T. Peters and G. Bennett (eds) *Bridging Science and Religion*, Minneapolis: Fortress Press.

Ramana Maharshi (1968a) *Day by Day with Bhagavan: From a Diary of A. Devaraja Mudaliar (Covering March 16, 1945 to January 4, 1947)*, Tiruvanamalai: Sri Ramanashramam.

— (1968b) *Talks with Sri Ramana Maharshi*, 4th edn, Tiruvanamalai: Sri Ramanashramam.

Rāmānuja (n.d.) *Śrī Rāmānuja Gītā Bhāṣya: With Text in Devanagari & English Rendering, and Index of First Lines of Verses*, trans. Svāmī Ādidevānanda, Mylapore, Madras: Sri Ramakrishna Math.

— (1985–1991) *Śrībhāṣyam: Śārīrakamīmāmsābhāṣyam*, 4 vols, Melkote: Academy of Sanskrit Research.

Ramatirtha (1930–1932) *In the Woods of God-Realization: Or, Complete Works of Swami Rama Tirtha*, 5th edn, 8 vols, Lucknow: Rama Tirtha Publication League.

Rambachan, A. (1987) "The Place of Reason in the Quest for *Moksha*—Problems in Vivekananda's Conceptualization of *Jñānayoga*," *Religious Studies*, 23: 279–88.

— (1994) "Redefining the Authority of Scripture: The Rejection of Vedic Infallibility by the Brahmo Samaj," in L.L. Patton (ed.) *Authority, Anxiety, and Canon: Essays in Vedic Interpretation*, Albany: State University of New York Press.

— (2006) *The Advaita Worldview: God, World, and Humanity*, Albany: State University of New York Press.

Rammohan Roy (1978) *The English Works of Raja Rammohun Roy, with an English Translation of "Tuhfatul Muwahhiddin,"* New York: AMS Press.

Ranade, M.G. (1902) "Philosophy of Indian Theism," in M.B. Kolasker (ed.) *Religious & Social Reform: A Collection of Essays and Speeches*, Bombay: Gopal Narayen and Co.

Ranganathananda, M. (1999) "Our Religion is Based on Verifiable Truths!" *Hinduism Today*, May: 52.

Rao, G. [Utukuru, P.] (2001a) "A Two Dimensional Model of Evolution," Part 1, *Global Spiral*. Available at http://www.metanexus.net/magazine/tabid/68/id/8532/Default.aspx (accessed 15 March 2011).

— (2001b) "A Two Dimensional Model of Evolution," Part 2, *Global Spiral*. Available at http://www.metanexus.net/magazine/tabid/68/id/4610/Default.aspx (accessed 15 March 2011).

Ray, A.K. (1976) *The Religious Ideas of Rammohun Roy: A Survey of His Writings on Religion, Particularly in Persian, Sanskrit and Bengali*, New Delhi: Kanak Publications.

Raychaudhuri, T. (1988) *Europe Reconsidered: Perceptions of the West in Nineteenth Century Bengal*. Delhi: Oxford University Press.

— (1999) *Perceptions, Emotions, Sensibilities: Essays on India's Colonial and Post-Colonial Experiences*, New Delhi: Oxford University Press.

Reid, T. (1915) "Knowledge and Reality," in G.A. Johnston (ed.) *Selections from the Scottish Philosophy of Common Sense*, Chicago: Open Court Publishing Company.

Richards, R.J. (1992) *The Meaning of Evolution: The Morphological Construction and Ideological Reconstruction of Darwin's Theory*, Chicago: University of Chicago Press.

Riepe, D. (1961) T*he Naturalistic Tradition in Indian Thought*, Seattle: University of Washington Press.

Roday, P.R. and Purohit, M.K. (2010) "Joseph G. Medlicott and the Line of the Narmada and Son Valleys," *Current Science*, 98, no. 8: 1134–8.

Ruse, M. (1999) *The Darwinian Revolution: Science Red in Tooth and Claw*, 2nd edn, Chicago: University of Chicago Press.

— (2003) *Darwin and Design: Does Evolution Have a Purpose?*, Cambridge, Massachusetts: Harvard University Press.

Sabbatini, R.M.E. (1997) "Franz Joseph Gall (1758–1828): A Brief Biography," *Brain and Mind*. Available at http://www.cerebromente.org.br/n01/frenolog/frengall.htm (accessed 14 October 2010).

Sangwan, S. (1991a) *Science, Technology and Colonisation: An Indian Experience, 1757–1857*, Delhi: Anamika Prakashan.

Śaṅkara (n.d.) *Srīmad Bhagavad Gītā Bhāṣya of Sri Sāṁkarācārya, with Text in Devanagiri & English Rendering, and Index of First Lines of Verses*, trans. A.G.K. Warrier, Madras: Sri Ramakrishna Math.

— (1983a) *Brahmasutra Bhashya*, rev. edn, vol. 7 in *Complete Works of Sri Sankaracharya in the Original Sanskrit*, Madras: Samata Books.

— (1983b) *Vivekachudamani of Sri Sankaracharya; Text with English Translation, Notes and Index*, 3rd edn, Madhavananda ed. Mayavati: Advaita Ashrama.

Sarkar, H.C. (1928) "Introduction," in *Brahmo Dharma of Maharshi Debendranath Tagore*, trans. H.C. Sarkar, Calcutta: H.C. Sarkar.

Scott, E.C. (2004) *Evolution vs. Creationism: An Introduction*, Berkeley: University of California Press.

Sellon, E.B. and Weber, R. (1992) "Theosophy and the Theosophical Society," in A. Faivre and J. Needleman (eds) *Modern Esoteric Spirituality*, New York: Crossroad.

Sen, A. (2005) *The Argumentative Indian: Writings on Indian History, Culture and Identity*, New York: Farrar, Straus and Giroux.

Sethna, K.D. (1981) *The Spirituality of the Future: A Search Apropos of R.C. Zaehner's Study in Sri Aurobindo and Teilhard de Chardin*, Rutherford: Fairleigh Dickinson University Press; London: Associated University Presses.

Sharma, B.N.K. (1986) *The Brahmasūtras and Their Principal Commentaries: A Critical Exposition*, 2nd edn, 2 vols, New Delhi: Munshiram Manoharlal.

Sharma, C. (1960) *A Critical Survey of Indian Philosophy*, Delhi: Motilal Banarsidass.

Shastri, M.H. (1982) *Lokāy[a]ta and Vratya*, Naihati: Haraprasad Shastri Gaveshana Kendra.

Shishya (2006–2009a) [Post on Theory of Evolution] *Arya Samaj Forum*. Now defunct. Formerly available at http://aryasamaj.forumwise.com/aryasamaj-thread284-0-asc-0.html (accessed 3 December 2009).

— (2006–2009b) [Post on Avataric Evolutionism] *Arya Samaj Forum*. Now defunct. Formerly available at http://aryasamaj.forumwise.com/aryasamaj-thread897.html (accessed 3 December 2009).

Shubin, N. (2008) *Your Inner Fish: A Journey into the 3.5-Billion-Year History of the Human Body*, New York: Pantheon.

Shukla, R. (2005) *India Science Report*, New Delhi: National Council of Applied Economic Research. Available at http://www.insaindia.org/India%20Science%20report-Main.pdf (accessed 29 January 2011).

Singh, C. (1971) *Life & Teachings of Swami Dayanand Saraswati*, 2nd edn, New Delhi: Jan Gyan Prakashan.

Singleton, M. (2007) "Yoga, Eugenics, and Spiritual Darwinism in the Early Twentieth Century," *International Journal of Hindu Studies*, 11, no. 2: 125–45.

Sircar, M. (2003a) "On the Physiological Basis of Psychology," in A.K. Biswas (ed.) *Collected Works of Mahendralal Sircar, Eugene Lafont and the Science Movement (1860–1910)*, Kolkata: The Asiatic Society.

— (2003b) "On the Desirability of Cultivation of the Sciences by the Natives of India," in A.K. Biswas (ed.) *Collected Works of Mahendralal Sircar, Eugene Lafont and the Science Movement (1860–1910)*, Kolkata: The Asiatic Society.

— (2003c) "Mahendra Lal Sircar's Proposal for the Indian Association for the Cultivation of Science (IACS)," in A.K. Biswas (ed.) *Collected Works of Mahendralal Sircar, Eugene Lafont and the Science Movement (1860–1910)*, Kolkata: The Asiatic Society.

Smart, N. (1964) *Doctrine and Argument in Indian Philosophy*, London: George Allen and Unwin.

Spencer, H. (1958) *First Principles*, 4th edn, New York: De Witt Revolving Fund.

Srivastava, R.S. (1968) *Sri Aurobindo and the Theories of Evolution (A Critical and Comparative Study of the Indian and Western Theories of Evolution with Special Reference to Sri Aurobindo's Philosophy of Evolution)*, Varanasi: Chowkhamba Sanskrit Series Office.

Sugirtharajah, S. (2003) *Imagining Hinduism: A Postcolonial Perspective*, London: Routledge.

Sullivan, B.M., Wiist, B. and Wayment, H. (2010) "The Buddhist Health Study: Meditation on Love and Compassion as Features of Religious Practice," *Crosscurrents*, 60, no. 2: 185–207.

Tagore, D. (1928) *Brahmo Dharma of Maharshi Debendranath Tagore*, trans. H.C. Sarkar, Calcutta: H.C. Sarkar.

— (1994) *The Auto-Biography of Maharshi Devendranath Tagore*, trans. S. Tagore and I. Devi, Calcutta: S.C. Sarkar & Sons.

Tagore, R. (1931) *The Religion of Man, being the Hibbert Lectures for 1930*, New York: Macmillan.

— (1969) *Our Universe*, trans. I. Dutt, Bombay: Jaico Publishing House.

Tagore, S. (1994) "Introduction 1," in *The Auto-Biography of Maharshi Devendranath Tagore*, trans. S. Tagore and I. Devi, Calcutta: S.C. Sarkar & Sons.

Teilhard de Chardin, P. (1975) *The Phenomenon of Man*, trans. B. Wall, New York: Harper & Row.

Thibaut, G. (trans.) (1890) *The Vedānta-Sūtras with the Commentary by Śaṅkarācārya*, part I, Oxford: Clarendon Press.

— (1904) *The Vedānta-Sūtras with the Commentary by Śaṅkarācārya*, part II, Oxford: Clarendon Press.

— (1966) *The Vedānta-Sūtras with the Commentary by Rāmānuja*, part III, 2nd edn, Delhi: Motilal Banarsidass.

Thistlethwayte, L.E.L. (1998) "The Role of Science in the Hindu–Christian Encounter," in G.A. Oddie (ed.) *Religious Traditions in South Asia: Interaction and Change*, Richmond, Surrey: Curzon.

Thompson, R. (1981) *Mechanistic and Nonmechanistic Science: An Investigation into the Nature of Consciousness and Form*, Los Angeles: Bhaktivedanta Book Trust.

Thompson, R. (Sadaputa dasa), Drutakarma dasa, and Bhutatma dasa (1984) "Higher Dimensional Science," in Sadaputa dasa (ed.) *Origins: Higher Dimensions in Science*, Los Angeles: Bhaktivedanta Institute.

Trautmann, T.R. (2004) *Aryans and British India*, New Delhi: Yoda Press.

Udayana (1995) *Ātmatattvaviveka by Udayanācārya, with Translation, Explanation and Analytical-Critical Survey* [with Sanskrit text], trans. N.S. Dravid, Shimla: Indian Institute of Advanced Study.

— (1996) *Nyāya-Kusumāñjali of Udayanācārya, with Translation and Explanation* [with Sanskrit text], trans. N.S. Dravid, vol. 1, New Delhi: Indian Council of Philosophical Research.

Uddyotakara (1984) *Nyāya-Vārṭika*, in *The Nyāya-Sūtras of Gauṭama with the Bhāṣya of Vāṭsyāyana and the Vārṭika of Uḍḍyoṭakara*, vol. 4, trans. M.G. Jhā, Delhi: Motilal Banarsidass.

— (1985) *Nyāya-Vārṭika*, in T. Nyaya-Tarkatirtha and A. Tarkatirtha (eds) *Nyāyadarśanam with Vātsyāyana's Bhāṣya, Udyotkara's Vārrtika, Vācaspati Miśra's Tātparyaṭīkā & Viśvanāthas Vṛtti, with an Introduction by Narendra Chandra Vedantatirtha*, 2nd edn, New Delhi: Munshiram Manorharlal Publishers Private Limited.

Van Buitenen, J.A.B. (1968) *Rāmānuja on the Bhagavad-Gītā: A Condensed Rendering of His Gītābhāṣya with Copious Notes and an Introduction*, 2nd edn, Delhi: Motilal Banarsidass.

Varadachari, V. (1977) "Ātmatattvaviveka," in K.H. Potter (ed.) *Encyclopedia of Indian Philosophies: Indian Metaphysics and Epistemology: The Tradition of Nyāya-Vaiśeṣika up to Gaṅgeśa*, Princeton: Princeton University Press.

Vatsyayana, Mallanaga (2002) *Kamasutra: A New, Complete English Translation of the Sanskrit Text with Excerpts from the Sanskrit* Jayamangala *Commentary of Yashodhara Indrapada, the Hindi* Jaya *Commentary of Devadatta Shastri, and Explanatory Notes by the Translators*, trans. W. Doniger and S. Kakar, Oxford: Oxford University Press.

Vivekananda (2003) *The Complete Works of Swami Vivekananda*, 9 vols, in *Swami Vivekananda: Life, Works & Research* (multimedia CD), Calcutta: Advaita Ashrama.

Ward, K. (1996) *Religion and Creation*, Oxford: Clarendon Press.

Ward, W. (1985) *History, Literature, and Mythology of the Hindoos: Including a Minute Description of Their Manners and Customs, and Translations from Their Principal Works*, 3rd edn, 4 vols, Delhi: B.R. Publishing Corp.

Williams, G.M. (1974) *The Quest for Meaning of Svāmī Vivekānanda: A Study of Religious Change*, Chico, California: New Horizons Press.

— (1981) "Swami Vivekananda: Archetypal Hero or Doubting Saint?" in R.D. Baird (ed.) *Religion in Modern India*, New Delhi: Manohar.

Yogananda (1955) *Autobiography of a Yogi*, Los Angeles: Self-Realization Fellowship.

Županov, I.G. (2005) *Missionary Tropics: The Catholic Frontier in India (16th–17th Centuries)*, Ann Arbor: University of Michigan Press.

Index

For Product Safety Concerns and Information please contact our EU representative GPSR@taylorandfrancis.com Taylor & Francis Verlag GmbH, Kaufingerstraße 24, 80331 München, Germany